Selected Titles in This Series

202 **V. P. Maslov and G. A. Omel′yanov,** Geometric asymptotics for nonlinear PDE. I, 2001
201 **Shigeyuki Morita,** Geometry of differential forms, 2001
200 **V. V. Prasolov and V. M. Tikhomirov,** Geometry, 2001
199 **Shigeyuki Morita,** Geometry of characteristic classes, 2001
198 **V. A. Smirnov,** Simplicial and operad methods in algebraic topology, 2001
197 **Kenji Ueno,** Algebraic geometry 2: Sheaves and cohomology, 2001
196 **Yu. N. Lin′kov,** Asymptotic statistical methods for stochastic processes, 2001
195 **Minoru Wakimoto,** Infinite-dimensional Lie algebras, 2001
194 **Valery B. Nevzorov,** Records: Mathematical theory, 2001
193 **Toshio Nishino,** Function theory in several complex variables, 2001
192 **Yu. P. Solovyov and E. V. Troitsky,** C^*-algebras and elliptic operators in differential topology, 2001
191 **Shun-ichi Amari and Hiroshi Nagaoka,** Methods of information geometry, 2000
190 **Alexander N. Starkov,** Dynamical systems on homogeneous spaces, 2000
189 **Mitsuru Ikawa,** Hyperbolic partial differential equations and wave phenomena, 2000
188 **V. V. Buldygin and Yu. V. Kozachenko,** Metric characterization of random variables and random processes, 2000
187 **A. V. Fursikov,** Optimal control of distributed systems. Theory and applications, 2000
186 **Kazuya Kato, Nobushige Kurokawa, and Takeshi Saito,** Number theory 1: Fermat's dream, 2000
185 **Kenji Ueno,** Algebraic Geometry 1: From algebraic varieties to schemes, 1999
184 **A. V. Mel′nikov,** Financial markets, 1999
183 **Hajime Sato,** Algebraic topology: an intuitive approach, 1999
182 **I. S. Krasil′shchik and A. M. Vinogradov, Editors,** Symmetries and conservation laws for differential equations of mathematical physics, 1999
181 **Ya. G. Berkovich and E. M. Zhmud′,** Characters of finite groups. Part 2, 1999
180 **A. A. Milyutin and N. P. Osmolovskii,** Calculus of variations and optimal control, 1998
179 **V. E. Voskresenskiĭ,** Algebraic groups and their birational invariants, 1998
178 **Mitsuo Morimoto,** Analytic functionals on the sphere, 1998
177 **Satoru Igari,** Real analysis—with an introduction to wavelet theory, 1998
176 **L. M. Lerman and Ya. L. Umanskiy,** Four-dimensional integrable Hamiltonian systems with simple singular points (topological aspects), 1998
175 **S. K. Godunov,** Modern aspects of linear algebra, 1998
174 **Ya-Zhe Chen and Lan-Cheng Wu,** Second order elliptic equations and elliptic systems, 1998
173 **Yu. A. Davydov, M. A. Lifshits, and N. V. Smorodina,** Local properties of distributions of stochastic functionals, 1998
172 **Ya. G. Berkovich and E. M. Zhmud′,** Characters of finite groups. Part 1, 1998
171 **E. M. Landis,** Second order equations of elliptic and parabolic type, 1998
170 **Viktor Prasolov and Yuri Solovyev,** Elliptic functions and elliptic integrals, 1997
169 **S. K. Godunov,** Ordinary differential equations with constant coefficient, 1997
168 **Junjiro Noguchi,** Introduction to complex analysis, 1998
167 **Masaya Yamaguti, Masayoshi Hata, and Jun Kigami,** Mathematics of fractals, 1997
166 **Kenji Ueno,** An introduction to algebraic geometry, 1997
165 **V. V. Ishkhanov, B. B. Lur′e, and D. K. Faddeev,** The embedding problem in Galois theory, 1997

(*Continued in the back of this publication*)

Geometric Asymptotics for Nonlinear PDE. I

Translations of
MATHEMATICAL MONOGRAPHS

Volume 202

Geometric Asymptotics for Nonlinear PDE. I

V. P. Maslov
G. A. Omel′yanov

Translated by
Dmitrii Chibisov

American Mathematical Society
Providence, Rhode Island

В. П. Маслов, Г. А. Омельянов

ГЕОМЕТРИЧЕСКИЕ АСИМПТОТИКИ ДЛЯ НЕЛИНЕЙНЫХ УРАВНЕНИЙ В ЧАСТНЫХ ПРОИЗВОДНЫХ

Translated from the Russian manuscript by Dmitrii Chibisov

2000 *Mathematics Subject Classification.* Primary 35Qxx, 76L05.

ABSTRACT. This book sets forth a relatively simple new method for constructing asymptotic solutions for nonintegrable nonlinear systems of partial differential equations. Many examples originating from problems of hydrodynamics, nonlinear optics, plasma physics, continuum mechanics, and the theory of phase transitions (problems with a free boundary) are treated. Many figures illustrating the various physical effects are included.

This book can be used by researchers and graduate students working in partial differential equations, applied mathematics, and various areas in physics.

Library of Congress Cataloging-in-Publication Data

Maslov, V. P. (Viktor Pavlovich), 1930–

[Geometricheskii asimptotiki dlia nelineinykh uravnenii v chastnykh proizvodnykh. English]

Geometric asymptotics for nonlinear PDE / V. P. Maslov, G. A. Omel′yanov ; [translated by Dmitrii Chibisov]

p. cm. — (Translations of mathematical monographs, ISSN 0065-9282 ; v. 202)

Includes bibliographical references.

ISBN 0-8218-2109-1 (alk. paper)

1. Differential equations, Nonlinear—Asymptotic theory. I. Omel′ianov, G. A. (Georgii Aleksandrovich), 1950– . II. Title. III. Series.

QA377.M2837 2001

515/.355—dc21 2001040045

Visit the AMS home page at URL: `http://www.ams.org/`

10 9 8 7 6 5 4 3 2 1 06 05 04 03 02 01

Contents

Introduction

This book is intended to set out the asymptotic theory of nonlinear waves with *localized fast variation*, i.e., the waves that, at each time t, vary "slowly" outside a small neighborhood of the front $x = \varphi(t)$ and "fast" near the front.

The best known example of a function with localized fast variation is given by a "smoothed shock wave" $u = u(x, t, \varepsilon)$ (see Figure 0.1), which is a regularization of the discontinuous function

$$\bar{u}(x,t) = \begin{cases} u^+(x,t) & \text{for} \quad x > \varphi_1(t), \\ u^-(x,t) & \text{for} \quad x < \varphi_1(t), \end{cases} \tag{0.1}$$

so that $u(x, t, \varepsilon)$ equals $u^\pm(x, t)$ outside a small "transition zone" (a neighborhood of the front $x = \varphi_1(t)$) and varies from $u^-|_{x=\varphi_1(t)-0}$ to $u^+|_{x=\varphi_1(t)+0}$ as a smooth function inside the transition zone. We will characterize the width of the transition zone by a quantity $\varepsilon \ll 1$ and treat it as a small parameter, assuming that ε is much less than any positive constant. One of the possible mathematical expressions for a smoothed shock wave is

$$u = \frac{1}{2}(u^+ + u^-) + \frac{1}{2}(u^+ - u^-) \tanh\left(\beta_1 \frac{x - \varphi_1(t)}{\varepsilon}\right), \tag{0.2}$$

where the function $\beta_1 = \beta_1(t) > 0$ describes the variation of the width of the transition zone.

Another characteristic example of a function with localized fast variation is a *"distorted solitary wave"* (soliton) (see Figure 0.2), which is a narrow (of width $\sim \varepsilon$) bell-shaped function moving along a smooth "background" $u_0(x, t)$. "Distortion" of the solitary wave consists in the variation of its amplitude and velocity with time. Such a wave can be described, e.g., by the formula

$$u = u_0 + A \cosh^{-2}\left(\beta_2 \frac{x - \varphi_2(t)}{\varepsilon}\right), \tag{0.3}$$

where the function $\beta_2 = \beta_2(t) > 0$ has the same meaning as β_1 in (0.2).

We assume that the amplitudes A and $u^+|_{x=\varphi_1(t)+0} - u^-|_{x=\varphi_1(t)-0}$ have magnitude $O(1)$ independent of the small parameter ε. Thus in ε-neighborhoods of the fronts the functions (0.2) and (0.3) vary by $O(1)$. We will refer to such a behavior of the wave as *fast* variation. If a function varies on an interval $O(\varepsilon)$ only by $O(\varepsilon)$, we will say that it is of *slow* variation. Here and in what follows we say that a function $f(x, t, \varepsilon)$ is $O(\varepsilon^k)$ or $O(1) = O(\varepsilon^0)$ if for $\varepsilon \in (0, \varepsilon_0]$ with some $\varepsilon_0 > 0$

$$\sup_{x,t \in Q} |f(x, t, \varepsilon)| \leq c\varepsilon^k,$$

where Q is the domain of f and the constant $c > 0$ does not depend on ε.

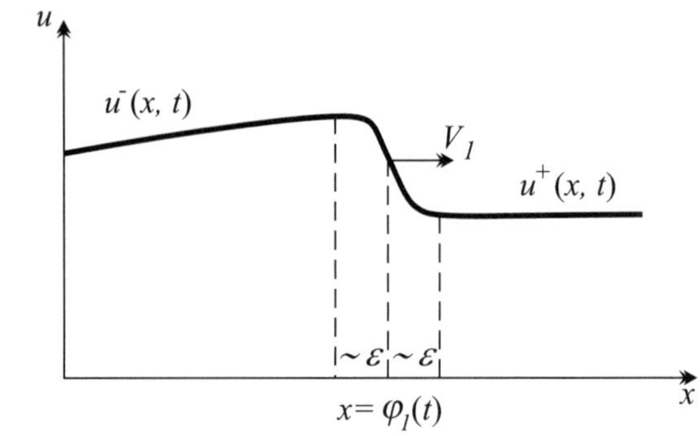

FIGURE 0.1. Typical spatial profile of a smoothed shock wave. $x = \varphi_1(t)$ is the front position, $V_1 = \dot{\varphi}_1(t)$ is the front velocity, $u^+(x,t)$ is the nonperturbed state ahead of the wave front, and $u^-(x,t)$ is the resulting state behind the wave front.

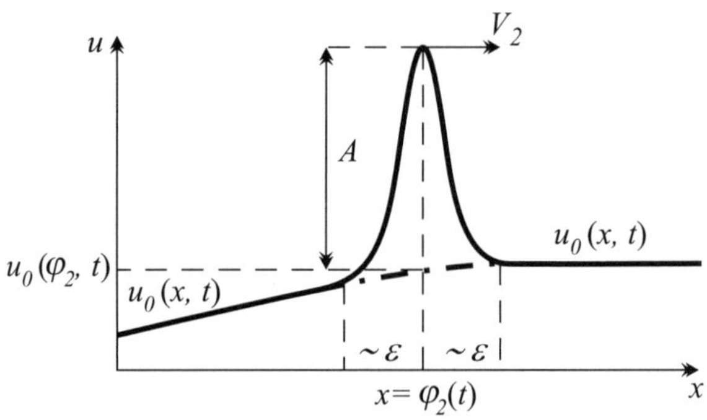

FIGURE 0.2. Typical spatial profile of a soliton type wave. $x = \varphi_2(t)$ is the front position, $V_2 = \dot{\varphi}_2(t)$ is the front velocity, and $u_0(x,t)$ is a smooth background.

When u_0, A, and $u^\pm$ are constant quantities, the smoothed shock wave (0.2) is an exact solution to the Burgers equation with small viscosity

$$\frac{\partial u}{\partial t} + u\frac{\partial u}{\partial x} = \varepsilon\frac{\partial^2 u}{\partial x^2}, \tag{0.4}$$

and the solitary wave (0.3) is an exact solution to the Korteweg–de Vries equation with small dispersion

$$\frac{\partial u}{\partial t} + 3u\frac{\partial u}{\partial x} + \varepsilon^2\frac{\partial^3 u}{\partial x^3} = 0. \tag{0.5}$$

In this case

$$\beta_1 = \frac{u^- - u^+}{4}, \quad \beta_2 = \frac{\sqrt{A}}{2}, \tag{0.6}$$

and the waves (0.2), (0.3) move with constant velocities

$$V_1 = \dot{\varphi}_1 = \frac{u^+ + u^-}{2} \quad \text{and} \quad V_2 = \dot{\varphi}_2 = A + 3u_0. \tag{0.7}$$

It is clear that the initial values of u for $t = 0$ specified by formulas (0.2), (0.3) along with relations (0.6), (0.7) completely determine the dynamics of these waves for all $t > 0$. However for variable $u_0, A, u^\pm$ the relations (0.7), to be rewritten now as

$$\dot{\varphi}_1 = \frac{u^+(x,t) + u^-(x,t)}{2}\bigg|_{x=\varphi_1(t)}, \qquad \dot{\varphi}_2 = (A + 3u_0(x,t))\big|_{x=\varphi_2(t)},$$

do not completely determine the dynamics of the waves, since the functions u_0, A, and $u^\pm$ are not yet specified. Accordingly, for variable $u_0, A, u^\pm$ the functions (0.2), (0.3) do not solve the equations (0.4), (0.5) exactly, but satisfy these equations within $O(1)$. This means that when we substitute (0.2), (0.3) into (0.4), (0.5) the terms of order ε^{-1} (which arise from differentiation of (0.2), (0.3)) cancel, but there remains a discrepancy of magnitude $O(1)$. One may naturally assume that in this case the functions (0.2), (0.3) are merely the leading terms of approximate solutions of equations (0.4), (0.5). These are "two-scale" solutions, since they involve two types of dependence on independent variables, namely, the fast dependence determined by "fast" arguments $\tau_i = (x - \varphi_i(t))/\varepsilon$ of the functions $\tanh \beta_1\tau_1$ and $\cosh^{-2} \beta_2\tau_2$ describing the profile of the wave in a neighborhood of the front, and the slow dependence determined by "slow" arguments x, t of functions $u_0 = u_0(x,t)$, $u^\pm = u^\pm(x,t)$ (as well as functions $A = A(t)$ and $\beta_i = \beta_i(t)$) describing the profile of the wave outside the neighborhood of the front.

For constant $u_0, A, u^\pm$ the exact solutions of (0.2), (0.3) can be easily found by integrating (0.4), (0.5) with the change of variables $\tau = (x - Vt)/\varepsilon$. The relations (0.6), (0.7) arise automatically as the conditions for existence of a solution having the form of a traveling wave, i.e., a solution with argument $(x - Vt)/\varepsilon$. Similar formal calculations for variable $u_0, A, u^\pm$ also lead to formulas (0.2), (0.3) and (0.6), (0.7) if we seek a solution as a distorted traveling wave and require the function $u = u(x, t, (x - \varphi(t))/\varepsilon)$ to satisfy the corresponding equation up to $O(1)$. However the fact that the relations (0.7) are not closed motivates the development of a systematic method of constructing an asymptotic solution with any given accuracy.

A similar situation arises when constructing rapidly oscillating asymptotic solutions of equations with a small parameter. For example, the function

$$u = \mu + \nu \operatorname{cn}^2\left(\beta\frac{S(x,t)}{\varepsilon}, k\right), \tag{0.8}$$

where $\operatorname{cn}(u,k)$ is the elliptic Jacobi function, for constant values of the parameters μ, ν, β, k, and $S = x - Vt$, $V = \text{const}$, is an exact oscillating solution to the Korteweg–de Vries equation (0.5) (the so-called cnoidal wave). For variable parameters μ, ν, β, k and nonlinear phase S the function (0.8) approximately describes the "distorted" cnoidal wave, and the problem arises of obtaining additional relations for its parameters. This problem was solved by Whitham [**259–262**] for a wide class of equations. The method he developed is widely known now as the "Whitham method of averaging", or "two-scale method". Recall that this method represents the rapidly oscillating solution of a partial differential equation in the form of the asymptotic expansion

$$u = F_0\Big(\frac{S(x,t)}{\varepsilon}, x, t\Big) + \varepsilon F_1\Big(\frac{S(x,t)}{\varepsilon}, x, t\Big) + \dots, \tag{0.9}$$

where the phase $S(x,t)$ and the functions $F_j(\tau,x,t)$, periodic in τ, are to be determined. Substituting (0.9) into the equation and equating the coefficients at powers of ε to zero, one obtains a chain of ordinary (in τ) differential equations, the equation for $F_0(\tau,x,t)$ being nonlinear, and those for the $F_i(\tau,x,t)$, $i \geq 1$, being nonhomogeneous linear. The conditions for the existence of τ-periodic solutions of these equations lead to additional relations (which form a system of nonlinear partial differential equations) that enable one to determine the phase $S(x,t)$ and the dependence of the functions $F_j(\tau,x,t)$ on the "slow" variables x and t.

We have briefly described here the idea of the Whitham method as developed by Luke [**133**]. Note that (0.9) is a natural extension to the nonlinear case of the classical WKB expansion (with $F_j(\tau,x,t) = \varphi_j(x,t)\exp(i\tau)$, $i = \sqrt{-1}$) for a solution to a linear equation (see, e.g., [**141, 149**]). For nonlinear ordinary differential equations, a representation of the solution in a form similar to (0.9) was apparently first proposed by Kuzmak [**122**]; its idea goes back to work by Poincaré and Krylov–Bogolyubov (for more details see, e.g., the books [**21, 22, 43, 182, 254**]).

It appears natural to use the expansion (0.9) also for constructing asymptotic solutions with localized fast variation. However, due to the qualitative difference between the solutions of type (0.2), (0.3) and the rapidly oscillating solutions, the Whitham method cannot be applied directly in our case, but has to be essentially modified. The matter is that if $F(\tau,x,t)$ stabilizes exponentially fast as $\tau \to \pm\infty$ on smooth functions $F^{\pm}(x,t)$, i.e., $F(\tau,x,t)$ converges exponentially fast to smooth functions $F^{\pm}(x,t)$ as $\tau \to \pm\infty$, and if $S(x,t) \in C^\infty$, then any smooth variation of the phase S outside its zero level line $\Gamma = \{(x,t),\ S(x,t) = 0\}$ changes the function $F(S(x,t)/\varepsilon, x, t)$ by $O(\varepsilon)$. Moreover, for soliton-type functions of the form (0.2) one can replace the amplitude $A(x,t)$ by any smooth function $\tilde{A}(x,t)$ which is equal to A on Γ without affecting the leading term of the solution. This leads to indeterminacy in specification of initial values of the phase and amplitude and to a certain freedom in determination of S and A for $t > 0$; actually, in order to construct a solution of the form (0.2) it suffices to determine the zero level line of the phase and to find $A|_\Gamma$. At the same time, by a formal application of the Whitham scheme we obtain a system of nonlinear equations of first order for determination of $S(x,t)$ and $A(x,t)$ for all $x \in \mathbb{R}^1$ and $t > 0$. It is well known that in general the classical solution of such a system of equations exists only within some time t^* which depends on the initial values $S(x,0)$ and $A(x,0)$.

Next, if we continue to determine the solution with localized fast variation following the Whitham scheme, the first correction which we obtain involves functions of a smoothed shock wave type $u_1(x,t) - u_1^-(S,A)(\tanh(S/\varepsilon) - 1)/2$ with coefficient u_1^- completely determined by $S(x,t)$ and $A(x,t)$. However it is easy to verify directly that the function $u_0(x,t) + \varepsilon\big(u_1(x,t) + u_1^-(S,A)\big)$ constructed in this way, i.e., the tentative solution in the domain $\{(x,t),\ t > 0,\ S(x,t) < 0\}$, in general does not satisfy the original equation with accuracy $O(\varepsilon^2)$. (In **1.1.3** we will exemplify this by the Korteweg–de Vries equation.) When repeating mechanically the standard Whitham scheme, this property requires one to construct separately the solutions in a small neighborhood of the wave front Γ and outside this neighborhood with subsequent matching of these solutions (see, e.g., [**2, 76**]).

In the modification of the Whitham method proposed here we take into account from the outset the specific character of functions with localized fast variation, namely, rather than determining the phase S (as well as the amplitude for soliton-type solutions) we find only its zero level line Γ (and $A|_\Gamma$ for solutions of the type (0.3)). Consequently, instead of solving an auxiliary system of nonlinear partial differential equations of first order, we only need to find a single characteristic of this system where $S \equiv 0$. Furthermore, when looking for a correction, we obtain $u_1^-(S, A)|_\Gamma$ rather than $u_1^-(S, A)$, and the freedom in extending $u_1^-(S, A)|\Gamma$ outside the front Γ allows us to avoid the contradiction mentioned above. As a result, we obtain a very simple constructive method for determining an asymptotic solution with any prescribed accuracy. Methodologically our approach is similar to the method of complex germs [**144**] for constructing rapidly oscillating localized solutions.

This preliminary discussion makes it clear that the asymptotics we discuss are closely related to a geometric object, namely, the curve (or a surface of codimension one in the multidimensional case) in whose vicinity the fast variation occurs. Hence the method we develop here is closely related to geometric asymptotics for linear equations. Recall that the method of geometric optics, or the ray method, known also as the WKB (Wentzel–Kramers–Brillouin) method, which first appeared in physical literature, associates with quasiclassical (short-wave) asymptotics of a linear equation a certain geometric object, namely rays, i.e., the projections of bicharacteristics of the Hamilton system on the configuration (x, t)-space. Formally, this correspondence can be easily obtained by substituting the fast oscillating exponent $\exp(iS(x, t)/\varepsilon)$ into the equation; then the terms of higher order in ε^{-1} are equated to zero to yield the Hamilton–Jacobi equation (eikonal); and finally, the corresponding Hamilton system is derived. (For more details on the WKB method see, e.g., [**13, 119, 134, 141, 142, 149, 248**].) This construction allowed for explanation of many physical phenomena, but it is intrinsically local because it is valid only in the domain where the rays do not intersect. A qualitatively new stage in the development of the theory was opened up by the work of Maslov (see [**141**]), who established that the global geometric objects here are the bicharacteristics and, accordingly, the Lagrangian manifold in the extended (phase) space. The subsequent extensive study of this object resulted in elaborating the method of geometric asymptotics in the modern sense, see [**79, 141, 149, 173**], among others. Substantial progress based on systematic use of this approach was achieved not only in constructing global short-wave asymptotics [**79, 141, 142, 148, 173, 248**] (see also [**13, 119**] for development of the method of standard equations), but also in other directions, especially in the theory of linear equations, quantum mechanics, symplectic geometry, etc., see [**79, 128, 246, 248**].

For nonlinear equations, the theory nowadays seems to be staying on the same level as the linear theory in the beginning of 20th century: the geometric object is considered in the configuration space, and all constructions are of a local nature. Nevertheless, comprehension of the geometric meaning of our construction allows for better understanding of the reasons for its locality and may guide one in passing to the global description.

The method proposed here was initially designed for smoothed shock waves [**166**], taking as an example the equations of gas dynamics, and for soliton-type solutions of equations with small dispersion and variable coefficients [**150**]. This method was further developed in papers dealing with localized solutions of equations with small viscosity [**46**], problems with a free boundary [**49, 196–201, 216**], and other related problems [**151, 152, 155, 161–163, 224–229, 238**].

The problem of constructing asymptotic soliton-type solutions attracted the attention of many mathematicians and physicists in the seventies and eighties, after the the discovery of the method of the inverse scattering transform (or the method of the inverse scattering problem), see, e.g., [**2, 37, 58, 126, 129, 263, 265**]. The first results in this direction were obtained in the papers [**1, 92, 135, 202, 267**], which developed the so-called "direct" asymptotic methods based primarily on the Whitham method. These ideas were advanced further in [**2, 75–77, 116, 117, 178, 179, 203, 204**].

In a parallel way in [**108, 110–112**] the asymptotic theory based on the method of the inverse scattering transform was elaborated. This theory resulted in obtaining fine, important results, in particular, results describing the collision of solitons and the appearance of the so-called radiation (see, e.g., [**87, 98–105, 107–109, 114, 139, 140, 169, 220**]). However, this theory has a rather narrow range of applicability, because the specific methods require the equations to have a specific structure, namely, the equation in its leading term must be reducible by a suitable change of variables to an equation with constant coefficients that can be integrated by the method of the inverse scattering problem. We do not describe these results here, because we consider mostly multidimensional equations and systems with variable coefficients, to which the method of the inverse problem is inapplicable.

The problem of constructing asymptotic solutions of smoothed shock wave type was treated by many mathematicians. To the authors' knowledge, the first essential results were obtained in [**85, 86, 188–191**] in the context of the problem on the relationship between the discontinuous solution to the Hopf equation and the solution to an equation of Burgers type with viscosity ε as $\varepsilon \to 0$. Asymptotic methods were elaborated, in particular, in [**21, 89, 214, 239, 240**] for one-dimensional problems. A detailed description of the procedure based on the matching method for finding solutions to Burgers' type equations with small viscosity, including the description of the process of forming a smoothed shock wave, was given by Il′in [**88**].

We should also mention the papers which can be regarded as conceptual predecessors of our method, namely, those dealing with asymptotics of the boundary layer type, and most importantly, with the method of Vishik and Lyusternik, see, in particular, [**27, 28, 250, 252, 253**].

When writing this book we tried to present the method in the simplest and most transparent way. For this reason we avoided elaborating general constructions, presenting instead the ideas and techniques of the method as applied to a number of particular problems. These problems were selected so that all of them are physically meaningful and each subsequent problem exhibits some new mathematical aspects as compared to the previous ones.

Now we will briefly describe the arrangement of the material in the book. The first two sections of Chapter 1 contain a very detailed exposition of the basic ideas of our method as applied to the Korteweg–de Vries equation (soliton-type solutions) and to the Korteweg–de Vries–Burgers equation (solutions of the smoothed shock wave type). These equations, being very simple and thoroughly investigated, are commonly used as basic model examples. They allow us to expound the ideas and the technical aspects of the method in the simplest way. In the subsequent sections of Chapter 1 we consider somewhat more complicated problems, to demonstrate that the same method can be successfully applied in the case of variable coefficients and equations of more complicated structure.

Chapter 2 deals with multidimensional equations and systems. The basic model example here is the Kadomtsev–Petviashvili equation. As is well known, it has an exact solution in the form of a crest (called also a two-dimensional soliton) whose cross-sections have the form of a solitary wave. The front of the crest is a straight line, and the amplitude and the velocity are constant in time. With reference to the Kadomtsev–Petviashvili equation, we describe in §2.1 a natural modification of our method for multidimensional problems. Note that in the multidimensional case distorted solitons arise not only due to the variable background, variable coefficients, or perturbing right-hand side as is in one-dimensional problems, but also due to a curvilinear wave front or a variable amplitude of the wave along the front. In the same section we consider the Kadomtsev–Petviashvili equation with variable coefficients.

The second basic example in Chapter 2 is the multidimensional sine-Gordon equation considered in §2.2. These two examples differ by the fact that the multidimensional sine-Gordon equation is isotropic with respect to the spatial variables, whereas the Kadomtsev–Petviashvili equation is inherently nonisotropic.

The main points of our method for constructing soliton-type asymptotics of multidimensional systems of equations are examined in §2.3 as applied to the Boussinesq system of equations. One more example of a physically meaningful problem which can be easily solved by our method is treated in §§2.5 and 2.6, where we obtain asymptotics of the smoothed shock wave type for three-dimensional equations of gas dynamics.

Finally, to illustrate the application of the method to related problems, we consider in §2.7 the construction of the boundary layer asymptotics for equations of gas dynamics and derive, as a consequence, the Prandtl equations.

In Chapters 3 and 4 the method is applied to problems of a different structure from those considered above. In Chapter 3 we consider a pseudodifferential equation and two dynamical systems which we also reduce to pseudodifferential equations. In Chapter 4 we consider systems of equations which qualitatively differ from those studied above in that the small parameter enters asymmetrically. The averaging of these systems results in problems with a free boundary.

In Chapter 5 we derive model equations describing the dynamics of the parameters of the leading term of the asymptotics in the multi-phase case. This amounts virtually to computation of the weak limit of original equations for a special class of functions.

In Chapter 6, written jointly with A. I. Shafarevich and based on his papers [**224–228**], an asymptotic solution of a stretched vortex type for stationary Navier–Stokes equations is constructed. These results elaborate the ideas presented above in two directions.

First, an asymptotic soliton-type solution to the Navier–Stokes equation is constructed. For simplified models describing surface and interior waves (the Korteweg–de Vries, Kadomtsev–Petviashvili, Whitham equations, the Boussinesq equation and system of equations, and so on) the soliton solutions have been known for a long time. Hence one could naturally expect that the fundamental equations of hydrodynamics, namely, the Navier–Stokes system, also possess solutions of this type. However, for a long time the attempts to find such solutions failed. This gave rise to the opinion that the Navier–Stokes equations could not have soliton solutions due to their parabolic structure (similarly to the Burgers equation, which is a "one-dimensional analog" of the Navier–Stokes equations). Nevertheless, detailed experimental studies on the turbulence problem revealed the existence of localized structures (vortex films, strings, and single vortexes; see, e.g., [**15, 69, 70, 177, 184**]), and it appeared natural that they can be identified with soliton solutions. The interest in verifying this conjecture has particularly grown after the 1994 discovery, by Moffatt, Kida, and Ohkitani [**177**], of a solution of a stationary solitary wave type in a certain specific setup. Namely, the leading term of this solution to the three-dimensional Navier–Stokes equations is a superposition of a linear outer flow (a smooth background) and a soliton-type function concentrated in a neighborhood of a straight line Γ_0.

Obviously, in the case of nonlinear outer flow the front becomes curvilinear. This brings up the question of stability of the solution [**177**] with respect to "geometric" perturbations. Attacks on this problem encountered fundamental rather than technical difficulties. They were overcome only recently in pioneering papers by A. I. Shafarevich [**224–228**]. Using a completely new approach to this problem, which will be discussed in more detail in Chapter 6, he established the existence of soliton-type solutions for any, however complicated, geometry of the outer flow V and the front Γ. It is important that in the simplest case $V = V_{\text{lin}}$, $\Gamma = \Gamma_0$ Shafarevich's solution turns into the solution by Moffatt, Kida, and Ohkitani.

The second completely new point is as follows. The common property of usual soliton solutions (both exact and asymptotic) is that their front Γ has codimension one. This means that for each fixed t the front of the soliton solution in the one-dimensional case is a point, in the two-dimensional case it is a curve, and so on. Clearly, for $x \in \mathbb{R}^n$ and $n = 1$ the case $\operatorname{codim}\Gamma = 1$ is exhaustive, whereas for $n \geq 2$ the cases $\operatorname{codim}\Gamma = 2, 3, \ldots, n$ are also possible. There is a well-known example of the solution with $\operatorname{codim}\Gamma = 2$. These are rational solitons (also called lamps), which were found by the method of the inverse scattering problem [**137**] for the Kadomtsev–Petviashvili equation. They have the form of a bell-shaped function (a hill) concentrated in a neighborhood of a moving point. Like all exact solutions to equations integrable by the method of the inverse scattering problem, a lamp has a constant amplitude and its trajectory is a straight line. It is clear that a perturbation will distort the trajectory of the lamp and make its amplitude variable. However, the problem of constructing such asymptotics remains open.

Another well-known example where $\operatorname{codim}\Gamma = 2$ was examined by I. A. Molotkov and S. A. Vakulenko [**179**]. Dealing with a nonlinear Helmholtz equation ($x \in \mathbb{R}^3$), they constructed a soliton-type solution concentrated near a curve under an additional assumption of a special kind of symmetry (similar to radial symmetry).

The special symmetry requirements on the solution in [**179**], as well as in [**177**], actually allow for the reduction of the problem to a single-phase one. Progress in constructing essentially two-phase soliton-type solutions with $\operatorname{codim}\Gamma = 2$ under minimal additional assumptions (of general position type) was achieved only in [**224–228**].

Finally we discuss an aspect related to the physical interpretation of the results. An equation with a small parameter in the terms containing higher order derivatives can be regarded as a model giving a "microscopic" description of a physical process. In this respect, a typical problem is to pass to the limit as $\varepsilon \to 0$ and to derive an averaged equation describing the process at the "macroscopic" level. For Burgers type equations with small viscosity this problem was solved in [**85, 86, 188–191**] and others, where it was shown that averaging in the class of discontinuous limiting functions results in *Rankine–Hugoniot conditions* on the discontinuity front. Recall that the Rankine–Hugoniot conditions relate the amplitude of the jump to the velocity of the motion of the discontinuity front. We will derive a similar result from construction of an asymptotic solution of smoothed shock wave type for the Korteweg–de Vries–Burgers equation (§1.2) as well as for three-dimensional equations of gas dynamics (§§2.5, 2.6, 5.4). For equations with small dispersion, when constructing their soliton-type solutions, we will also obtain relations connecting the amplitude and the velocity of the distorted soliton. By analogy with shock waves these relations, which remain valid in the limit as $\varepsilon \to 0$, will be called *Hugoniot type conditions.*

For problems with a single spatial variable the Hugoniot type conditions have the form of a system of ordinary differential equations. In particular, for the simplest version of the problem on the motion of a distorted solitary wave in a basin of variable depth we obtain the following Hugoniot type conditions (see §1.3):

$$\frac{d\varphi}{dt} = \sqrt{gH(\varphi)} + \frac{1}{2}A\sqrt{\frac{g}{H(\varphi)}}, \tag{0.10}$$

$$\zeta\left(1 + \frac{53}{60}\zeta\right)^{-23/53} = E(H/H_0)^{-2}, \tag{0.11}$$

where $H = H(x) > 0$ is the nonperturbed depth, $A = A(\varphi) > 0$ is the amplitude of the distorted soliton, $x = \varphi(t)$ is its position at time t, $\zeta = A(\varphi)/H(\varphi)$, $H_0 = H(\varphi(0))$, the constant E is determined by $A(\varphi(0))$ and H_0, and g is the acceleration of gravity.

For problems with multidimensional spatial variables the Hugoniot type conditions have the form of partial differential equations considered on the wave front. For some physical models (see §4.1 and [**198**]) the Hugoniot type conditions take the form of the so-called "mean curvature problem"" i.e., the problem of finding a surface Γ of codimension one whose motion is described by the equation

$$V_\nu = \kappa\mathcal{K} + f. \tag{0.12}$$

Here V_ν is the normal velocity of the motion of the surface Γ, $\mathcal{K}$ is its mean curvature, f is the external force, and κ is a constant. The mean curvature problem is widely known and has been thoroughly investigated by geometric methods (see, e.g., [**24, 41, 61–64**]). At the same time a number of problems describing different physical phenomena, such as waves in a two-dimensional basin (§2.3), waves in multidimensional Josephson junctions (§2.2), and waves in a plasma (§5.1), give rise to Hugoniot type conditions of the form

$$V_\nu = \kappa_1\mathcal{K} + \kappa_2 a + f, \tag{0.13}$$

where a is the acceleration of the surface Γ, while κ_1 and κ_2 are functions. The equation (0.13) is a natural extension of (0.12), and apparently it plays no less important role than the latter. Unfortunately we are unaware of any papers where equations of the form (0.13) are discussed.

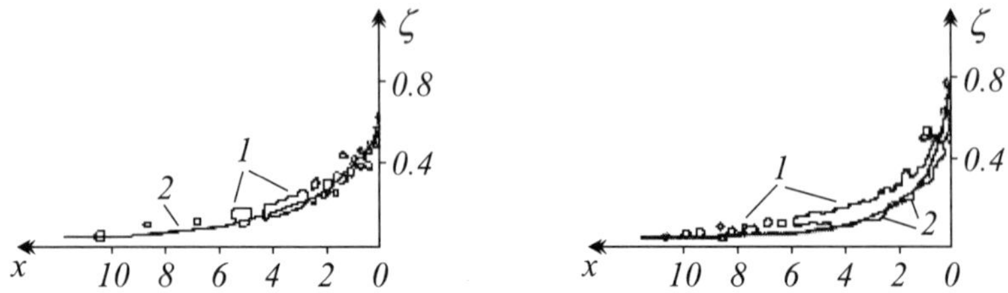

FIGURE 0.3. Comparison of the theory with experimental data. $\zeta = A(x)/H(x)$, $A(x)$ is the amplitude of the soliton type wave at the point x, $H(x)$ is the depth, x is the distance to the break-down point, 1 are experimental data, and 2 is the theoretical curve.

To conclude, we compare the amplitude of the distorted solitary wave in a basin of variable depth as computed by the Hugoniot type conditions (0.10), (0.11) with the measurements in actual experiments [**38**].

As is seen from these graphs, the theoretical curve fits well to the experimental data.

The authors express their sincere gratitude to A. I. Shafarevich, who read the manuscript and made a number of valuable comments. The authors are also thankful to K. Omel'yanov, and to D. Kulagin, N. Ordinartseva, A. Titov, and V. Trushkov, students at the Moscow State Institute of Electronics and Mathematics, who carried out all numerical computations. This book was written under partial support from the Russian Foundation for Basic Research, Grant No. 99–01–01074, and the INTAS Foundation, Grant No. 94–2763.

CHAPTER 1

Waves in One-dimensional Nonlinear Media

1.1. Dynamics of a distorted solitary wave in media with small dispersion: Korteweg–de Vries equation

1.1.1. Construction of the asymptotic solution. In this section we describe in detail the method of constructing asymptotic soliton-type solutions to any degree of accuracy. We present the method as applied to the simplest model equation

$$\frac{\partial u}{\partial t} + 3u\frac{\partial u}{\partial x} + \varepsilon^2 \frac{\partial^3 u}{\partial x^3} = 0, \qquad x \in \mathbb{R}^1, \tag{1.1}$$

where $\varepsilon \to 0$ is a small parameter characterizing the magnitude of dispersion. The standard form of the Korteweg–de Vries equation (with $\varepsilon = 1$) can be obtained from (1.1) by the substitution $t = t'\varepsilon$, $x = x'\varepsilon$, so that the equation (1.1) can be viewed as an equation describing the behavior of "large-scale" waves ($x' \sim 1/\varepsilon$) for "large" time $t' \sim 1/\varepsilon$.

It is well known that the Korteweg–de Vries equation, derived in 1895, plays a fundamental role in the theory of waves in dispersive media. This equation arises in the context of description of long waves on the surface of a fluid, and of waves in a plasma and in solid media, as well as in many other physical problems (see, e.g., [**14, 26, 94, 97, 123, 127, 148, 172, 174, 213, 221, 262**]). This equation is renowned also because in the course of its investigation Gardner, Green, Kruskal, and Miura [**72**] discovered the method of the inverse scattering problem. The fundamental character of the Korteweg–de Vries equation and the remarkable properties it possesses motivated extensive study of this equation as well as of its closely related analogs. In particular, the equation (1.1) was used by many authors as the basic model example in elaborating the perturbation theory for solitons (see, e.g., [**2, 76, 92, 108, 111, 150, 178, 202**]). Following this tradition and aiming at simplification of exposition, we introduce our proposed method of obtaining asymptotics with localized fast variation by the simplest example, viz., by constructing the distorted solitary wave for the Korteweg–de Vries equation (1.1).

To simplify the presentation, it will be expedient to introduce notation for the classes of functions to be used in the sequel.

We will denote by $\mathcal{S}$ the set of infinitely differentiable functions $f(\tau, x, t)$ such that f itself and each of its derivatives tends to zero as $\tau \to \pm\infty$ uniformly in x, t faster than any positive power of $1/|\tau|$. For an example of such a function one can take $f = A(x,t)/(\cosh\tau)^k$, $k > 0$, with any bounded smooth $A(x,t)$, as well as any function $f = f(\tau)$ from the Schwartz space.

Denote by $\mathcal{H}$ the set of infinitely differentiable functions $g(\tau, x, t)$ such that $\partial g(\tau, x, t)/\partial\tau \in \mathcal{S}$ for any t. It is clear that any function $g \in \mathcal{H}$ converges as

$\tau \to \pm\infty$ at an exponential rate to "constants" (in τ) to be denoted by $g^{\pm}(x,t)$, and in general $g^{+} \neq g^{-}$. As an example of a function from $\mathcal{H}$ one can take

$$g(\tau,x,t) = \frac{1}{2}(g^{+}(x,t) + g^{-}(x,t)) + \frac{1}{2}(g^{+}(x,t) - g^{-}(x,t)) \tanh \tau.$$

Henceforth we mean by τ the "fast" variable $S(x,t)/\varepsilon$ with a smooth phase S. Clearly, the behavior of $f(\tau,x,t)$ for $\tau \to +\infty$ ($\tau \to -\infty$) corresponds to the behavior of the function $f(S(x,t)/\varepsilon, x, t)$ as $\varepsilon \to 0$ and $S \geq c\varepsilon^{1-\delta}$ with some $c > 0$, $\delta \in (0,1)$ ($S \leq -c\varepsilon^{1-\delta}$), i.e., outside a small neighborhood of the set of zeroes of the phase S.

It is not hard to prove that for any functions $f(\tau,x,t) \in \mathcal{S}$, $g(\tau,x,t) \in \mathcal{H}$, and any smooth function $S(x,t)$ such that $\partial S/\partial x|_{\Gamma} \neq 0$, where $\Gamma = \{(x,t),\ S(x,t) = 0\}$, the following relations hold:

$$(1.2)\qquad \begin{aligned} f\Big(\frac{S}{\varepsilon},x,t\Big) &= f_1\Big(\frac{x-\varphi(t)}{\varepsilon},t\Big) + \varepsilon\tilde{f}\Big(\frac{x-\varphi(t)}{\varepsilon},t,\varepsilon\Big),\\ g\Big(\frac{S}{\varepsilon},x,t\Big) &= g_1\Big(\frac{x-\varphi(t)}{\varepsilon},x,t\Big) + \varepsilon\tilde{g}\Big(\frac{x-\varphi(t)}{\varepsilon},t,\varepsilon\Big), \end{aligned}$$

where $x = \varphi(t)$ is the equation of the zero level line of the phase $S(x,t)$,

$$\begin{aligned} f_1\Big(\frac{x-\varphi}{\varepsilon},t\Big) &= f\bigg(\frac{\partial S}{\partial x}\bigg|_{x=\varphi}\frac{x-\varphi}{\varepsilon},\varphi,t\bigg),\\ g_1\Big(\frac{x-\varphi}{\varepsilon},x,t\Big) &= g\bigg(\frac{\partial S}{\partial x}\bigg|_{x=\varphi}\frac{x-\varphi}{\varepsilon},x,t\bigg), \end{aligned}$$

and $\tilde{f}(\tau,t,\varepsilon)$, $\tilde{g}(\tau,t,\varepsilon)$ are functions from $\mathcal{S}$ which regularly depend on ε. The proof of these properties, though rather obvious, can be found in [**150**].

Now we proceed to the construction of an asymptotic soliton-type solution of equation (1.1) with accuracy $O(\varepsilon^{N+1})$. Due to (1.2), without loss of generality this solution can be written as

$$(1.3)\qquad u(x,t,\varepsilon) = u_0(x,t) + V_0\Big(\frac{S}{\varepsilon},t\Big) + \sum_{j=1}^{N+1} \varepsilon^j\Big(u_j(x,t) + V_j\Big(\frac{S}{\varepsilon},x,t\Big)\Big),$$

where $S = x - \varphi(t)$, $u_i(x,t) \in C^{\infty}$, $V_0(\tau,t) \in \mathcal{S}$, $V_j(\tau,x,t) \in \mathcal{H}$ are functions to be determined.

The regular expansion

$$u_{\mathrm{reg}}(x,t,\varepsilon) = \sum_{i=0}^{N+1} \varepsilon^i u_i(x,t)$$

describes the "smooth background" over which the wave

$$u_{\mathrm{sing}}(x,t,\varepsilon) = \sum_{i=0}^{N+1} \varepsilon^i V_i\Big(\frac{S}{\varepsilon},x,t\Big)$$

with localized fast variation propagates.

The assumption $V_0 \in \mathcal{S}$ singles out soliton-type solutions from the class of solutions with localized fast variation. In general, higher order corrections to V_0 do not vanish outside a small neighborhood of the front; hence we assume only that $V_j \in \mathcal{H}$ for $j \geq 1$. Moreover, outside a small neighborhood of the wave front (i.e., for $\tau \to \pm\infty$) we have $u_j + V_j \sim u_j(x,t) + V_j^{\pm}(x,t)$. Since all functions in (1.3) are arbitrary, we can redefine u_j (e.g., by letting $u_j := u_j + V_j^+$ or $u_j := u_j + V_j^-$), so that we can set one of the limiting values V_j^+ or V_j^- equal to zero. We will assume, for definiteness, that

$$V_j^+ = 0, \quad j = 1, \dots, N+1. \tag{1.4}$$

Note that the limiting values of the expansion (1.3) outside a small neighborhood of the wave front,

$$u = u_0(x,t) + \sum_{j=1}^{N+1} (u_j(x,t) + V_j^-(x,t)) \qquad \text{for } x \leq \varphi(t) - c\varepsilon^{1-\delta_1},$$

$$u = \sum_{j=0}^{N+1} \varepsilon^j u_j(x,t) \qquad \text{for } x \geq \varphi(t) + c\varepsilon^{1-\delta_1}, \quad \delta_1 \in (0,1),$$

can be regarded as the "outer" expansion in the matching method and the expansion (1.3) for $|x - \varphi(t)| \leq c\varepsilon^{1-\delta_2}$, $1 > \delta_2 > \delta_1$, as the "inner" expansion. In contrast to the matching method, our representation is valid uniformly in both "outer" and "inner" domains.

Now we turn to determination of the coefficients in expansion (1.3). For that we substitute (1.3) into (1.1) and collect the coefficients of equal powers of ε. Taking into account the identities

$$\begin{aligned}
\frac{\partial}{\partial x} V\Big(\frac{x-\varphi}{\varepsilon}, x, t\Big) &= \Big(\frac{1}{\varepsilon}\frac{\partial}{\partial \tau} + \frac{\partial}{\partial x}\Big) V(\tau, x, t)\Big|_{\tau=(x-\varphi)/\varepsilon}, \\
\frac{\partial}{\partial t} V\Big(\frac{x-\varphi}{\varepsilon}, x, t\Big) &= \Big(-\frac{1}{\varepsilon}\varphi_t\frac{\partial}{\partial \tau} + \frac{\partial}{\partial t}\Big) V(\tau, x, t)\Big|_{\tau=(x-\varphi)/\varepsilon}, \\
\varepsilon^2 \frac{\partial^3}{\partial x^3} V\Big(\frac{x-\varphi}{\varepsilon}, x, t\Big) &= \Big(\frac{1}{\varepsilon}\frac{\partial^3}{\partial \tau^3} + 3\frac{\partial^3}{\partial \tau^2 \partial x} + 3\varepsilon\frac{\partial^3}{\partial \tau \partial x^2} + \varepsilon^2\frac{\partial^3}{\partial x^3}\Big) \\
&\quad \times V(\tau, x, t)\Big|_{\tau=(x-\varphi)/\varepsilon},
\end{aligned}$$

we obtain the relation

$$\begin{aligned}
&\bigg[\frac{1}{\varepsilon}\Big\{-\varphi_t V_{0\tau} + 3(u_0(x,t) + V_0)V_{0\tau} + V_{0\tau\tau\tau}\Big\} \\
&\quad + \Big\{-\varphi_t V_{1\tau} + 3\frac{\partial}{\partial \tau}((u_0(x,t) + V_0)V_1) + V_{1\tau\tau\tau} + f_1\Big\} \\
&\quad + \sum_{j=2}^{N+1} \varepsilon^{j-1}\Big\{\varphi_t V_{j\tau} + 3\frac{\partial}{\partial \tau}((u_0(x,t) + V_0)V_j) + V_{j\tau\tau\tau} + f_j\Big\}\bigg]\bigg|_{\tau=(x-\varphi)/\varepsilon} \\
&= O(\varepsilon^{N+1}).
\end{aligned} \tag{1.5}$$

Here $f_j = f_j(u_0, V_0, \dots, V_{j-1}, u_j)$; in particular,

$$f_1 = V_{0t} + 3(u_{0x}V_0 + u_1 V_{0\tau}) + u_{0t} + 3u_0 u_{0x},$$

$$f_2 = V_{1t} + 3\Big(\frac{\partial}{\partial x}((u_0 + V_0)V_1) + u_{1x}V_0 + (u_1 + V_1)V_{1\tau} + u_2 V_{0\tau}\Big) \tag{1.6}$$

$$+ 3V_{1\tau\tau x} + u_{1t} + 3\frac{\partial}{\partial x}(u_0 u_1);$$

moreover, $V_\tau = \partial V(\tau, x, t)/\partial\tau$, $V_x = \partial V(\tau, x, t)/\partial x$, $V_t = \partial V(\tau, x, t)/\partial t$, and the differentiation is conducted as if τ, x, and t were independent variables.

Consider the first expression in braces in (1.5). Since $V_{0\tau}(\tau, t)$ decreases exponentially fast as $\tau \to \pm\infty$, the jth term in the expansion

$$u_0(x,t)V_{0\tau}\Big(\frac{x-\varphi}{\varepsilon}, t\Big) = u_0(\varphi, t)V_{0\tau}\Big(\frac{x-\varphi}{\varepsilon}, t\Big) + (x-\varphi)u_{0x}(\varphi, t)V_{0\tau}\Big(\frac{x-\varphi}{\varepsilon}, t\Big) + \dots + \frac{1}{j!}(x-\varphi)^j \frac{\partial^j u_0}{\partial x^j}\Big|_{x=\varphi} V_{0\tau}\Big(\frac{x-\varphi}{\varepsilon}, t\Big) + \dots$$

has magnitude $O(\varepsilon^j)$, $j = 0, 1, \dots$. Therefore, using the equality $x - \varphi = \varepsilon\tau$, the last relation can be rewritten as

$$u_0(x,t)V_{0\tau}\Big(\frac{x-\varphi}{\varepsilon}, t\Big) = \Big\{\breve{u}_0 V_{0\tau}(\tau, t) + \varepsilon\tau\breve{u}_{0x}V_{0\tau}(\tau, t) + \dots + \frac{1}{N!}\varepsilon^N\tau^N\Big(\frac{\partial^N u_0}{\partial x^N}\Big|_{x=\varphi}\Big)V_{0\tau}(\tau, t)\Big\}\Big|_{\tau=(x-\varphi)/\varepsilon} + O(\varepsilon^{N+1}).$$

Here and later we denote

$$\breve{g} = g(\tau, x, t)\big|_{x=\varphi(t)}. \tag{1.7}$$

In the above formula we used the fact that for any function $f \in \mathcal{S}$ and any $k \geq 0$, a bound $|\tau^k f(\tau, x, t)| \leq c_k(x, t)$ holds for all $\tau \in \mathbb{R}^1$, where c_k is a smooth uniformly bounded function.

After a similar transformation of $u_0(x,t)V_{j_\tau}$, $j = 1, 2, \dots, N+1$, the relation (1.5) can be rewritten as

$$\begin{aligned}\Bigg[&\frac{1}{\varepsilon}\{-\varphi_t V_{0\tau} + 3(\breve{u}_0 + V_0)V_{0\tau} + V_{0\tau\tau\tau}\} \\ &+ \Big\{-\varphi_t V_{1\tau} + 3\frac{\partial}{\partial\tau}((\breve{u}_0 + V_0)V_1) + V_{1\tau\tau\tau} + f_1 + 3\tau\breve{u}_{0x}V_{0\tau}\Big\} \\ &+ \sum_{j=2}^{N+1}\varepsilon^{j-1}\Big\{-\varphi_t V_{j_\tau} + 3\frac{\partial}{\partial\tau}((\breve{u}_0 + V_0)V_j) + V_{j_{\tau\tau\tau}} \\ &+ 3\sum_{i=1}^{j}\frac{1}{i!}\tau^i\Big(\frac{\partial^i u_0}{\partial x^i}\Big|_{x=\varphi}\Big)\frac{\partial}{\partial\tau}V_{j-i} + f_j\Big\}\Bigg]\Bigg|_{\tau=(x-\varphi)/\varepsilon} = O(\varepsilon^{N+1}).\end{aligned} \tag{1.8}$$

It is clear that for this relation to hold for all $N \geq 0$ and $\varepsilon \to 0$ the expression in the first braces must be equal to zero. This leads to the equation

$$\frac{\partial}{\partial \tau}\left\{\frac{\partial^2 V_0}{\partial \tau^2} + \left(3\check{u}_0 + \frac{3}{2}V_0 - \varphi_t\right)V_0\right\} = 0.$$

On integrating with respect to τ we obtain

$$\frac{\partial^2 V_0}{\partial \tau^2} + \left(3\check{u}_0 + \frac{3}{2}V_0 - \varphi_t\right)V_0 = 0, \tag{1.9}$$

since the "constant" of integration equals zero by virtue of the condition $V_0 \in \mathcal{S}$. Now we solve the ordinary differential equation (1.9) to obtain

$$V_0(\tau, t) = A\cosh^{-2}(\beta(\tau + \varphi_1)) \in \mathcal{S},$$

where $\beta = \sqrt{A}/2$, $\varphi_1 = \varphi_1(t)$ is the "constant" of integration, and the amplitude $A = A(t) > 0$ is related to the velocity of the wave φ_t and the background on the front by the equation

$$\frac{d\varphi}{dt} = A + 3u_0(\varphi, t). \tag{1.10}$$

Now we return to (1.8). Since V_0 and $V_{1\tau}$ decrease exponentially fast outside a small neighborhood of the front, for (1.8) to hold with accuracy $O(\varepsilon)$ uniformly in τ, x, t it is necessary that $f_j \in \mathcal{S}$. This condition and the explicit form (1.6) of f_1 imply the equation

$$u_{0t} + 3u_0u_{0x} = 0. \tag{1.11}$$

Recall that the solution of a first order equation can be found by the method of characteristics (see, e.g., [**44, 219**]). For the Hopf equation (1.11) completed with the initial condition

$$u_0\big|_{t=0} = u^0(x), \tag{1.12}$$

this method leads to the characteristic equations

$$\frac{dX}{dt} = 3U_0, \quad \frac{dU_0}{dt} = 0, \qquad X\big|_{t=0} = x^0, \quad U_0\big|_{t=0} = u^0(x^0).$$

Hence $U_0 = u^0(x^0)$ is the value of the function u_0 on the characteristic $X(t, x^0) = 3u^0(x^0)t + x^0$, where $x^0 \in \mathbb{R}^1$ is a parameter. Solving the equation

$$x = X(t, x^0) \tag{1.13}$$

for x^0, we obtain the solution of the Cauchy problem (1.11), (1.12):

$$u_0 = u^0(x^0(x, t)), \tag{1.14}$$

where $x^0(x,t)$ is the solution of (1.13). Clearly the solution (1.14) exists only if the Jacobian does not vanish,

$$J(x^0,t) = 3u_x^0(x^0)t + 1 > 0. \tag{1.15}$$

We will assume that the Hopf equation (1.11) has a smooth solution for $x \in \mathbb{R}^1$ and $t \in [0,T]$.

Now all the terms in the second braces in (1.8) belong to $\mathcal{S}$, and we can apply to them the Taylor expansion around $x = \varphi(t)$. Using the equality $x - \varphi = \varepsilon\tau$ again and taking (1.9) into account, we transform (1.8) to the following form:

$$\begin{aligned}\Bigg[&\Big\{ - A\check{V}_{1\tau} + 3\frac{\partial}{\partial\tau}(V_0\check{V}_1) + \check{V}_{1\tau\tau\tau} + \check{f}_1 + 3\tau\check{u}_{0x}V_{0\tau}\Big\}\\ &+ \sum_{j=2}^{N+1}\varepsilon^{j-1}\Big\{ - AV_{j\tau} + 3\frac{\partial}{\partial\tau}(V_0V_j) + V_{j\tau\tau\tau} + f_j\\ &+ 3\sum_{i=1}^{j}\frac{1}{i!}\tau^i\frac{\partial^i u_0}{\partial x^i}\bigg|_{x=\varphi}\frac{\partial}{\partial\tau}V_{j-i} + \frac{1}{(j-1)!}\tau^{j-1}\\ &\times\left(\frac{\partial}{\partial\tau}L\frac{\partial^{j-1}V_1}{\partial x^{j-1}} + \frac{\partial^{j-1}f_1}{\partial x^{j-1}}\right)\bigg|_{x=\varphi}\Big\}\Bigg]\Bigg|_{\tau=(x-\varphi)/\varepsilon} = O(\varepsilon^{N+1}).\end{aligned} \tag{1.16}$$

Here we have used (1.10), (1.7), and the notation

$$L = \frac{\partial^2}{\partial\tau^2} + 3V_0 - A$$

for the variation of the operator in (1.9).

Now we obtain from (1.16) the equation

$$\frac{\partial}{\partial\tau}L\check{V}_1 = -\check{f}_1 - 3\tau\check{u}_{0x}V_{0\tau}. \tag{1.17}$$

Taking (1.11) and the explicit form (1.6) of $\check{f}_1$ into account, we obtain by integrating (1.17)

$$L\check{V}_1 = -F_1, \tag{1.18}$$

with

$$F_1 = \frac{\partial}{\partial t}\Big(\frac{A}{\beta}\tanh\beta(\tau+\varphi_1)\Big) + 3\check{u}_1V_0 + 3\tau\check{u}_{0x}V_0 + c_1,$$

where $c_1 = c_1(t)$ is the “constant” of integration.

It is not hard to establish that $v_1 = V_{0\tau}$ is a solution of the equation $Lv = 0$, while the second solution

$$v_2 = \left\{\frac{15}{\cosh^2 x}(x\tanh x - 1) + 2\sinh^2 x + 7\right\}\bigg|_{x=\beta(\tau+\varphi_1)}$$

grows exponentially as $\tau \to \pm\infty$. Using this fact, one can easily prove (see, e.g., [**78, 150**]) that for solvability of (1.18) in the class $\mathcal{H}$, i.e., in the class of functions stabilizing at $\pm\infty$, it is necessary and sufficient that

$$\int_{-\infty}^{\infty} F_1(\tau, t) V_{0\tau}\, d\tau = 0. \tag{1.19}$$

The same fact can be established by direct integration of the ordinary equation (1.18) making use of standard formulas (see, e.g., [**81**]).

By simple calculations, (1.19) can be rewritten as

$$\frac{dA}{dt} + 2u_{0x}(\varphi, t)A = 0. \tag{1.20}$$

To transform (1.20), we will compute the total derivative of $u_0(\varphi(t), t)$. By (1.10) and (1.11) we have

$$\frac{du_0(\varphi, t)}{dt} = \left\{ \frac{\partial u_0(x,t)}{\partial t} + \frac{\partial u_0(x,t)}{\partial x}(A + 3u_0(x,t)) \right\}\bigg|_{x=\varphi} = A\check{u}_{0x}.$$

Therefore the equation (1.20) can be easily integrated to yield

$$A(t) + 2u_0(\varphi, t) = E, \tag{1.21}$$

where E is a constant determined by the initial data.

Thus we have completed the construction of the leading term of the asymptotic formula (1.3) (up to a shift φ_1 of the "fast" variable τ). Namely, we obtained the Hopf equation (1.11) for the smooth background u_0 and the equations (1.10) and (1.21) for the amplitude A and the phase φ of the distorted solitary wave.

Now we return to equation (1.18), assuming condition (1.21). It is clear that the assumptions $V_1 \in \mathcal{H}$, $V_1^+ = 0$ imply $\check{V}_1 \in \mathcal{H}$, $\check{V}_1^+ = 0$. Therefore, by letting τ tend to $+\infty$, we find the "constant" $c_1 = -A_t/\sqrt{A}$. Next, letting $\tau \to -\infty$ in (1.18), we obtain the equality

$$\check{V}_1^- = -2\frac{A_t}{A^{3/2}}. \tag{1.22}$$

Now the equation (1.18) allows us to find the function V_1 on the front $x = \varphi(t)$,

$$\check{V}_1(\tau, \varphi, t) = z_1(\tau, t) - \frac{\varphi_2(t) V_{0\tau}(\tau, t)}{2A}, \tag{1.23}$$

where z_1 is a particular solution of the nonhomogeneous equation, which is given by the formula (see, e.g., [**81**])

$$z_1(\tau, t) = V_{0\tau}(\tau, t) \int_0^{\tau} V_{0\tau}^{-2}(\tau_1, t) \int_{-\infty}^{\tau_1} F_1(\tau', t) V_{0\tau}(\tau', t)\, d\tau'\, d\tau_1, \tag{1.24}$$

and φ_2 is the "constant" of integration. We introduced the normalizing coefficient $-1/2A$ in (1.23) in order to simplify the subsequent formulas.

Using the formula (1.24) and the explicit form of F_1, and taking (1.20) into account, one can write down the function $\check{V}_1$ explicitly:

$$\check{V}_1(\tau,\varphi,t) = -\frac{1}{2\beta}\frac{\partial^2}{\partial\tau^2}\{(\varphi_2+\alpha\tau+\check{u}_{0x}\tau^2)(\tanh(\beta(\tau+\varphi_1))-1)\}, \tag{1.25}$$

where $\alpha = (3\check{u}_1\beta + 3\check{u}_{0x} + (\beta\varphi_1)_t)/\beta$.

It will be convenient to rewrite $\check{V}_1$ in the form

$$\check{V}_1(\tau,\varphi,t) = \omega_1(\tau,t) + \nu_1(t)\chi(\tau,t),$$

where $\chi = \big(1-\tanh(\beta(\tau+\varphi_1))\big)/2$,

$$\omega_1 = -\frac{(\varphi_2+\alpha\tau+\check{u}_{0x}\tau^2)V_{0\tau}}{2A} - \frac{(\alpha+2\tau\check{u}_{0x})V_0}{A}, \qquad \nu_1 = \check{V}_1^-. \tag{1.26}$$

We have used here the equality $\check{u}_{0x} = -A_t/2A$.

Since $\omega_1 \in \mathcal{S}$, $\chi^+ = 0$, and $\chi^- = 1$, it is natural to seek the extension of (1.22) outside the front $x = \varphi(t)$ in the form of a smoothed shock wave

$$V_1(\tau,x,t) = \omega_1(\tau,t) + u_1^-(x,t)\chi(\tau,t), \tag{1.27}$$

where u_1^- is a smooth function to be determined, such that

$$u_1^-\big|_{x=\varphi(t)} = \nu_1(t), \quad t\in[0,T]. \tag{1.28}$$

To determine the shift $\varphi_1(t)$ and the function $u_1^-(x,t)$, consider the terms of order ε in the left-hand side of (1.16). Since V_0, $V_{1\tau}$, and $V_{2\tau}$ decrease exponentially as $\tau\to\pm\infty$, a necessary condition for (1.16) to hold within $O(\varepsilon^2)$ is that $f_2\to 0$ as $\tau\to\pm\infty$. From this condition and the explicit form of f_2 we obtain the equations

$$\frac{\partial u_1}{\partial t} + 3\frac{\partial}{\partial x}(u_0u_1) = 0, \tag{1.29}$$

$$\frac{\partial u_1^-}{\partial t} + 3\frac{\partial}{\partial x}(u_0u_1^-) = 0. \tag{1.30}$$

We consider the linearized Hopf equation (1.29) for all $x\in\mathbb{R}^1$ and $t\in(0,T]$. It is clear that the existence of a smooth solution of the Hopf equation (1.11) implies the existence of a smooth solution of (1.29) if $u_1|_{t=0}\in C^\infty$.

Next, formula (1.27) for $\tau = (x-\varphi)/\varepsilon$ shows that it suffices to determine the smooth function u_1^- only in the domain $\Omega_T^- = \{(x,t),\ x<\varphi(t),\ t\in(0,T]\}$, since in the strip $\Omega_T^\delta = \{(x,t),\ \varphi(t)<x<\varphi(t)+c\varepsilon^{1-\delta},\ t\in(0,T)\}$, $c>0$, $0<\delta<1$, for sufficiently small ε one can always define an infinitely differentiable extension of the function u_1^-, while for $x>\varphi(t)+c\varepsilon^{1-\delta}/2$ the term $u_1^-\chi$ is negligibly small. Therefore we will consider the equation (1.30) in the domain Ω_T^-, completing it with boundary condition (1.28) and initial condition

$$u_1^-\big|_{t=0} = u_{10}^-(x), \quad x\le\varphi(0). \tag{1.31}$$

This is a well-posed problem, since the velocity of the solitary wave front $d\varphi/dt$ is greater than the velocity dX/dt of the motion along the characteristics (1.13).

This problem can also be solved by the method of characteristics. In the domain $\{(t,x),\ x \leq 3u^0(\varphi(0))t + \varphi(0)\}$ that corresponds to the characteristics $x = X(t, x^0)$ originating at the points $x^0 \leq \varphi(0)$, the solution of (1.30), (1.31) has the form

$$u_1^-(x,t) = \frac{u_{1,0}^-(x^0)}{J(x^0,t)}\bigg|_{x^0=x^0(x,t)}, \tag{1.32}$$

where $x^0(x,t)$ is the solution of equation (1.13) and the Jacobian J is given by (1.15).

In the domain bounded by the curves $x = X(t, \varphi(0))$ and $x = \varphi(t)$ the solution of (1.30), (1.28) is given by the formula

$$u_1^-(x,t) = \frac{\nu_1(t^0)J(x_\varphi^0, t^0)}{J(x_\varphi^0, t-t^0)}\bigg|_{t^0=t^0(x,t)} = 4\frac{u_x^0(x_\varphi^0)}{\sqrt{A(t^0)}J(x_\varphi^0, t-t^0)}\bigg|_{t^0=t^0(x,t)}, \tag{1.33}$$

where $x_\varphi^0 = x_\varphi^0(t^0)$ is the solution to the equation

$$3u^0(x_\varphi^0)t^0 + x_\varphi^0 = \varphi(t^0)$$

and $t^0 = t^0(x,t)$ solves the equation

$$x = 3u_0(\varphi(t^0), t^0)(t - t^0) + \varphi(t^0).$$

It is not hard to see that the conditions (1.15) and $A > 0$ are sufficient for solvability of these equations.

Now to construct the extension of u_1^- into the strip Ω_T^δ we can use (1.32), selecting the points (x,t) lying in Ω_T^δ. It is easily seen that for sufficiently small ε this procedure is correct. As a result, (1.26) and (1.32), (1.33), along with equation (1.29), determine the first correction term in the asymptotic expansion of the solution (1.3) up to the "constant" of integration $\varphi_2(t)$.

Recall that the shift $\varphi_1(t)$ of the argument τ of the distorted solitary wave has not been determined yet. Now we determine it. The main question here is whether we can write down an equation for φ_1 which would not involve other unknown functions. In other words, whether the Hugoniot type conditions for solitary waves form triangular systems of equations or they form, as is the case for smoothed shock waves, infinite chains of linked equations.

To answer this question, consider the terms of order ε in the left-hand side of (1.16). Applying the Taylor expansion about the point $x = \varphi(t)$ and taking (1.29) and (1.30) into account, we obtain the equation for $\check{V}_2 = V_2(\tau, x, t)|_{x=\varphi(t)}$ similar to (1.17):

$$\frac{\partial}{\partial \tau} L\check{V}_2 = -\mathcal{F}_2(\tau, t), \tag{1.34}$$

where

$$\begin{aligned}\mathcal{F}_2 = {} & \frac{\partial}{\partial t}\check{V}_1 + 3(\check{u}_1 + \check{V}_1 + \tau\check{u}_{0x})\check{V}_{1\tau} + 3\check{u}_{0x}\check{V}_1 \\ & + 3\Big(\check{u}_2 + \tau\check{u}_{1x} + \frac{1}{2}\tau^2\check{u}_{0xx}\Big)V_{0\tau} + 3(\check{u}_{1x} + \tau\check{u}_{0xx})V_0 \\ & + 3\check{u}_{1x}^-\Big((\tau V_{0\tau} + V_0)\chi - \frac{\beta}{4A}(2V_{0\tau} + \tau V_0^2)\Big) - (\nu_{1t} + 3\nu_1\check{u}_{0x})\chi.\end{aligned}$$

In the computation of the right-hand side of (1.34) we have used (1.27), (1.29), (1.30), and the equality $\chi_\tau = -\beta V_0/2A$.

When we integrate with respect to τ, (1.34) becomes

$$L\check{V}_2 = -F_2(\tau, t), \tag{1.35}$$

where

$$F_2(\tau, t) = \int_{-\infty}^{\tau} \mathcal{F}_2(\tau', t)\, d\tau' - \int_{-\infty}^{\infty} \mathcal{F}_2(\tau', t)\, d\tau',$$

and the "constant" of integration has been chosen according to the condition $\check{V}_2 \to 0$ as $\tau \to +\infty$.

Consider the condition

$$\int_{-\infty}^{\infty} F_2(\tau, t) V_{0\tau}(\tau, t)\, d\tau = 0 \tag{1.36}$$

of solvability of (1.35) in the class $\mathcal{H}$. To simplify the subsequent calculations we will use the auxiliary identity

$$\begin{aligned} 3\int_{-\infty}^{\infty} V_0 \check{V}_1 \check{V}_{1\tau}\, d\tau = \int_{-\infty}^{\infty} \check{V}_1 V_{0t}\, d\tau + 3\check{u}_{0x} \int_{-\infty}^{\infty} (V_0 + \tau V_{0\tau}) \check{V}_1\, d\tau \\ + 3\check{u}_1 \int_{-\infty}^{\infty} V_{0\tau} \check{V}_1\, d\tau + \frac{1}{2} A \nu_1^2, \end{aligned} \tag{1.37}$$

which can be easily obtained by multiplying (1.17) by $\check{V}_1$ and integrating with respect to τ. By means of (1.37) condition (1.36) can easily be rewritten as

$$\begin{aligned} \left\{\frac{d}{dt} + 3\check{u}_{0x}\right\} \int_{-\infty}^{\infty} V_0 \check{V}_1\, d\tau + \frac{3}{2}\left(\check{u}_{1x} + \frac{1}{2}\check{u}_{1x}^{-} - \check{u}_{0xx}\varphi_1\right) \int_{-\infty}^{\infty} V_0^2\, d\tau \\ - \frac{1}{2}(\nu_{1t} + 3\check{u}_{0x}\nu_1) \int_{-\infty}^{\infty} V_0\, d\tau + \frac{1}{2} A \nu_1^2 = 0. \end{aligned} \tag{1.38}$$

The explicit formula (1.25) for $\check{V}_1$ allows us to establish that

$$\int_{-\infty}^{\infty} V_0 \check{V}_1\, d\tau = \frac{1}{2}\nu_1 \int_{-\infty}^{\infty} V_0\, d\tau - \frac{3}{4A}(\alpha - 2\check{u}_{0x}\varphi_1) \int_{-\infty}^{\infty} V_0^2\, d\tau.$$

Now, taking integrals of V_0 and V_0^2, we rewrite (1.38) as

$$\begin{aligned} \frac{d}{dt}\{\sqrt{A}(\alpha - 2\check{u}_{0x}\varphi_1)\} - 2A^{3/2}\Big\{\check{u}_{1x} - \frac{1}{2}\check{u}_{1x}^{-} - \check{u}_{0xx}\varphi_1 \\ - \frac{3}{2A}\check{u}_{0x}(\alpha - 2\check{u}_{0x}\varphi_1)\Big\} = 0, \end{aligned} \tag{1.39}$$

where α is the same as in (1.25). Further transformations use the Hugoniot condition (1.20) and the equalities

$$\frac{d}{dt}\check{u}_{0x} = A\check{u}_{0xx} - 3\check{u}_{0x}^2, \quad \frac{d}{dt}\check{u}_1 = A\check{u}_{1x} - 3\check{u}_{0x}\check{u}_1,$$

which follow in an obvious way from (1.11), (1.29), and (1.10).

Finally, taking into account the explicit form of the function α (see (1.25)), we transform (1.39) into the differential equation

$$\frac{d}{dt}\Big\{\frac{d\varphi_1}{dt} - \check{u}_{0x}\varphi_1\Big\} = G(t), \tag{1.40}$$

where

$$G(t) = -\frac{d}{dt}\Big(\check{u}_1 + \frac{5}{2}\nu_1\Big) - \frac{3}{2}\nu_1^2\sqrt{A},$$

and $\nu_1 = \check{V}_1^-$ is given by (1.22).

Thus the Hugoniot type conditions for distorted solitary waves split up: the trajectory and the amplitude of the wave are determined by equations (1.10), (1.21), and the "shift" φ_1 by equation (1.40). Moreover, we have to find the "smooth" background, i.e., the functions $u_0(x,t)$ and $u_1(x,t)$ with accuracy $O(\varepsilon^2)$. However, unlike the problem on the smoothed shock wave, the equations (1.11) and (1.29) are considered for all $x \in \mathbb{R}^1$, so that they can be solved without any difficulties.

Note that if we eliminate the amplitude from (1.10) and (1.21), the equation for the phase φ can be rewritten as

$$\frac{d}{dt}\Big(\frac{d\varphi}{dt} - u_0(\varphi, t)\Big) = 0. \tag{1.41}$$

It is seen from this equation that the operator in the left-hand side of (1.40) is the variation of the operator in (1.41).

Construction of the subsequent terms of the asymptotic solution (1.3) follows the same lines: letting τ tend to $-\infty$, we find from (1.35) the limiting value $\nu_2 = \check{V}_2^-$:

$$\nu_2 = -\frac{1}{A}\int_{-\infty}^{\infty} \mathcal{F}_2(\tau', t)\, d\tau'. \tag{1.42}$$

Next, using a formula similar to (1.24), we calculate the solution of (1.35) up to the "constant" of integration φ_3, i.e.,

$$\check{V}_2(\tau, t) = \nu_2(t)\chi(\tau, t) + \omega_2(\tau, t),$$

where $\omega_2 = \widetilde{\omega}_2 + \varphi_3(t)V_{0\tau}$.

After we determine the extension of V_2 into the domain Ω_T^- in the form similar to (1.27),

$$V_2(\tau, x, t) = u_2^-(x,t)\chi(\tau, t) + \omega_2(\tau, t),$$

and consider the terms of order ε^2 in (1.16), we obtain the equations

$$u_{2t} + 3(u_0u_2)_x = -3u_1u_{1x} - u_{0xxx}, \quad x \in \mathbb{R}^1,\ t > 0, \tag{1.43}$$

$$u_{2t}^- + 3(u_0u_2^-)_x = -3u_1^-u_{1x}^-, \quad x < \varphi(t),\ t > 0. \tag{1.44}$$

The equation (1.43) must be completed with an initial condition for $t = 0$, $x \in \mathbb{R}^1$, and the equation (1.44) by an initial condition for $t = 0$, $x \leq \varphi(0)$ and the boundary condition

$$u_2\big|_{x=\varphi(t)} = \nu_2(t), \quad t > 0.$$

Finally, the condition

$$\int_{-\infty}^{\infty} F_3(\tau,t)V_{0\tau}\,d\tau = 0$$

of solvability in $\mathcal{H}$ of the equation

$$L\check{V}_3 = -F_3(\tau,t)$$

leads to an equation of the form (1.40) for the "constant" of integration $\varphi_2(t)$ involved in V_1.

1.1.2. The choice of the Cauchy data. We have constructed the so-called self-similar asymptotic solution for the Korteweg–de Vries equation. The self-similar solutions are characterized by the property that they preserve their shape during the period of their existence. There are many equations which have such solutions. For example, a solution of the linear equation

$$u_t + au_x = 0, \quad u\big|_{t=0} = u^0(x)$$

is self-similar,

$$u = u^0(x - at),$$

for any initial data $u^0 \in C^1$. However, in general the self-similarity condition imposes restrictions on the choice of initial data. For example, a solution of the wave equation

$$u_{tt} - a^2u_{xx} = 0, \qquad u\big|_{t=0} = u^0(x),\ u_t\big|_{t=0} = u^1(x),$$

will be self-similar only when $u^1 = \pm au^0_x$ (although the function $u^0 \in C^2$ may still be arbitrary). It is clear that for nonlinear equations these restrictions will be very strong. For example, the Cauchy problem for the Korteweg–de Vries equation has a self-similar solution only when the initial data is a particular exact solution (either an almost periodic solution or a soliton).

Asymptotic self-similarity of a solution to an equation with small parameter means self-similarity only with respect to the fast variable. Thus the class of initial data for which such an asymptotic solution exists is somewhat wider than for strictly self-similar solutions. We will describe the initial data which correspond to a self-similar soliton-type asymptotic solution. It follows from the above construction that the regular part of the solution $u_{\text{reg}} = \sum_{i=0}^{N} \varepsilon^i u_i(x,t)$ for $t = 0$ may be an arbitrary sufficiently smooth function. Moreover, it is clear that the "tail" $u^- = \sum_{i=1}^{N} \varepsilon^i u_i^-(x,t)$ which follows the distorted soliton can also be specified arbitrarily for $t = 0$ and $x < \varphi^0(0)$. However, at the point $x = \varphi(0)$ the initial and boundary (on the curve $x = \varphi(t)$) data must agree with each other. In particular, the equality

$$u_1^-\big|_{t=0,\,x=\varphi(0)} = \frac{4u_{0x}}{\sqrt{A}}\bigg|_{t=0,\,x=\varphi(0)} \tag{1.45}$$

must necessarily hold. Further, it is obvious that the position of the center $\varphi^0 = \varphi(0)$ of the initial soliton $V_0((x-\varphi^0)/\varepsilon + \varphi_1^0, 0)$, its amplitude $A^0 = A(0) > 0$, as well as the initial shift $\varphi_1^0 = \varphi_1(0)$ may be arbitrary. However the form of the solitary wave $V_0(\eta,0)$, i.e., its dependence on $\eta = (x-\varphi^0)/\varepsilon + \varphi_1^0$, is fixed, $V_0(\eta,0) = A^0/\cosh^2(\sqrt{A^0}\eta/2)$.

Consider the asymptotic solution of accuracy $O(\varepsilon^2)$. By (1.26), (1.27) we have for $t=0$

$$\tag{1.46} \begin{aligned} u\big|_{t=0} = u_0^0(x) + V_0(\eta,0) + \varepsilon\Big\{&u_1^0(x) + u_{1,0}^-(x)\chi(\eta,0) \\ &- \frac{\alpha^0}{2A^0}[\tau V_{0\tau}(\eta,0) + 2V_0(\eta,0)] \\ &- u_{0x}^0(x)\frac{\tau}{2A^0}[\tau V_{0\tau}(\eta,0) + 4V_0(\eta,0)] \\ &- \frac{\varphi_2(0)}{2A^0}V_{0\tau}(\eta,0)\Big\}\Big|_{\eta=\tau+\varphi_1^0} + O(\varepsilon^2), \end{aligned}$$

where $\tau = (x-\varphi^0)/\varepsilon$,

$$\alpha^0 = \varphi_{1t}\big|_{t=0} + 3u_1^0(\varphi^0) + u_{0x}^0(\varphi^0)(6-\sqrt{A^0}\varphi_1^0)/\sqrt{A^0},$$

and, as was pointed out above, u_0^0, u_1^0, and $u_{1,0}^-$ are arbitrary smooth functions satisfying (1.45).

Recall that we have obtained an equation of second order for the function $\varphi_1(t)$, while the specification of $u|_{t=0}$ up to $O(\varepsilon)$ determines a single initial condition $\varphi_1|_{t=0} = \varphi_1^0$. Therefore, since $\varphi_{1t}|_{t=0}$ is arbitrary, α^0 in (1.46) can also be chosen arbitrarily. Hence the choice of the free parameters α^0 and $\varphi_2(0)$ effects to some extent the profile of the fast varying correction of order ε. It is clear that a similar property holds for higher order terms of the asymptotic expansion as well; namely, the choice of $\varphi_j(0)$ and $d\varphi_{j-1}/dt|_{t=0}$ is determined by the profile of the fast varying correction of order ε^j.

Consider a simple example. Let

$$u\big|_{t=0} = u_0^0 + V_0(\tau,0) + \varepsilon u_1^0(x) + O(\varepsilon^2),$$

where $u_0^0 = \text{const}$, $\tau = (x-\varphi^0)/\varepsilon$. Then, by (1.46), $\varphi_1(0) = 0$, $\varphi_{1t}|_{t=0} = -3u_1^0(\varphi^0)$, and $\varphi_2(0) = 0$. Next, from the corresponding equations we find that

$$\begin{gathered} u_0 \equiv u_0^0, \qquad u_1^- \equiv 0, \qquad A \equiv A^0, \\ \varphi = (A^0 + 3u_0^0)t + \varphi^0, \qquad u_1 = u_1^0(x - 3u_0^0 t), \\ \varphi_1 = -\int_0^t u_1^0(A^0 t' + \varphi^0)\,dt' - 2u_1^0(\varphi^0)t. \end{gathered}$$

Therefore

$$\tag{1.47} \begin{aligned} u = u_0^0 + V_0(\eta) + \varepsilon\Big\{&u_1^0(x - 3u_0^0 t) \\ &- \frac{\alpha(t)}{2A^0}(\tau V_{0\tau}(\eta) + 2V_0(\eta)) - \frac{\varphi_2(t)}{2A^0}V_{0\tau}(\eta)\Big\}\Big|_{\eta=\tau+\varphi_1(t)} + O(\varepsilon^2), \end{aligned}$$

where

$$V_0(\eta) = \frac{A^0}{\cosh^2(\sqrt{A^0}\eta/2)}, \quad \tau = \frac{x - (A^0 + 3u_0^0)t - \varphi^0}{\varepsilon}, \quad \alpha = 2\big(u_1^0(A^0 t + \varphi^0) - u_1^0(\varphi^0)\big).$$

Formula (1.47) also illustrates the following property of a soliton-type solution of the Korteweg–de Vries equation: small (of order $O(\varepsilon)$) smooth perturbations of the initial data affect the leading term of the solution, since the phase shift φ_1 depends on u_1^0. However this is a weak instability, since for finite t the distance between the curves $x = \varphi(t)$ and $x = \varphi(t) - \varepsilon\varphi_1(t)$ is of order $O(\varepsilon)$.

Note also that in this example the "tail" appears only in the correction terms of order ε^2, because in this case $\nu_1 \equiv 0$, but, as one can easily derive by formula (1.42), $\nu_2 = 4u_{1x}^0(A^0 t + \varphi^0)/\sqrt{A^0}$, so that $u_2^- \not\equiv 0$.

Let us emphasize once more that the partial arbitrariness in the specification of the initial profile of the fast varying correction discussed above is unique: if for $t = 0$ the fast varying correction of order ε^j, $j \geq 1$, does not have a special structure, then the correction terms of order ε^j in the solution cannot be found in the form of a one-phase expansion (1.3). This question has been studied in detail by Kalyakin and Il′in [**88, 98–105**].

1.1.3. Comparison with the standard two-scales method. Let us try to apply the standard two scales method to construction of the asymptotic soliton-type solution. According to the usual scheme [**133, 162**] we will seek the solution in the form of expansion

$$u = \sum_{j=0}^{N} \varepsilon^j \left\{ u_j(x,t) + U_i\left(\frac{S}{\varepsilon}, x, t\right) \right\} + O(\varepsilon^{N+1}), \tag{1.48}$$

where S, u_j are smooth functions, $U_0(\tau, x, t) \in \mathcal{S}$, $U_i(\tau, x, t) \in \mathcal{H}$, $i = 1, \dots, N$. Substituting (1.48) into equation (1.1) and equating to zero the coefficients of ε^{-1} and ε^0, we obtain the equation (1.11) as well as the equalities

$$S_t \frac{\partial U_0}{\partial \tau} + 3S_x(u_0 + U_0)\frac{\partial U_0}{\partial \tau} + S_x^3 \frac{\partial^3 U_0}{\partial \tau^3} = 0, \tag{1.49}$$

$$\frac{\partial}{\partial \tau} \mathcal{L} U_1 = -\mathcal{F}_1(\tau, x, t), \tag{1.50}$$

where

$$\begin{aligned} \mathcal{L} &= S_x^3 \frac{\partial^2}{\partial \tau^2} + 3S_x(u_0 + U_0) + S_t, \\ \mathcal{F}_1 &= \frac{\partial U_0}{\partial t} + 3\frac{\partial}{\partial x}\left(\left(u_0 + \frac{1}{2}U_0\right)U_0\right) + 3S_x u_1 \frac{\partial U_0}{\partial \tau} \\ &\quad + 3S_x^2 \frac{\partial^3 U_0}{\partial \tau^2 \partial x} + 3S_x S_{xx} \frac{\partial^2 U_0}{\partial \tau^2}. \end{aligned}$$

It is not difficult to establish that (1.49) has a solution in the class of exponentially decreasing functions only if $S_x \neq 0$. For definiteness, let $S_x > 0$. Then by simple calculations we obtain from (1.49) the explicit formula

$$U_0 = \frac{A}{\cosh^2(\beta(\tau + S_1))},$$

with $\beta = \sqrt{A}/2S_x$ and $S_1 = S_1(x,t)$ being the "constant" of integration, as well as the equation

$$S_t + (3u_0(x,t) + A)S_x = 0. \tag{1.51}$$

Next, integrating (1.50) and choosing the "constant" of integration from the condition $U_1 \to 0$ as $\tau \to \infty$, we obtain

$$\mathcal{L}U_1 = -F_1(\tau, x, t), \tag{1.52}$$

where

$$\begin{aligned} F_1 = & -4\{(S_x\sqrt{A})_t + 3(S_x\sqrt{A}u_0)_x + (S_xA^{3/2})_x(3-2\chi)\chi\}\chi \\ & + \beta^{-1}U_0\{[\beta_t + 3(u_0 + \tfrac{1}{2}U_0)\beta_x]\tau + (\beta S_1)_t + 3(u_0 + \tfrac{1}{2}U_0)(\beta S_1)_x\} \\ & + 3S_xu_1U_0 + 3S_x^2\frac{\partial^2 U_0}{\partial\tau\partial x} + 3S_xS_{xx}\frac{\partial U_0}{\partial\tau}, \\ \chi = & \frac{1}{2}(1 - \tanh\beta(\tau + S_1)). \end{aligned}$$

The condition for solvability of (1.52) in the class $\mathcal{H}$ similar to (1.19) yields the second Hugoniot type condition:

$$A_t + (3u_0(x,t) + A)A_x + 2Au_{0x} = 0. \tag{1.53}$$

Finally, after some algebra we find from (1.52) that

$$U_1 = U_1^-(x,t)\chi + \omega_1(\tau, x, t), \tag{1.54}$$

where $U_1^- = 4u_{0x}/\sqrt{A}$ and ω_1 is a function exponentially decreasing as $\tau \to \pm\infty$.

Consider the Hugoniot type conditions (1.51), (1.53). In order for this system of equations to have a unique solution we need to specify initial data for any $x \in \mathbb{R}^1$. Let

$$S\big|_{t=0} = S^0(x), \quad A\big|_{t=0} = A^0(x),$$

where S^0 and A^0 are smooth functions such that $S^0(\varphi^0) = 0$ and $S_x^0 > 0$, $A^0 > 0$ in a neighborhood of the point $x = \varphi^0$. The characteristic system for the problem (1.51), (1.53), (1.55) has the form

$$\begin{aligned} &\frac{dq}{dt} = a(t, q^0) + 3u_0(q,t), && q\big|_{t=0} = q^0, \\ &\frac{ds}{dt} = 0, && s\big|_{t=0} = S^0(q^0), \\ &\frac{da}{dt} = -2au_{0x}(q,t), && a\big|_{t=0} = A^0(q^0). \end{aligned} \tag{1.56}$$

Let the Hopf equation (1.11) have a solution u_0 for $t < T$ and $x \in \mathbb{R}^1$ which belongs to C^∞. Then (1.56) also has a solution for $t < T$, and, provided the Jacobian does not vanish,

$$J = \frac{\partial q}{\partial q^0} \neq 0, \tag{1.57}$$

the solution of the initial problem is given by the formulas

$$S = S^0(q^0(x,t)), \quad A = a(t, q^0(x,t)),$$

where $q^0(x,t)$ is the solution of the equation $x = q(t, q^0)$.

Observe that

$$\frac{du_0(q,t)}{dt} = au_{0x}(q,t).$$

Hence the last equation in (1.56) can be easily integrated,

$$a(q,t) + 2u_0(q,t) = E(q^0), \tag{1.58}$$

where $E = A^0(q^0) + 2u_0^0(q^0)$, while the first equation in (1.56) can be rewritten as

$$\frac{dq}{dt} = u_0(q,t) + E(q^0). \tag{1.59}$$

The last equation shows that the choice of various initial values of the amplitude $A^0(x)$ affects the dependence of the function $q = q(t, q^0)$ on the parameter q^0. As a consequence, a "bad" choice of A^0 may result in the violation of condition (1.57) for $t < T$. Let us illustrate this by a simple example. Let $\varphi^0 = 0$, $u_0^0 = 1 - x$, $S^0 = x$, and $A^0 = 2(\alpha - \gamma x)$, $\alpha, \gamma > 0$. Then the solution of equation (1.11) can be easily found in an explicit form,

$$u_0 = \frac{1-x}{1-3t},$$

and it exists for $t < T = 1/3$. Next we solve (1.59) to find the characteristics

$$q = (1-3t)^{1/3}\left\{q^0 - \tfrac{1}{2}E(q^0)[(1-3t)^{2/3} - 1] + (1-3t)^{-1/3} - 1\right\},$$

whence

$$J = (1+\gamma)(1-3t)^{1/3}\left\{(1-3t)^{2/3} - \frac{\gamma}{1+\gamma}\right\}.$$

Thus condition (1.57) fails at time $t^* = \{1 - (\gamma/(1+\gamma))^{3/2}\}/3 < 1/3$.

Consider now the function (1.54). It is clear that outside a small neighborhood of the front $\Gamma = \{(x,t),\ S(x,t) = 0\}$ of the solitary wave we have, up to exponentially small terms, $U_1 = U_1^-(x,t)$ for $S < 0$ and $U_1 = 0$ for $S > 0$. Hence in the domain $\Omega_T^- = \{(x,t),\ S(x,t) < 0\}$ the function $U_1^-(x,t)$ must satisfy the linearized Hopf equation. However, it is easily seen that

$$\frac{\partial U_1^-}{\partial t} + 3\frac{\partial}{\partial x}(u_0 U_1^-) = \frac{2u_{0x}(A_x + 2u_{0x})}{\sqrt{A}} \not\equiv 0$$

for variable u_0 and A. This means that in the domain Ω_T^- the expansion (1.48) does not provide an asymptotic solution, and so we have to use the procedure of matching.

As was pointed out in the Introduction, the method presented above avoids both problems indicated here. Moreover, we virtually replace the partial differential equations (1.51), (1.53) by the characteristic equations (1.56) and consider only one characteristic which originates at the initial wave front $x = \varphi^0$. Accordingly, the only natural restriction on the time of existence of the asymptotic solution which remains is the condition for existence of a solution to the Hopf equation.

1.2. Waves in media with viscosity and dispersion: The KdV–Burgers equation

In this section we will show that the ideas presented above can be easily applied for the description of the wave motion in a medium with viscosity. If the viscosity effects are prevalent, instead of soliton-type solutions we obtain so-called solutions of smoothed shock wave type, or kinks. The physical interpretation of such solutions is obvious — for vanishing viscosity they go over into shock waves.

As an example we will consider the simplest model of a visco-elastic medium, given by the Korteweg–de Vries–Burgers equation

$$u_t + 3uu_x + \varepsilon_1 u_{xxx} = \varepsilon_2 u_{xx}, \tag{1.60}$$

where ε_1 and ε_2 are small parameters which characterize the magnitude of the dispersion and viscosity respectively.

We will treat this equation in two cases: for $\varepsilon_1 \sim \varepsilon_2$, i.e., when the effects of dispersion are prevalent, and for $\varepsilon_1 \ll \varepsilon_2$, i.e., when the prevalent effects are due to viscosity.

Let $\varepsilon_1 \sim \varepsilon_2 \ll 1$. Rewrite (1.60) as

$$u_t + 3uu_x + \varepsilon^2 u_{xxx} = \delta_1 \varepsilon^2 u_{xx}, \tag{1.61}$$

where $\delta_1 \geq 0$ is a constant and $\varepsilon \to 0$ is a small parameter. It is not hard to see that the solution of equation (1.61) does not differ qualitatively from the solution of the KdV equation. Indeed, writing the solution in the form of expansion (1.3) we obtain the equations (1.11) and (1.29) for the coefficients u_0, u_1 of the asymptotic expansion for the background, and the equation (1.9), which entails (1.10) and the formula $V_0 = A\cosh^{-2}(\sqrt{A}(\tau + \varphi_1)/2)$, for the leading term of the rapidly varying component of the solution. Further, for constructing the correction V_1 we obtain an equation of the form (1.17) with additional term $\delta_1 V_{0\tau\tau}$ in the right-hand side. Accordingly, the solvability condition (1.19) results, instead of (1.20), in the Riccati equation

$$\frac{d}{dt}(A + 2u_0(\varphi, t)) + \frac{4}{15}\delta_1 A^2 = 0, \tag{1.62}$$

while the orthogonality condition for the higher order correction entails the equation

$$A^{-1}\varphi_{1tt} + \alpha_1\varphi_{1t} + \alpha_2\varphi_1 = G,$$

where

$$\alpha_1 = -A^{-2}(2A_t + 3\dot{u}_0), \quad \alpha_2 = -(A^{-2}\dot{u}_0)_t - \frac{28}{5}\delta_1 A^{-1}\dot{u}_0 - 2\check{u}_{0xx},$$
$$G = -0.1A^{-2}\{A^3(A^{-5/2})_{tt} + A(5\check{u}_1 - 32\delta_1\sqrt{A})_t + 3\check{u}_1(25\dot{u}_0 - 6\delta_1 A^2)\},$$
$$\dot{u}_0 = du_0(\varphi, t)/dt.$$

In the particular case $u_0 = u_1 = 0$ the last equation can be easily solved in an explicit form,

$$\varphi_1 = -30\sqrt{A} + C_1 A + C_2,$$

where C_1, C_2 are the constants of integration. The subsequent constructions are carried through along the same lines as in §1.1.

Therefore, if the dispersion effects are prevalent, the KdV–Burgers equation has a soliton-type solution (in the class we are interested in). The viscosity effects result only in the dissipation of the solution. In particular, for $u_0 = \text{const}$ the amplitude of the solitary wave decreases inverse proportionally to time,

$$A = \frac{A^0}{1 + \frac{4}{15}\delta_1 A^0 t}.$$

Note that (1.61) is a particular case of the weakly perturbed KdV equation, i.e., an equation of the form

$$u_t + 3uu_x + \varepsilon^2 u_{xxx} = F\Big(x, t, u, \varepsilon\frac{\partial u}{\partial x}, \dots, \varepsilon^k \frac{\partial^k u}{\partial x^k}\Big), \tag{1.63}$$

where $F(x, t, u, p_1, \dots, p_k)$ is a smooth function, $k < \infty$. It is not hard to see that the solution with localized fast variation of this equation is of soliton type, while the right-hand side F affects only the dynamics of the distorted soliton. Note also that the presence of the perturbation F results in the fact that the parameters of the solitary wave vary with time even for constant background u_0.

Now we consider the case where the viscosity effects are prevalent. Rewrite (1.60) as

$$u_t + 3uu_x = \varepsilon u_{xx} - \varepsilon^3 \delta_2 u_{xxx}, \tag{1.64}$$

where δ_2 is a constant. In particular, for $\delta_2 = 0$ (1.64) becomes the Burgers equation. This equation leads us to another class of physically meaningful solutions with localized fast variation, viz., solutions of smoothed shock wave type.

We will represent the solution of (1.64) in the form generalizing (1.3),

$$u = \sum_{i=0}^{N+1} \varepsilon^i \Big(u_i(x, t) + V_i\Big(\frac{x - \varphi(t)}{\varepsilon}, x, t\Big)\Big) + O(\varepsilon^{N+2}), \tag{1.65}$$

where $u_i \in C^\infty$ and all the functions $V_i(\tau, x, t)$ are assumed to stabilize at infinity with respect to τ, i.e., $V_i(\tau, x, t) \in \mathcal{H}$ and $V_i \to 0$ as $\tau \to \infty$. (Formula (1.3) is obtained from (1.65) under the additional assumption $V_0 \in \mathcal{S} \subset \mathcal{H}$.) Substituting (1.65) into equation (1.64) and collecting the terms according to the powers of ε, we obtain

$$\begin{aligned}
\Bigg[&\frac{1}{\varepsilon}\{-\varphi_t V_{0\tau} + 3(u_0(x, t) + V_0)V_{0\tau} - V_{0\tau\tau}\} \\
&+ \Big\{ - \varphi_t V_{1\tau} + 3\frac{\partial}{\partial\tau}((u_0(x, t) + V_0)V_1) - V_{1\tau\tau} + f_1 + \delta_2 V_{0\tau\tau\tau}\Big\} \\
&+ \sum_{j=2}^{N+1} \varepsilon^{j-1}\Big\{ - \varphi_t V_{j\tau} + 3\frac{\partial}{\partial\tau}((u_0(x, t) + V_0)V_j) - V_{j\tau\tau} \\
&+ f_j + \delta_2(V_{j\tau\tau\tau} + 3V_{j-1\,\tau\tau x} + 3V_{j-2\,\tau xx} \\
&+ (V_{j-3} + u_{j-3})_{xxx})\Big\}\Bigg]\Bigg|_{\tau=(x-\varphi)/\varepsilon} = O(\varepsilon^{N+1}),
\end{aligned} \tag{1.66}$$

where the functions f_j are computed by the same formulas as in (1.5).

The first expression in braces in (1.66) belongs to $\mathcal{S}$. Hence we can apply the Taylor series expansion around $x = \varphi(t)$ to the functions involved in it. On doing so and equating to zero the terms of order ε^{-1} we obtain the equation

$$-\varphi_t \check{V}_{0\tau} + 3(\check{u}_0 + \check{V}_0)\check{V}_{0\tau} = \check{V}_{0\tau\tau}. \tag{1.67}$$

Here and subsequently we again use the notation (1.7). Obviously, the integration yields

$$-\varphi_t \check{V}_0 + 3(\check{u}_0 + \tfrac{1}{2}\check{V}_0)\check{V}_0 = \check{V}_{0\tau}, \tag{1.68}$$

where the constant of integration is equal to zero in view of the normalizing condition $\check{V}_0 \to 0$ as $\tau \to \infty$. Now. letting τ tend to $-\infty$ and denoting $\nu_0 = \lim_{\tau\to-\infty} \check{V}_0$, we obtain from (1.68) the equality

$$\varphi_t \nu_0 = 3(\check{u}_0 + \tfrac{1}{2}\nu_0)\nu_0. \tag{1.69}$$

Taking (1.69) into account, the equation (1.68) can be easily integrated to yield the explicit formula

$$\check{V}_0 = \nu_0(t)\chi(\tau, t), \tag{1.70}$$

where $\chi = \frac{1}{2}\big(1-\tanh(\beta(\tau+\varphi_1))\big)$, $\beta = 3\nu_0/4$. Obviously, the condition that $\check{V}_0 \to 0$ as $\tau \to \infty$, for a nontrivial solution ($\nu_0 \neq 0$), implies $\nu_0 > 0$.

We will use the standard notation $[f]|_{x=\varphi}$ for the amplitude of the jump of a function $f(\tau, x, t)$ at the point $x = \varphi$,

$$[f]\big|_{x=\varphi} = \lim_{\varepsilon\to 0}\lim_{x\to\varphi+0} f\Big(\frac{x-\varphi}{\varepsilon}, x, t, \varepsilon\Big) - \lim_{\varepsilon\to 0}\lim_{x\to\varphi-0} f\Big(\frac{x-\varphi}{\varepsilon}, x, t, \varepsilon\Big). \tag{1.71}$$

Formulas (1.65) and (1.70) for the asymptotic solution of the KdV–Burgers equation imply that

$$[u]\big|_{x=\varphi} = -\nu_0, \tag{1.72}$$

with the right-hand limit of u at the jump point equal to $u_0|_{x=\varphi}$, and the left-hand limit equal to $u_0|_{x=\varphi} + \nu_0$. Thus we obtain the so-called entropy condition

$$[u]\big|_{x=\varphi} < 0,$$

which, as is well known, is the condition for uniqueness and stability of a discontinuous solution of the limiting (in the sense of $\mathcal{D}'$ as $\varepsilon \to 0$) Hopf equation ([**74, 188**], see also [**73, 96, 189–191**]). Further, for a nontrivial solution ($\nu_0 \neq 0$) of the KdV–Burgers equation the relation (1.69) can be rewritten as

$$\varphi_t = \frac{3}{2}\frac{[u^2]|_{x=\varphi}}{[u]|_{x=\varphi}}. \tag{1.73}$$

Thus the front of the smoothed shock wave moves in the same manner as for the discontinuous solution of the Hopf equation, namely, according to the Hugoniot condition (1.73).

As in §1.1, we will seek the extension V_0 of the function $\check{V}_0$ outside the curve $x = \varphi(t)$ in the form

$$V_0(\tau, x, t) = V_0^-(x, t)\chi(\tau, t), \tag{1.74}$$

where V_0^- is a function to be determined such that $V_0^-|_{x=\varphi} = \nu_0(t)$. To determine u_0 and V_0^-, consider the terms of order ε^0 in (1.66). Taking their limits as $\tau \to \pm\infty$ and denoting $u_0^\pm = \lim_{\varepsilon\to 0} u$ for $x > \varphi(t)$ and $x < \varphi(t)$ respectively, we obtain the Hopf equations

$$\begin{aligned} u_{0\,t}^+ + 3u_0^+ u_{0\,x}^+ &= 0, \qquad t > 0, \quad x \in \Omega_t^+, \\ u_{0\,t}^- + 3u_0^- u_{0\,x}^- &= 0, \qquad t > 0, \quad x \in \Omega_t^-, \end{aligned} \tag{1.75}$$

where $\Omega_t^+ = \{x;\, x > \varphi(t)\}$, $\Omega_t^- = \{x;\, x < \varphi(t)\}$.

Thus we have arrived at the classical setting of the problem on the motion of a shock wave for the Hopf equation: for three unknown functions $u_0^+(x,t)$, $u_0^-(x,t)$, and $\varphi(t)$ we have equations (1.75) related by the condition (1.73) on the curve $x = \varphi(t)$. Note that (1.73) is virtually an ordinary differential equation,

$$\frac{d\varphi}{dt} = \frac{3}{2}\big(u_0^-(\varphi, t) + u_0^+(\varphi, t)\big). \tag{1.76}$$

Denote by $\widetilde{u}_0^\pm$ infinitely differentiable extensions of $u_0^\pm$ into the domains $\Omega_t^\mp$, and let the corresponding Hopf equations hold in the $\varepsilon^{1-\delta}$-neighborhood of the curve $x = \varphi(t)$. Then, setting

$$u_0(x,t) = \widetilde{u}_0^+(x,t), \quad V_0^- = \widetilde{u}_0^-(x,t) - \widetilde{u}_0^+(x,t),$$

we see that we have constructed the leading term of the asymptotic formula (1.65) up to a phase shift φ_1.

Now we return to the computation of the first correction in the expansion (1.65) of the solution. Taking (1.75) into account, we expand the terms of order ε^0 in (1.66) by the Taylor formula about the point $x = \varphi(t)$. Then, equating to zero the terms of order ε^0, we obtain the equation

$$-\varphi_t \check{V}_{1\tau} + 3\frac{\partial}{\partial\tau}((\check{u}_0 + \check{V}_0)\check{V}_1) - \check{V}_{1\tau\tau} = -\check{F}_1,$$

where

$$\check{F}_1 = \Big(\frac{(\beta\varphi_1)_t}{\beta} + 3\check{u}_1\Big)\check{V}_{0\tau} + \frac{3}{4\beta}\check{V}_{0x}^-\tau\check{V}_{0\tau\tau} + \delta_2\check{V}_{0\tau\tau\tau}.$$

On integrating we get

$$\check{V}_{1\tau} - 3\nu_0\Big(\chi - \frac{1}{2}\Big)\check{V}_1 = \mathcal{F}_1(\tau, t), \tag{1.77}$$

where

$$\mathcal{F}_1 = \Big(\frac{(\beta\varphi_1)_t}{\beta} + 3\check{u}_1\Big)\check{V}_{0\tau} + \frac{3}{4\beta}\check{V}_{0x}^-(\tau\check{V}_{0\tau} - \check{V}_0) + \delta_2\check{V}_{0\tau\tau}.$$

Hence, letting $\tau \to -\infty$ and taking (1.69) into account, we obtain the relation

$$\varphi_{1t} - 3\Big(\check{u}_{0x} + \frac{1}{2}\check{V}_{0x}^-\Big)\varphi_1 + 3\Big(\check{u}_1 + \frac{1}{2}\check{V}_1^-\Big) = -\frac{\check{V}_{0x}^-}{\nu_0}. \tag{1.78}$$

Observing that the functions φ and φ_1 enter with opposite signs in the phase $(x - \varphi(t) + \varepsilon\varphi_1)/\varepsilon$, we see that (1.78) is a nonhomogeneous equation in variations with respect to the Hugoniot condition (1.69).

The solution of equation (1.77) can be easily found explicitly by the formula

$$\check{V}_1(\tau, t) = \varphi_2\chi_\tau + \int_{-\infty}^{\tau} \mathcal{F}_2(\tau', t)\exp\Big(\int_{\tau'}^{\tau} q(\tau'', t)\,d\tau''\Big)\,d\tau',$$

where $q = 3\nu_0(\chi - \frac{1}{2})$ and $\varphi_2 = \varphi_2(t)$ is the "constant" of integration. Since

$$\exp\Big(\int_{\tau'}^{\tau} q(\tau'', t)\,d\tau''\Big) = \Big(\frac{\cosh\beta(\tau' + \varphi_1)}{\cosh\beta(\tau + \varphi_1)}\Big)^2,$$

we get after some algebra that

$$\check{V}_1(\tau, t) = \nu_1(t)\chi + \omega_1(\tau, t), \tag{1.79}$$

where $\nu_1 = \check{V}_1^-$ denotes the limiting value of $\check{V}_1$ as $\tau \to -\infty$,

$$\begin{aligned} \omega_1(\tau, t) &= \Big(\varphi_2 - \nu_1\tau + \frac{1}{2}\check{V}_{0x}^-\tau^2\Big)\chi_\tau + \delta_2\big((\tau + \varphi_1)\chi_{\tau\tau} - 2\zeta\chi_\tau\big), \\ \zeta &= \log\cosh\eta - \eta\tanh\eta\big|_{\eta = \beta(\tau + \varphi_1)} \in \mathcal{S}. \end{aligned} \tag{1.80}$$

As before, we will seek the extension $V_1(\tau, x, t)$ of the function $\check{V}_1(\tau, t)$ in the form

$$V_1(\tau, x, t) = V_1^-(x, t)\chi(\tau, t) + \omega_1(\tau, t), \tag{1.81}$$

where V_1^- is a smooth function satisfying the condition $V_1^-(\varphi, t) = \nu_1(t)$. Set $u_1^+(x, t) = u_1(x, t)$ for $x > \varphi(t)$ and $u_1^-(x, t) = u_1(x, t) + V_1^-(x, t)$ for $x < \varphi(t)$. Then, collecting the terms of order ε in (1.66) and taking the limits as $\tau \to \pm\infty$, we obtain the equations

$$u_{1\,t}^{\pm} + 3(u_0^{\pm}u_1^{\pm})_x = u_{0\,xx}^{\pm}, \qquad t > 0, \quad x \in \Omega_t^{\pm}. \tag{1.82}$$

Note that (1.78) can be rewritten in terms of $u_0^{\pm}, u_1^{\pm}$ as

$$\frac{d\varphi_1}{dt} - \frac{3}{2}(u_{0\,x}^+ + u_{0\,x}^-)\big|_{x=\varphi}\varphi_1 = -\frac{3}{2}(u_1^+ + u_1^-)\big|_{x=\varphi} - \frac{[u_{0\,x}^{\pm}]|_{x=\varphi}}{[u_0^{\pm}]|_{x=\varphi}}. \tag{1.83}$$

Thus to determine the three functions u_1^+, u_1^-, and φ_1 we have obtained the equations (1.82), (1.83), which are linearized equations (1.75), (1.77).

Next, denote by $\widetilde{u}_1^{\pm}$ the smooth extensions of $u_1^{\pm}$ into the domain $\Omega_t^{\mp}$ such that in the $\varepsilon^{1-\delta}$-neighborhood of the curve $x = \varphi(t)$ the equations (1.82) hold. By setting

$$u_1(x, t) = \widetilde{u}_1^+(x, t), \quad V_1^-(x, t) = \widetilde{u}_1^-(x, t) - \widetilde{u}_1^+(x, t),$$

we complete the construction of the correction $u_1 + V_1$ up to the "constant" of integration φ_2. The equation (1.83) enables us to find the phase shift φ_1, and hence to complete the construction of the leading term of the asymptotic expansion.

The construction of the higher order terms of the asymptotic expansion is carried out in a similar way.

Let us remark in conclusion that for the leading term of the solution of smoothed shock wave type of the KdV–Burgers equation (1.64) we have obtained an auxiliary problem (1.73), (1.75), which has exactly the same form as the problem on the motion of a shock wave for the limiting (as $\varepsilon \to 0$) Hopf equation. The dispersion effects in (1.64) affect only the profile of the soliton component of the correction of order ε.

It should be emphasized that the limiting problems (1.9), (1.11) and (1.73), (1.75) for dispersive and dissipative media are qualitatively different: the equations (1.9), (1.62) describing the motion of the front of the distorted soliton split off from the equation (1.11) for the background, and the smooth solution of (1.11) can be easily found by the method of characteristics. On the contrary, the equation (1.77) describing the dynamics of the front of the smoothed shock wave can be solved only jointly with equations (1.75) which describe the state of the medium ahead of and behind the front of the wave. Even in the simplest case of the Hopf equation such a problem presents certain technical difficulties.

The relationship between the solution of the equation

$$u_t + (\varphi(u))_x = \varepsilon u_{xx} \tag{1.84}$$

for viscosity ε tending to 0 and the discontinuous solution of the limiting equation

$$v_t + (\varphi(v))_x = 0 \tag{1.85}$$

is well known. In particular, this relationship forms the basis for Oleinik's method of vanishing viscosity [**190**], which treats the generalized solutions of (1.85) by passage to the regularized equation (1.84) with subsequent passage to the limit as $\varepsilon \to 0$. We have discussed here the KdV–Burgers equation (1.64) only as one of the simplest examples where the solutions of smoothed shock wave type exist, in order to make more transparent the exposition of the general method of construction of such solutions.

1.3. Soliton-type waves in nonhomogeneous media

In this section we will show that the method presented above can be easily applied to the analysis of wave processes in nonhomogeneous media. As an example, we will consider the problem of constructing the asymptotic soliton-type solution of the KdV equation with variable coefficients,

$$u_t + (\rho_1 + 3\rho_2 u)u_x + \varepsilon^2 \rho_3 u_{xxx} + \rho_4 u = 0, \tag{1.86}$$

where $\rho_j = \rho_j(x,t)$ are smooth functions, $\rho_2 > 0$, $\rho_3 > 0$. An actual physical example where an equation of the form (1.86) arises is the problem of wave motion on a shelf of variable depth $H(x)$, which was considered in [**203**]. In this case

$$\rho_1 = \sqrt{gH}, \quad \rho_2 = \frac{1}{2}\sqrt{\frac{g}{H}}, \quad \rho_3 = \frac{1}{6}\sqrt{gH^5}, \quad \rho_4 = \frac{1}{2}\rho_{1x}, \tag{1.87}$$

where g is the acceleration of gravity and $H > 0$ is a smooth function.

A similar equation in the case

$$\rho_1 = 0, \quad \rho_4 = 0, \quad \rho_2 = \rho_2(t), \quad \rho_3 = \rho_3(t) \tag{1.88}$$

was obtained in modeling some processes of hydrodynamics and plasma physics [**92, 116**]. Finally, the equation (1.86) with

$$\rho_1 = 0, \quad \rho_2 = 1, \quad \rho_3 = 1, \quad \rho_4 = 1/t, \tag{1.89}$$

referred to as the spherical KdV equation, appears in plasma physics [**213**]. Since the solution of this equation has a singularity at $t = 0$, the Cauchy problem in the case (1.89) is considered for $t > t_0 > 0$. Other examples of physical problems where the KdV equation with variable coefficients arises are investigated in [**115, 135, 221**].

Now we will set out the formal scheme of constructing the asymptotic solution; we present here a version which will enable us to simplify the necessary computations as much as possible.

We will seek the asymptotic soliton-type solution in the form (1.3), i.e.,

$$u(x,t,\varepsilon) = \sum_{j=0}^{N+1} \varepsilon^j \Big(u_j(x,t) + V_j\Big(\frac{x-\varphi}{\varepsilon}, x, t\Big)\Big) + O(\varepsilon^{N+2}), \tag{1.90}$$

where $u_j \in C^\infty$, $V_j(\tau, x, t) \in \mathcal{H}$ with $V_0 = V_0(\tau, t) \in \mathcal{S}$. As before, we assume that $V_j(\tau, x, t) \to 0$ as $\tau \to \infty$ and denote $V_j^-(x,t) = \lim_{\tau\to-\infty} V_j(\tau, x, t)$.

Substituting (1.90) into equation (1.86) and collecting the terms according to the powers of ε, we obtain the relation

$$\begin{aligned} &\frac{1}{\varepsilon}\{-\varphi_t V_{0\tau} + (\rho_1 + 3\rho_2(u_0 + V_0))V_{0\tau} + \rho_3 V_{0\tau\tau\tau}\} \\ &\quad + \sum_{j=1}^{N+1} \varepsilon^{j-1}\Big\{ -\varphi_t V_{j\tau} + \frac{\partial}{\partial\tau}(\rho_1 + 3\rho_2(u_0+V_0))V_j \\ &\qquad + \rho_3 V_{j\tau\tau\tau} + f_j\Big\}\Big|_{\tau=(x-\varphi)/\varepsilon} = O(\varepsilon^{N+1}), \end{aligned} \tag{1.91}$$

where the f_j are polynomials in $u_0, V_0, \dots, V_{j-1}, u_j$ and their derivatives. In particular,

$$\begin{aligned} f_1 &= u_{0t} + V_{0t} + (\rho_1 + 3\rho_2(u_0+V_0))u_{0x} + 3\rho_2 u_1 V_{0\tau} + \rho_4(u_0 + V_0), \\ f_2 &= u_{1t} + V_{1t} + (\rho_1 + 3\rho_2(u_0+V_0))(u_{1x} + V_{1x}) + 3\rho_2 u_2 V_{0\tau} \\ &\quad + 3\rho_2(u_1 + V_1)(u_{0x} + V_{1\tau}) + 3\rho_3 V_{1\tau\tau x} + \rho_4(u_1 + V_1). \end{aligned}$$

Let us derive equations for the solution outside the front $x = \varphi$. To this end we consider (1.91) for $\tau \to \pm\infty$. Equating to zero the coefficients of ε^j, $j = 0, 1, \dots$, we obtain

$$u_{0t} + (\rho_1 + 3\rho_2 u_0)u_{0x} + \rho_4 u_0 = 0, \tag{1.92}$$

$$u_{jt} + \rho_1 u_{jx} + 3\rho_2(u_0 u_j)_x + \rho_4 u_j = g_j(x,t), \tag{1.93}$$

$$V_{jt}^- + \rho_1 V_{jx}^- + 3\rho_2(u_0 V_j^-)_x + \rho_4 V_j^- = g_j^-(x,t), \tag{1.94}$$

where $j = 1, 2, \dots, N+1$ and g_j, g_j^- are polynomials in $u_0, \dots, u_{j-1}, V_{j-1}^-$ and their derivatives. In particular,

$$g_1 = g_1^- = 0, \quad g_2 = -3\rho_2 u_1 u_{1x} - \rho_3 u_{0xxx}, \quad g_2^- = -3\rho_2 V_1^- V_{1x}^- .$$

We will assume that for $t \leq T$ there exists a smooth solution of (1.92).

Now we find the solution in a small neighborhood of the front $x = \varphi$. For this, taking (1.92)–(1.94) into account, we expand the functions involved in (1.91) by the Taylor formula around $x = \varphi(t)$. Using the notation $\check{g} = g|_{x=\varphi}$ and the equality $x - \varphi = \varepsilon\tau$, we collect similar terms with respect to their order in ε. Now, equating to zero the coefficients of ε^k, $k = -1, 0, 1, \dots$, we obtain the equations

$$-\varphi_t V_{0\tau} + (\check{\rho}_1 + 3\check{\rho}_2(\check{u}_0 + V_0))V_{0\tau} + \check{\rho}_3 V_{0\tau\tau\tau} = 0, \tag{1.95}$$

$$\frac{\partial}{\partial\tau} L\check{V}_j + \mathcal{F}_j(\tau, t) = 0, \tag{1.96}$$

where

$$L = -\varphi_t + \check{\rho}_1 + 3\check{\rho}_2(\check{u}_0 + V_0) + \check{\rho}_3 \frac{\partial^2}{\partial\tau^2}$$

and the $\mathcal{F}_j$ are polynomials in $\tau, \check{u}_0, V_0, \dots, \check{V}_{j-1}$ and their derivatives. In particular,

$$\begin{aligned} \mathcal{F}_1 &= V_{0t} + 3\check{\rho}_2(\check{u}_1 V_{0\tau} + \check{u}_{0x} V_0) + \check{\rho}_4 V_0 \\ &\quad + \tau(\check{\rho}_{1x} + 3\check{\rho}_2 \check{u}_{0x} + 3\check{\rho}_{2x}(\check{u}_0 + V_0))V_{0\tau} + \tau\check{\rho}_{3x} V_{0\tau\tau\tau}. \end{aligned} \tag{1.97}$$

The soliton-type solution (1.95) can easily be found explicitly:

$$V_0 = A(t) \cosh^{-2} \beta(\tau + \varphi_1), \tag{1.98}$$

where $\beta = \frac{1}{2}\sqrt{\check{\rho}_2 A/\check{\rho}_3}$ and $\varphi_1 = \varphi_1(t)$ is the "constant" of integration. As a by-product of computing (1.98) we obtain also the first Hugoniot type condition, viz., the relationship between the amplitude $A > 0$ and the velocity of the solitary wave

$$\varphi_t = \rho_1(\varphi, t) + \rho_2(\varphi, t)(A(t) + 3u_0(\varphi, t)). \tag{1.99}$$

Consider the equations (1.96). The necessary and sufficient conditions for their solvability in the class of functions stabilizing at infinity are

$$\mathcal{F}_j \in \mathcal{S}, \tag{1.100}$$

$$\int_{-\infty}^{\infty} \mathcal{F}_j(\tau, t) V_0(\tau, t)\, d\tau = 0. \tag{1.101}$$

The conditions (1.100) are fulfilled by (1.92)–(1.94). The orthogonality conditions (1.101) enable us to find the second Hugoniot type condition, as well as to compute the integration "constants" which arise when solving (1.96).

If condition (1.101) is fulfilled, the solution $\check{V}_j \in \mathcal{H}$ of equation (1.96) can be found explicitly by means of the formula (1.24). It is not hard to see that this solution has the form

$$\check{V}_j(\tau, t) = \nu_j(t)\chi(\tau, t) + \omega_j(\tau, t), \tag{1.102}$$

where $\chi = \frac{1}{2}(1 - \tanh\beta(\tau + \varphi_1))$, $\omega_j = \varphi_{j+1}(t)V_{0\tau} + \omega_{j,1}(\tau, t)$ with $\omega_{j,1}$ a function from $\mathcal{S}$, and φ_{j+1} the "constant" of integration. The limiting value ν_j of the function $\check{V}_j$ as $\tau \to -\infty$ can be easily found by integration of (1.96) with respect to τ:

$$\nu_j = -\frac{1}{A\check{\rho}_2}\int_{-\infty}^{\infty} \mathcal{F}_j(\tau, t)\,d\tau. \tag{1.103}$$

In particular, we easily derive from (1.97) and (1.103) that

$$\nu_1 = -\frac{4}{A\check{\rho}_2}\left\{\frac{d}{dt} + \check{\rho}_4 - \check{\rho}_{1x} - \check{\rho}_{2x}(A + 3\check{u}_0)\right\}\sqrt{A\frac{\check{\rho}_3}{\check{\rho}_2}}. \tag{1.104}$$

We will determine the extension of the function $\check{V}_j$ outside the curve $x = \varphi(t)$ by the formula

$$V_j(\tau, x, t) = V_j^-(x, t)\chi(\tau, t) + \omega_j(\tau, t). \tag{1.105}$$

Then for finding V_j^- we obtain the equation (1.94) for $t > 0$ and $x < \varphi(t)$, which is completed with the boundary condition

$$V_j^-\big|_{x=\varphi(t)} = \nu_j(t) \tag{1.106}$$

and with initial condition for $t = 0$ and $x < \varphi(0)$. Obviously, the curve $x = \varphi(t)$ is not a characteristic $x = X(t)$ of equation (1.94), and, moreover, $\dot{\varphi}_t > X_t$ when $\rho_2 A > 0$. Therefore such a problem is well posed.

Now we will derive the second Hugoniot type condition. To this end we consider (1.101) for $j = 1$. Using the explicit form of the function $\mathcal{F}_1$, this condition can be rewritten in the form of the equality

$$\begin{aligned}&\left\{\frac{d}{dt} + 2\check{\rho}_4 + 3\check{\rho}_2\check{u}_{0x} - 3\check{\rho}_{2x}\check{u}_0 - \check{\rho}_{1x}\right\}\int_{-\infty}^{\infty} V_0^2\,d\tau\\ &\qquad + 3\check{\rho}_{3x}\int_{-\infty}^{\infty}(V_{0\tau})^2\,d\tau - 2\check{\rho}_{2x}\int_{-\infty}^{\infty} V_0^3\,d\tau = 0.\end{aligned} \tag{1.107}$$

Computing the integrals in (1.107) and using the equality

$$\frac{d\check{u}_0}{dt} = A\check{\rho}_2\check{u}_{0x} - \check{\rho}_4\check{u}_0,$$

we can transform (1.107) into the following Riccati equation:

$$\frac{dA}{dt} + q_1 A + q_2 A^2 = q_3, \tag{1.108}$$

where

$$\begin{aligned}q_1 &= \frac{1}{3}\frac{d}{dt}\log\frac{\check{\rho}_3}{\check{\rho}_2} + \frac{2}{3}(2\check{\rho}_4 - \check{\rho}_{1x} - 3\check{\rho}_{2x}\check{u}_0),\\ q_2 &= \left(\frac{4}{5}\frac{\check{\rho}_2}{\check{\rho}_3}\rho_{3x} - \frac{16}{15}\rho_{2x}\right)\bigg|_{x=\varphi}, \qquad q_3 = -2\frac{d\check{u}_0}{dt} - 2\check{\rho}_4\check{u}_0.\end{aligned}$$

It is easily seen that the equations (1.92), (1.99), and (1.108) form a closed system that allows us to determine the smooth background u_0, as well as the front $\varphi(t)$ and the amplitude A of the distorted soliton V_0.

To complete our determination of the leading term of the asymptotic expansion, it remains to find the phase shift φ_1. Consider the condition (1.101) for $j = 2$. The function $\mathcal{F}_2$ is given by the formula

$$\begin{aligned}\mathcal{F}_2 = \Big\{&\check{V}_{1t} + (3\rho_2 u_{0x} + \rho_4)V_1 - (\nu_{1t} + \nu_1(3\rho_2 u_{0x} + \rho_4))\chi \\ &+ 3\rho_2(u_1 + V_1)V_{1\tau} + 3\rho_2(u_{1x} + V_{1x})V_0 + 3\rho_3 V_{1\tau\tau x} \\ &+ 3\rho_2 u_2 V_{0\tau} + (\rho_{4x} + 3(\rho_2 u_{0x})_x)\tau V_0 + 3(\rho_2 u_1)_x \tau V_{0\tau} \\ &+ \frac{1}{2}\tau^2\big((\rho_{1xx} + 3(u_0\rho_2)_{xx})V_{0\tau} + 3\rho_{2xx}V_0V_{0\tau} + \rho_{3xx}V_{0\tau\tau\tau}\big) \\ &+ \tau\big((\rho_{1x} + 3(u_0\rho_2)_x)V_{1\tau} + 3\rho_{2x}(V_0V_1)_\tau + \rho_{3x}V_{1\tau\tau\tau} \\ &+ 3\rho_2(V_0V_{1x})_\tau + \rho_3 V_{1\tau\tau\tau x} - A\rho_2 V_{1\tau x}\big)\Big\}\Big|_{x=\varphi}.\end{aligned}$$

We will show how an equation for φ_1 can be derived without obtaining $\check{V}_1$ explicitly. We will use the identity

$$\begin{aligned}&\int_{-\infty}^{\infty} V_{0t}\check{V}_1\,d\tau + 3\check{\rho}_2\int_{-\infty}^{\infty}\{\check{u}_{0x}V_0 + \check{u}_1 V_{0\tau} + (V_0\check{V}_1)_\tau\}\check{V}_1\,d\tau \\ &\quad + \check{\rho}_4\int_{-\infty}^{\infty} V_0\check{V}_1\,d\tau + (\rho_{1x} + 3(\rho_2 u_0)_x)\big|_{x=\varphi}\int_{-\infty}^{\infty}\tau V_{0\tau}\check{V}_1\,d\tau \\ &\quad + 3\check{\rho}_{2x}\int_{-\infty}^{\infty}\tau V_0 V_{0\tau}\check{V}_1\,d\tau + \check{\rho}_{3x}\int_{-\infty}^{\infty}\tau V_{0\tau\tau\tau}\check{V}_1\,d\tau + \frac{1}{2}A\check{\rho}_2\nu_1^2 = 0,\end{aligned}$$

which can be easily obtained by multiplying the equation (1.96) for $j = 1$ by $\check{V}_1$ and integrating with respect to τ. Then the condition (1.101) for $j = 2$ can be rewritten as

$$\begin{aligned}&\left\{\frac{d}{dt} + 3(\check{\rho}_2\check{u}_{0x} - \check{\rho}_{2x}\check{u}_0) + 2\check{\rho}_4 - \check{\rho}_{1x}\right\}\int_{-\infty}^{\infty} V_0\check{V}_1\,d\tau \\ &\qquad - 3\rho_{2x}\int_{-\infty}^{\infty} V_0^2\check{V}_1\,d\tau + 3\rho_{3x}\int V_{0\tau}\check{V}_{1\tau}\,d\tau + G(t,\varphi_1) = 0,\end{aligned} \tag{1.109}$$

where

$$\begin{aligned}G(t,\varphi_1) &= G_0(t)\varphi_1 + G_1(t), \\ G_0 &= -\frac{3}{2}\check{\rho}_{3xx}\int_{-\infty}^{\infty} V_{0\ \tau}^{2}\,d\tau \\ &\qquad - \left\{\frac{3}{2}(\rho_2 u_{0xx} - \rho_2 u_{0xx}) + \rho_{4x} - \frac{1}{2}\rho_{1x}\right\}\Big|_{x=\varphi}\int_{-\infty}^{\infty} V_0^2\,d\tau, \\ G_1 &= \frac{3}{2}\left\{\rho_2 u_{1x} - \rho_{2x}u_1 + \frac{1}{2}\rho_2 V_{1\ x}^{-}\right\}\Big|_{x=\varphi}\int_{-\infty}^{\infty} V_0^2\,d\tau + \frac{1}{2}A\check{\rho}_2\nu_1^2 \\ &\qquad - \frac{1}{2}\{\nu_{1t} + (3\check{\rho}_2\check{u}_{0x} + \rho_4)\nu_1\}\int_{-\infty}^{\infty} V_0\,d\tau, \\ \nu_1 &= -\frac{2}{\beta\rho_2}\left\{\frac{d}{dt}\ln(A/\beta) + \rho_4 - \rho_{1x} - \rho_{2x}(A + 2u_0)\right\}\Big|_{x=\varphi}.\end{aligned}$$

In order to further transform (1.109), observe that (1.95) and (1.99) imply the equality

$$A\check{\rho}_2 V_0 = \frac{3}{2}\check{\rho}_2 V_0^2 + \check{\rho}_3 V_{0\tau\tau}. \tag{1.110}$$

Multiplying (1.110) by $\check{V}_1$ and integrating with respect to τ, we arrive at the auxiliary identity

$$3\check{\rho}_2 \int_{-\infty}^{\infty} V_0^2 \check{V}_1\, d\tau = 2A\check{\rho}_2 \int V_0 \check{V}_1\, d\tau + 2\check{\rho}_3 \int V_{0\tau}\check{V}_{1\tau}\, d\tau.$$

Therefore we can rewrite (1.109) as

$$\left\{\frac{d}{dt} + q_4\right\} \int_{-\infty}^{\infty} V_0 \check{V}_1\, d\tau + q_5 \int_{-\infty}^{\infty} V_{0\tau}\check{V}_{1\tau}\, d\tau + G(t, \varphi_1) = 0, \tag{1.111}$$

where

$$q_4 = \left\{3(\rho_2 u_{0x} - \rho_{2x} u_0) + 2\rho_4 - \rho_{1x} - 2A\rho_{2x}\right\}\Big|_{x=\varphi},$$
$$q_5 = \left\{3\rho_{3x} - \frac{2\rho_3 \rho_{2x}}{\rho_2}\right\}\Big|_{x=\varphi}.$$

We will compute the integrals in the left-hand side of (1.111) without obtaining the explicit form of $\check{V}_1$. Denote the function $\check{\rho}_3/\check{\rho}_2$ by q. Then (1.110) and (1.96) for $j = 1$ can be rewritten as

$$-AV_0 + \frac{3}{2}V_0^2 + qV_{0\tau\tau} = 0, \tag{1.112}$$
$$\hat{L}_1 \check{V}_1 = \check{\rho}_2^{-1}\Phi_1(\tau, t), \tag{1.113}$$

where

$$\Phi_1(\tau, t) = -\int_{-\infty}^{\tau} \mathcal{F}_1(\tau', t)\, d\tau' + \int_{-\infty}^{\infty} \mathcal{F}_1(\tau', t)\, d\tau',$$
$$\hat{L}_1 = -A + 3V_0 + q\frac{\partial^2}{\partial\tau^2}.$$

Now, as is easily seen by differentiating (1.112) with respect to the parameter A, we have $V_0 = \hat{L}_1 V_{0A}$. Using this equality, we obtain from (1.113)

$$\int_{-\infty}^{\infty} V_0 \check{V}_1\, d\tau = \int_{-\infty}^{\infty} \check{V}_1 \hat{L}_1 V_{0A}\, d\tau = \frac{1}{\check{\rho}_2} \int_{-\infty}^{\infty} V_{0A}\Phi_1\, d\tau. \tag{1.114}$$

In a similar way, differentiating (1.112) with respect to the parameter q, we obtain $V_{0\tau\tau} = -\hat{L}_1 V_{0q}$. This equality and (1.113) imply that

$$\int_{-\infty}^{\infty} V_{0\tau}\check{V}_{1\tau}\, d\tau = \frac{1}{\check{\rho}_2} \int_{-\infty}^{\infty} V_{0q}\Phi_1\, d\tau. \tag{1.115}$$

The solution of (1.112) depending on the parameters A and q has the form (1.98), where $\beta = \sqrt{A/q}/2$. Therefore $V_{0A} = (V_0 + \frac{1}{2}\tau V_{0\tau})/A$, $V_{0q} = -\tau V_{0\tau}/2q$. These formulas and the explicit expression for Φ_1,

$$\Phi_1 = 2\alpha_1\chi - 2\alpha_2\chi^2(3-2\chi) - \alpha_3 V_0 - \alpha_4\tau V_0 - \frac{3}{2}\check\rho_{2x}\tau V_0^2 - \check\rho_{3x}(\tau V_{0\tau\tau} - V_{0\tau}),$$

where

$$\alpha_1 = \Big\{\Big(\frac{A}{\beta}\Big)_t + \frac{A}{\beta}(\rho_4 - \rho_{1x} - 3\rho_{2x}u_0)\Big\}\Big|_{x=\varphi}, \quad \alpha_2 = \frac{\check\rho_{2x}A^2}{\beta},$$

$$\alpha_3 = 2\check\rho_2\check u_1 + \frac{(\beta\varphi_1)_t}{\beta}, \quad \alpha_4 = \Big\{\rho_{1x} + 3(\rho_2 u_0)_x + \frac{\beta_t}{\beta}\Big\}\Big|_{x=\varphi},$$

allow us to evaluate the integrals

$$\int_{-\infty}^{\infty}\Phi_1 V_{0A}\,d\tau = \frac{2}{3}\frac{A}{\beta}\Big\{-\frac{1}{2}\frac{d\varphi_1}{dt} + q_6\varphi_1\Big\} + q_7,$$

$$\int_{-\infty}^{\infty}\Phi_1 V_{0q}\,d\tau = -\frac{A}{q}\int_{-\infty}^{\infty}\Phi_1 V_{0A}\,d\tau. \tag{1.116}$$

Here

$$q_6 = \Big\{3\rho_2 u_{0x} + \rho_4 + \frac{A_t}{A} - \frac{\beta_t}{2\beta} + \frac{2}{5}A\Big(\frac{\rho_2\rho_{3x}}{\rho_3} - \rho_{2x}\Big)\Big\}\Big|_{x=\varphi},$$

$$q_7 = \frac{1}{\beta}\Big\{\Big(\frac{A}{\beta}\Big)_t + \frac{A}{\beta}(\rho_4 - \rho_{1x} - 3\rho_{2x}u_0) - \rho_{2x}\frac{A^2}{\beta} - A\rho_2 u_1\Big\}\Big|_{x=\varphi}.$$

Finally, (1.111) together with (1.114)–(1.116) implies the required equation

$$\frac{d^2\varphi_1}{dt^2} - q_8\frac{d\varphi_1}{dt} - q_9\varphi_1 = q_{10}, \tag{1.117}$$

where

$$q_8 = \frac{d}{dt}\log(A\check\rho_2) + \Big\{3(\rho_2 u_0)_x + \rho_{1x} + \frac{4}{5}A\Big(\frac{19\rho_2\rho_{3x}}{4\rho_3} - \rho_{2x}\Big)\Big\}\Big|_{x=\varphi},$$

$$q_9 = \Big\{2q_6\Big(\frac{d}{dt}\log\frac{Aq_6}{\beta\check\rho_2} + 3(\rho_2 u_{0x} - \rho_{2x}u_0) + 2\rho_4 - \rho_{1x} - 3\frac{A\rho_2\rho_{3x}}{\rho_3}\Big) + 3\frac{\beta\rho_2 G_0}{A}\Big\}\Big|_{x=\varphi},$$

$$q_{10} = 3\frac{\beta}{A}\Big\{\check\rho_2\frac{d}{dt}\frac{q_7}{\check\rho_2} + q_7\Big(q_4 - \frac{A\check\rho_2 q_5}{\check\rho_3}\Big) + \check\rho_2 G_1\Big\}.$$

Consider the particular cases (1.87)–(1.89) of the Korteweg–de Vries equation (1.86). Let the coefficients be computed according to (1.87). Then the equations (1.99) and (1.108) become

$$\frac{d}{dt}(A + 2u_0(\varphi,t)) + H^{-1}(\varphi)\Big\{A + \frac{u_0(\varphi,t)(A+H(\varphi)) + (23/15)A^2}{A + 2H(\varphi) + 3u_0(\varphi,t)}\Big\}\frac{dH(\varphi)}{dt} = 0,$$

$$\frac{d\varphi}{dt} = \frac{1}{2}\sqrt{g}H^{-1/2}(\varphi)\big(A + 2H(\varphi) + 3u_0(\varphi,t)\big). \tag{1.118}$$

When there is no background, i.e., when $u_0 \equiv 0$, the first equation (1.118) can be integrated. This yields the equality

$$AH(\varphi)\Big(1+\frac{\mu_1 A}{H(\varphi)}\Big)^{-\mu_2}=E, \tag{1.119}$$

where $\mu_1 = 53/60$, $\mu_2 = 23/53$, $E = \text{const}$. This equality is a refinement of the relation $A \sim E/H$ obtained in [**203**] for waves of small amplitude. The relationship (1.119) fits well with the experimental data [**38**] (see Figure 0.3 in the Introduction).

Figures 1.1–1.4 demonstrate the leading term of the asymptotics computed numerically by the formulas (1.119) and (1.118) for $u_0 = 0$ and the bottom profile

$$H(x)=\frac{7}{4}-\frac{1}{4}\tanh(2(x-3))-\frac{1}{2\cosh(2(x-3))}$$

(all numerical data are rounded off).

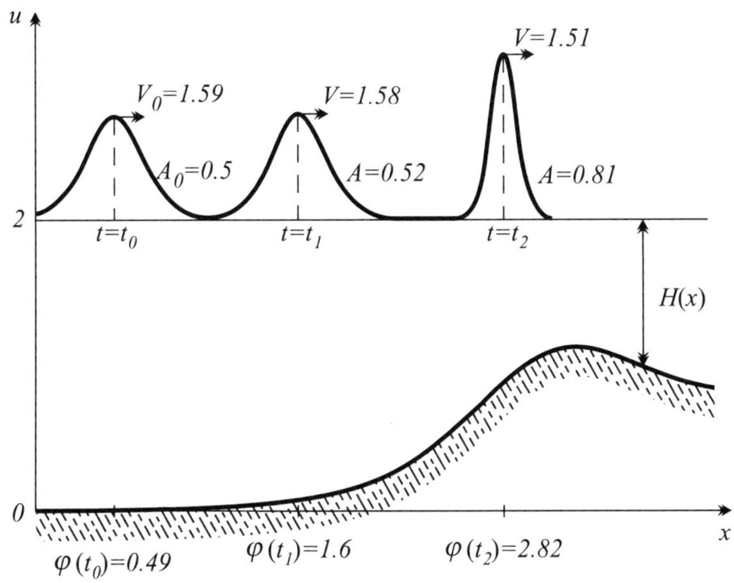

FIGURE 1.1. Dynamics of a solitary wave along an uneven bottom at time $t_0 = 0.3$, $t_1 = 1$, and $t_2 = 1.8$. Here $x = \varphi(t)$ is the front position, $V = \dot{\varphi}(t)$ is the velocity, $A = A(t)$ is the amplitude of the wave, and $H(x)$ is the depth.

When there is no background, the formula (1.104) transforms into

$$\nu_1(t)=-\frac{2}{\sqrt{3}}\sqrt{\frac{H}{A}}\Big(3H+\frac{43}{30}A\Big)H_x\Big|_{x=\varphi(t)}.$$

Hence, assuming that $\varphi(0) = 0$, we obtain the following formula for the solution of the problem (1.94), (1.106) for $j = 1$:

$$V_1^-(X)\big|_{t_0=t_0(x,t)}=\nu_1(t_0)\Big(\frac{H(\varphi(t_0))}{H(X)}\Big)^{1/4}\Big|_{t_0=t_0(x,t)}$$
$$\text{for}\quad X^0(t)\le x\le\varphi(t),\ t\ge 0,$$
$$V_1^-(x,t)=0\qquad\text{for}\quad x\le X^0(t),\ t\ge 0.$$

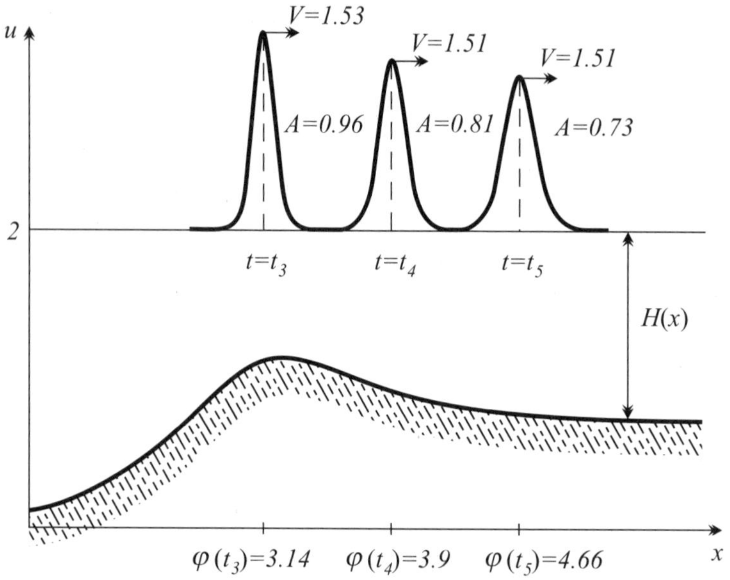

FIGURE 1.2. Dynamics of a solitary wave along an uneven bottom at time $t_3 = 2$, $t_4 = 2.5$, and $t_5 = 3$.

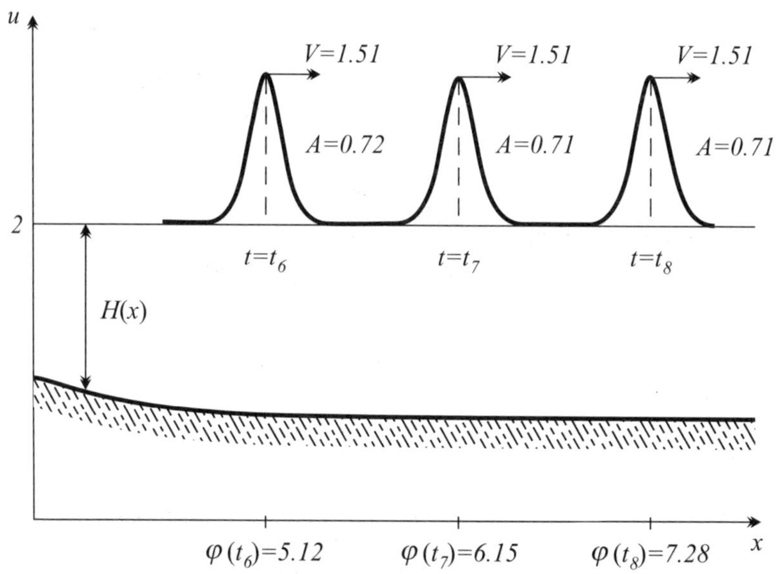

FIGURE 1.3. Dynamics of a solitary wave along an uneven bottom at time $t_6 = 3.3$, $t_7 = 4$, and $t_8 = 4.6$.

Here the characteristics $X = X(\xi, t_0)$ solve the problem

$$\frac{dX}{d\xi} = \sqrt{gH(X)}, \quad X\big|_{\xi=0} = \varphi(t_0)$$

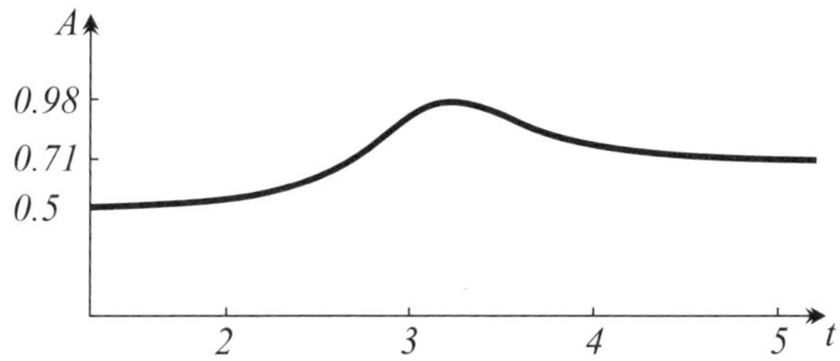

FIGURE 1.4. Time dependence of the amplitude of a solitary wave.

with parameter $t_0 \geq 0$; moreover, $\xi = t - t_0$, where $t_0 = t_0(x, t)$ is the solution of the equation

$$x = X(t - t_0, t_0),$$

and $X^0(t) = X(\xi, 0)$ is the characteristic originating at the point $x = \varphi(0) = 0$.

The numerical results are shown in Figures 1.5–1.7. It is seen that in the interval of bottom nonhomogeneity from $x \sim 0.96$ through $x \sim 4.56$ there is a "tail" V_1^-. It is not generated on the right of this interval (since $\nu_1 = 0$), and the wave V_1^- lags behind the soliton V_0.

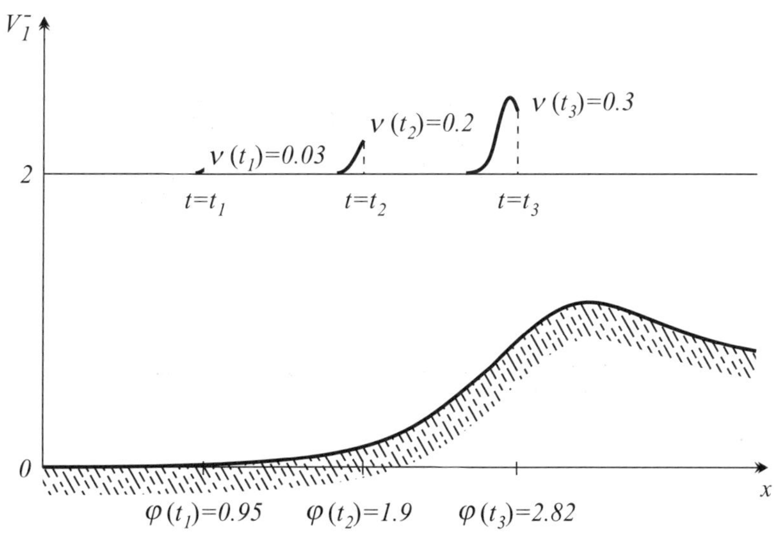

FIGURE 1.5. Generation and dynamics of the tail of a solitary wave due to an uneven bottom. Here $x = \varphi(t)$ are positions of the front of the solitary wave and $\nu(t)$ are boundary values of V_1^- on the front at time $t_1 = 0.6$, $t_2 = 1.2$, and $t_3 = 1.8$.

Consider now the case (1.88). It is easy to calculate that then the equations (1.99) and (1.108) can be rewritten as

$$\begin{aligned} &\frac{d}{dt}\big(A + 2u_0(\varphi, t)\big) + \frac{1}{3} A \frac{d}{dt} \log q = 0, \\ &\frac{d\varphi}{dt} = \rho_2\big(A + 3u_0(\varphi, t)\big), \end{aligned} \tag{1.120}$$

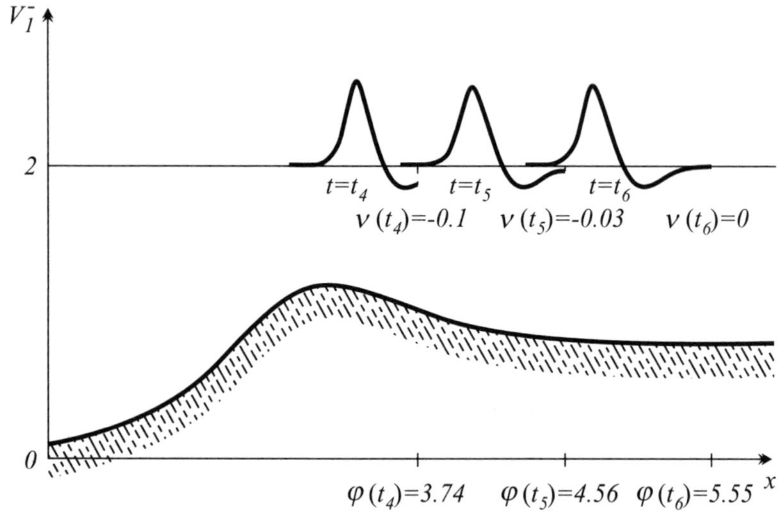

FIGURE 1.6. Generation and dynamics of the tail of a solitary wave due to an uneven bottom at time $t_4 = 2.4$, $t_5 = 3$, and $t_6 = 3.6$.

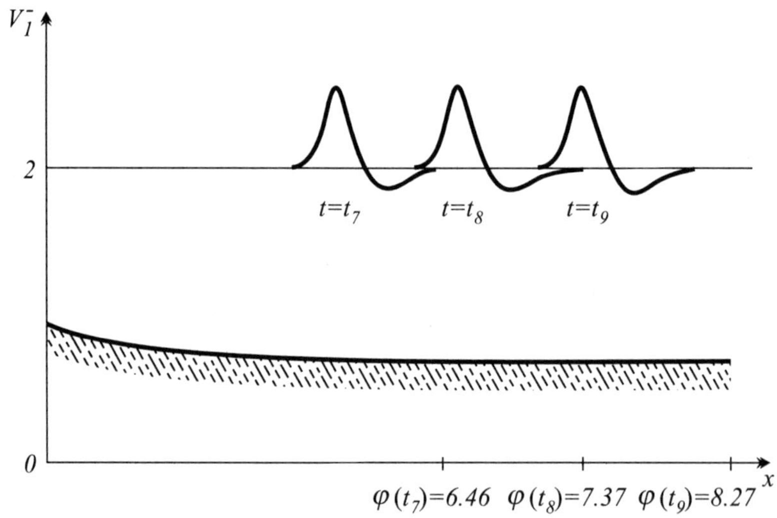

FIGURE 1.7. Dynamics of the tail of a solitary wave at time $t_7 = 4.2$, $t_8 = 4.8$, and $t_9 = 5.4$.

where $q = \rho_3/\rho_2$.

The first equation (1.120) with $u_0 \equiv 0$ can be easily integrated to yield

$$A(t) = E^2 \left(\frac{\rho_2(t)}{\rho_3(t)} \right)^{1/3}.$$

In this case (1.117) becomes

$$\frac{d}{dt}\left(\frac{q^{4/3}}{\rho_2} \frac{d}{dt} \frac{\varphi_1}{q} \right) = \frac{3}{2E}\left(\frac{d}{dt} \frac{q_7}{\rho_2} + G_1 \right).$$

Finally, we obtain the following Hugoniot type conditions for the coefficients (1.89):

$$\begin{aligned} &\frac{d}{dt}\{(t(A+2u_0(\varphi,t))\}+\frac{1}{3}A=0,\\ &\frac{d\varphi}{dt}=A+3u_0(\varphi,t). \end{aligned} \tag{1.121}$$

When $u_0=0$, $u_1=0$, the equations (1.117), (1.121) can be solved explicitly. It follows from (1.121) that

$$A=A_0t^{-4/3},\quad \varphi=-A_0t^{-1/3}+c_1,$$

where A_0 and c_1 are constants, and we recall that the initial condition is set for $t=t_0>0$.

Next, we find from (1.103) and (1.94) for $j=1$ that

$$\nu_1=-\frac{4}{3\sqrt{A_0}t^{1/3}},\quad V_1^-=-12\frac{A_0^{3/2}}{t(x-c_1)^2}.$$

Now, on computing the coefficients (1.117), we obtain

$$\varphi_1=5A_0^{-1/2}t^{2/3}+c_2t^{-1/3}+c_3.$$

1.4. Boussinesq equation

Here we consider the equation

$$u_{tt}-u_{xx}=\frac{3}{2}(u^2)_{xx}+\varepsilon^2u_{xxxx}, \tag{1.122}$$

which is a more general model for long-wave propagation than the Korteweg–de Vries equation. It was derived by Boussinesq in 1871 in the context of shallow water approximation for the rectangle geometry case. The Boussinesq equation was studied by many authors using, in particular, spectral methods. A detailed bibliography can be found in [**37**]. This equation is of interest for us mainly because it is of higher order and hence admits motion in both directions. This fact gives rise to some new points in the construction of the asymptotic solution; namely, two boundary conditions appear on the curve $x=\varphi(t)$, $t>0$, rather than a single condition as was the case for the KdV equation.

Let us construct an asymptotic solution describing the motion of a soliton over a smooth background. Writing down the solution in the form (1.90), collecting the terms according to order in ε, and equating to zero the coefficient of ε^{-2}, we obtain the model equation for the soliton $V_0(\tau,t)$:

$$\frac{\partial^2}{\partial\tau^2}\Big\{(\varphi_t^2-1-3\breve{u}_0)V_0-\frac{3}{2}V_0^2-V_{0\tau\tau}\Big\}=0. \tag{1.123}$$

Integrating (1.123), we obtain the formula

$$V_0=A\cosh^{-2}(\beta(\tau+\varphi_1)) \tag{1.124}$$

with $\beta = \sqrt{A}/2$, as well as the first Hugoniot type condition

$$\varphi_t^2 = A + 3u_0(\varphi, t) + 1. \tag{1.125}$$

This shows that the soliton $V_0((x-\varphi(t))/\varepsilon, t)$ can have only a positive amplitude, it can move either in the positive or in the negative direction along the x-axis, and in any case the condition

$$A + 3u_0(\varphi, t) + 1 > 0 \tag{1.126}$$

must be fulfilled. The relation (1.125) shows also that for $A > 0$ the trajectory of the soliton is different from the characteristics of the degenerate (as $\varepsilon = 0$) equation (1.122).

Next, denote by I_{-1} the terms of order ε^{-1} which arise when substituting the expansion (1.90) into (1.122),

$$\begin{aligned} &I_{-1}(\tau, x, t) \\ &\quad = \frac{\partial^2}{\partial\tau^2}\{(\varphi_t^2 - 1 - 3u_0(x,t))V_1 - 3V_0V_1 - V_{1\tau\tau} - 3u_1(x,t)V_0\} \\ &\quad - \frac{\partial}{\partial\tau}\Big\{\Big(2\varphi_t\frac{\partial}{\partial t} + \varphi_{tt}\Big)V_0 + 6u_{0x}(x,t)V_0\Big\} - 3\check{u}_{0x}\tau V_{0\tau\tau}\Big|_{\tau=(x-\varphi)/\varepsilon}. \end{aligned} \tag{1.127}$$

The last term in (1.127) is the coefficient in the Taylor expansion of the function $\varepsilon^{-2}3u_0(x,t)V_{0\tau\tau}$ around the point $x = \varphi(t)$. Observe that all the functions in (1.127) decrease exponentially as $\tau \to \pm\infty$. This enables us to expand them by the Taylor formula around $x = \varphi(t)$ as well, and to represent I_{-1} as

$$\begin{aligned} I_{-1}(\tau, x, t)\big|_{\tau=(x-\varphi)/\varepsilon} = \Big\{I_{-1}(\tau, \varphi, t) + \varepsilon\tau\frac{\partial I_{-1}}{\partial x}\Big|_{x=\varphi} \\ + \frac{1}{2}\varepsilon^2\tau^2\frac{\partial^2 I_{-1}}{\partial x^2}\Big|_{x=\varphi} + \dots\Big\}\Big|_{\tau=(x-\varphi)/\varepsilon}. \end{aligned}$$

Excluding from (1.127) the terms of order $O(\varepsilon)$, we arrive at the model equation for the first fast-varying correction on the front $x = \varphi(t)$:

$$\begin{aligned} &\frac{\partial^2}{\partial\tau^2}\{(\varphi_t^2 - 1 - 3\check{u}_0)\check{V}_1 - 3V_0\check{V}_1 - \check{V}_{1\tau\tau}\} - 3\check{u}_1V_{0\tau\tau} \\ &\quad - \frac{\partial}{\partial\tau}\Big\{\Big(2\varphi_t\frac{\partial}{\partial t} + \varphi_{tt}\Big)V_0 + 6\check{u}_{0x}V_0\Big\} - 3\check{u}_{0x}\tau V_{0\tau\tau} = 0. \end{aligned} \tag{1.128}$$

Hence, taking (1.125) into account, we obtain

$$\frac{\partial}{\partial\tau}L\check{V}_1 = F_1(\tau, t), \tag{1.129}$$

where

$$\begin{aligned} L &= A - 3V_0 - \frac{\partial^2}{\partial\tau^2}, \\ F_1 &= 3\check{u}_1V_{0\tau} + \Big\{2\varphi_t\frac{\partial}{\partial t} + \varphi_{tt} + 3\check{u}_{0x}\Big\}V_0 + 3\check{u}_{0x}\tau V_{0\tau}. \end{aligned}$$

The right-hand side of equation (1.129) belongs to $\mathcal{S}$. Hence for solvability of (1.129) in the class of bounded functions the following orthogonality condition is sufficient:

$$\int_{-\infty}^{\infty} F_1(\tau,t)V_0(\tau,t)\,d\tau = 0. \tag{1.130}$$

After some algebra, (1.130) yields the second Hugoniot type condition:

$$A_t + 3\sqrt{A}\frac{u_{0t} + u_{0x}\varphi_t(1+2\sqrt{A})}{6(A+3u_0+1)+\sqrt{A}} = 0. \tag{1.131}$$

Now we can integrate (1.129) to find using (1.24) that

$$\check{V}_1(\tau,t) = \nu_1(t)\chi + \omega_1(\tau,t), \tag{1.132}$$

where

$$\nu_1 = 4A^{-1/2}u_{0x}(\varphi,t) - \frac{1}{3}(A\varphi_t)^{-1}A_t, \tag{1.133}$$

$\chi = \frac{1}{2}(\tanh\beta(\tau+\varphi_1)-1)$, and ω_1 is a function exponentially decreasing as $\tau\to\pm\infty$, which contains the term $\varphi_2(t)V_{0\tau}$ with so far arbitrary "constant" of integration φ_2. We will seek the function $V_1(\tau,x,t)$ in the form

$$V_1(\tau,x,t) = V_1^-(x,t)\chi(\tau,t) + \omega_1(\tau,t),$$

where V_1^- satisfies the condition

$$V_1^-\big|_{x=\varphi} = \nu_1(t). \tag{1.134}$$

Consider now the terms I_0 of order ε^0 which arise when substituting the expansion (1.90) into (1.122):

$$I_0 = \frac{\partial^2}{\partial\tau^2}\{(\varphi_t^2 - 1 - 3u_0(x,t) - 3V_0)V_2 - V_{2\tau\tau}\} - \mathcal{F}_2(\tau,x,t), \tag{1.135}$$

where

$$\begin{aligned}\mathcal{F}_2 = {} & \frac{\partial^2}{\partial\tau^2}\Big\{3u_2V_0 + 3u_1V_1 + \frac{3}{2}V_1^2 - 4V_{1\tau x}\Big\} \\ & + \frac{\partial}{\partial\tau}\Big\{\Big(2\varphi_t\frac{\partial}{\partial t} + \varphi_{tt}\Big)V_1 + 2\frac{\partial}{\partial x}\big[(1+3(u_0+V_0))V_1 + 3u_1V_0\big]\Big\} \\ & + 3u_{0xx}V_0 - V_{0tt} + \frac{3}{2}\check{u}_{0xx}\tau^2V_{0\tau\tau} + 6\check{u}_{0xx}\tau V_{0\tau} \\ & - \tau\frac{\partial^2}{\partial\tau^2}\frac{\partial}{\partial x}\{LV_1 - 3u_1V_0\}\big|_{x=\varphi} + u_{0tt} - u_{0xx} - \frac{3}{2}(u_0^2)_{xx}.\end{aligned}$$

It is clear that the relation $I_0 = 0\,(\operatorname{mod} O(\varepsilon))$ can only be fulfilled under the condition that

$$u_{0tt} - u_{0xx} - \frac{3}{2}(u_0^2)_{xx} = 0. \tag{1.136}$$

Now we expand the functions involved in (1.135) by the Taylor formula and equate to zero the terms of order ε^0 to obtain the equation

$$\frac{\partial^2}{\partial\tau^2}L\check{V}_2 = \check{\mathcal{F}}_2.$$

On integrating we have

$$\frac{\partial}{\partial\tau}L\check{V}_2 = F_2(\tau,t). \tag{1.137}$$

Here

$$F_2 = \Big\{2\varphi_t\frac{\partial}{\partial t} + \frac{\partial}{\partial x}(1+3u_0(x,t)+\varphi_t^2)\Big\}V_1\Big|_{x=\varphi} + 2(\sqrt{A})_{tt}\tanh(\beta(\tau+\varphi_1)) + G_2(\tau,t) + c(t),$$

where G_2 is a function from $\mathcal{S}$ whose explicit form is immaterial for us, and c is the "constant" of integration.

It is clear that for solvability of (1.137) in the class $\mathcal{H}$ it is necessary that $F_2 \to 0$ as $\tau \to \pm\infty$. This condition implies the choice $c = -2(\sqrt{A})_{tt}$, as well as the necessity of the following equality:

$$\Big\{2\varphi_t\frac{\partial}{\partial t} + \frac{\partial}{\partial x}(1+3u_0(x,t)+\varphi_t^2)\Big\}V_1^-(x,t)\Big|_{x=\varphi} = 4(\sqrt{A})_{tt}. \tag{1.138}$$

Denote by ζ the vector with components $(1, \varphi_t - A/2\varphi_t)$ in the space of variables (t,x). Then, using (1.133), the equality (1.138) can be rewritten as

$$\frac{\partial V_1^-}{\partial\zeta}\Big|_{x=\varphi} = \sigma_1(t), \tag{1.139}$$

where

$$\sigma_1 = -\frac{(\varphi_{tt} + 3\check{u}_{0x})\nu_1 - 4(\sqrt{A})_{tt}}{2\varphi_t}$$

and $\partial/\partial\zeta$ is the derivative along ζ.

Next, the orthogonality condition of the form (1.130) implies an equation for the phase shift φ_1, and the integration of (1.137) leads to the function $\check{V}_2$ of the form (1.132) with coefficient

$$\nu_2 = -\frac{1}{A}\int_{-\infty}^{\infty} F_2(\tau,t)\,d\tau.$$

Finally, if we determine $V_2(\tau,x,t)$ similarly to (1.134) and consider the terms of order ε, we obtain in a similar way the equations

$$u_{1tt} - u_{1xx} - 3(u_0u_1)_{xx} = 0, \qquad x\in\mathbb{R}^1, \quad t>0, \tag{1.140}$$

$$V_1^-{}_{tt} - V_1^-{}_{xx} - 3(u_0V_1^-)_{xx} = 0, \qquad x<\varphi(t), \quad t>0. \tag{1.141}$$

It follows from the aforesaid that the equation (1.141) has to be completed with conditions (1.134) and (1.139) on the curve $x=\varphi(t)$, $t>0$. Of course, (1.140) and

(1.141) must be also completed with Cauchy conditions for $t=0$, $x\in\mathbb{R}^1$ and for $t=0$, $x\le\varphi(0)$, respectively.

We assume that there exists a solution of the Cauchy problem for the equation (1.136), which is a smooth function for $t\le T$. This implies solvability of the Cauchy problem for (1.140). Let, further, for $t\le T$ condition (1.126) be fulfilled and let $A>0$. Then it is not hard to verify that the curve $x=\varphi(t)$ is not a characteristic for (1.141) and the vector ζ is transversal to this curve. This makes equation (1.141) with conditions (1.134), (1.139) a well-posed problem.

The subsequent derivations follow the scheme set out above. For the corrections $u_j(x,t)$ to the background we obtain nonhomogeneous linear equations of the form (1.140), while for the "tail" $V_j^-(x,t)$ we obtain nonhomogeneous equations of the form (1.141) in the domain $\{(x,t),\ x<\varphi(t),\ t>0\}$ completed with conditions

$$V_j^-\big|_{x=\varphi}=\nu_j(t),\qquad \frac{\partial V_j^-}{\partial\zeta}\bigg|_{x=\varphi}=\sigma_j(t).$$

The integration "constants" $\varphi_j(t)$, $j\ge 1$, which are virtually the higher order corrections to the trajectory $x=\varphi(t)$ of the distorted soliton, are found from the orthogonality condition of the form (1.130).

1.5. Other examples

In this section we consider two equations important for physical applications, namely, the sine-Gordon equation and the Benjamin–Bona–Mahogany equation (the regularized long waves equation). There are no new methodological aspects in constructing their asymptotic solutions, so we briefly present only the construction of the first asymptotic term.

The various versions of the sine-Gordon equation

$$u_{\tau\tau}-u_{\xi\xi}+\sin u=0, \tag{1.142}$$

where the function u is interpreted as an angle, arise in many applications, such as differential geometry or nonlinear optics. In particular, solutions of (1.142) explain the phenomenon of self-induced transparency. Moreover, this equation has been extensively studied as a model of relativistic field theory. There is a vast literature on the sine-Gordon equation. Its discussion and further references can be found in [**14, 37, 129, 270**].

1. We will consider a perturbed sine-Gordon equation with small parameter which is related to large space and time scales of the process under study ($\xi=x/\varepsilon$, $\tau=t/\varepsilon$, $x,t\sim 1$).

The version

$$\varepsilon^2(u_{tt}-u_{xx})+\frac{\alpha^2}{\gamma}\sin\gamma u=0 \tag{1.143}$$

of the sine-Gordon equation with variable coefficients

$$\alpha=\alpha(x,t)>0,\quad \gamma=\gamma(x,t)>0$$

was treated in [**150**]. Skipping the calculations, we present here the leading term of the asymptotic solution to (1.143) obtained in [**150**]:

$$u = \frac{4}{\gamma(x,t)} \arctan\Big(\exp\Big(-\beta\frac{x-\varphi+\varepsilon\varphi_1}{\varepsilon}\Big)\Big) + O(\varepsilon), \tag{1.144}$$

where $\beta = \alpha(\varphi, t)/\sqrt{1-\varphi_t^2}$ and $\varphi = \varphi(t)$ is the solution of the Hamilton system

$$\frac{d\varphi}{dt} = \frac{\partial H}{\partial p}, \quad \frac{dp}{dt} = -\frac{\partial H}{\partial \varphi}, \quad H = \sqrt{\frac{\alpha^2(\varphi,t)}{\gamma^4(\varphi,t)} + p^2}, \tag{1.145}$$

with $p = \beta\gamma^{-2}(\varphi, t)\varphi_t$. The phase shift $\varphi_1 = \varphi_1(t)$ is the solution of the equation

$$\mu_1 \frac{d^2}{dt^2}\Big(\frac{\varphi_1}{\beta}\Big) + \mu_2 \frac{d}{dt}\Big(\frac{\varphi_1}{\beta}\Big) + \mu_3\varphi_1 = \frac{3}{4}\pi^2\gamma\Box\gamma^{-1}\big|_{x=\varphi}, \tag{1.146}$$

where

$$\begin{aligned}
&\mu_1 = \beta^2/\alpha(\varphi,t), \qquad \mu_2 = \gamma^4\frac{d}{dt}(\mu_1\gamma^{-4}(\varphi,t)) - \kappa\kappa_t\big|_{x=\varphi},\\
&\mu_3 = \gamma^2\frac{d}{dt}\Big(\mu_1\frac{d}{dt}\gamma^{-2}(\varphi,t)\Big) - \kappa^{-1}\Box\kappa + \alpha^{-1}\kappa^{-1}\langle\nabla_r\alpha, \nabla_r\kappa\rangle\\
&\qquad - 2\mu_1\varphi_{tt}\gamma^{-1}\gamma_x\big|_{x=\varphi},\\
&\kappa = \alpha(x,t)\gamma^{-2}(x,t),
\end{aligned}$$

$\nabla_r = (\partial/\partial t, -\partial/\partial x)$, and $\Box = \partial^2/\partial t^2 - \partial^2/\partial x^2$ is the d'Alembert operator.

Observe that the solution of (1.145) satisfies the condition $\varphi_t^2 < 1$, since

$$1 - \varphi_t^2 = 1 - (H_p)^2 = \frac{\kappa^2}{\kappa^2 + p^2} > 0.$$

In the particular case where κ does not depend on x the system (1.145) reduces to the equation

$$(\varphi_t)^2 = \frac{E_1}{\kappa^2(t) + E_1}, \qquad E_1 = \text{const} > 0.$$

If κ depends on the variable x only, (1.145) implies

$$(\varphi_t)^2 = 1 - E_2\kappa^2(\varphi), \qquad E_2 = \text{const} > 0.$$

Finally, if κ is a constant, i.e., $\alpha(x,t) = c\gamma^2(x,t)$, $c \neq 0$, then $\varphi_t = E_3$, where $E_3 \neq 0$ is a constant such that $|E_3| < 1$. Then the equation (1.146) becomes

$$\frac{d^2}{dt^2}\Big(\frac{\varphi_1}{\beta\gamma^2(\varphi,t)}\Big) = \frac{3}{4}\pi^2(1 - E_3^2)\gamma^{-1}\Box\gamma^{-1}\big|_{x=\varphi}.$$

It is easily seen that the solution of (1.144) is a smoothed shock wave with jump

$$[u] = \lim_{\varepsilon\to 0}\lim_{x\to\varphi+0} u - \lim_{\varepsilon\to 0}\lim_{x\to\varphi-0} u = -2\pi\gamma^{-1}(\varphi(t), t).$$

Since the limiting (as $\varepsilon \to 0$) equation (1.143) is a trigonometric equation $\sin \gamma u = 0$, we have no difficulties in finding the solution to the right and to the left of the wave front.

Note that in the literature on the inverse scattering problem such solutions (which are exact for constant α and γ) are often referred to as kinks, fluxons, or solitons. In order to emphasize the direction of the front motion, the solutions of the form (1.144) are often called antifluxons or antikinks, while the solutions with β replaced by $-\beta$ in (1.144) are called fluxons or kinks.

2. Now we consider in more detail another version of the perturbed sine-Gordon equation with small parameter,

$$\begin{aligned}\varepsilon^2(u_{tt} - u_{xx}) + \alpha^2 \sin u \\ + \varepsilon^3(\rho_1(x) + 3\rho_2(x)\cos u)|u_t|u_t - \eta(x)\varepsilon^4 u_{xxt} = \varepsilon\delta(x,t),\end{aligned} \tag{1.147}$$

where $\alpha > 0$, $\eta \geq 0$, ρ_1, ρ_2, δ are smooth functions. Such an equation arises in the theory of long Josephson junctions when taking into consideration the dissipative effects and when the displacement current (i.e., energy pumping) is present [**205**]. Then the meaning of the function u is the relative phase difference between the macroscopic quantum wave functions of two superconductors, the normalized coefficients $\alpha, \rho_1, \rho_2, \delta, \eta$ are certain functions of the temperature (which we assume to be a specified function of x), and the normalized function δ describes the displacement current. The arrangement of the parameters in (1.147) corresponds to the case of small dissipation and small energy pumping.

We will seek the asymptotic solution of smoothed shock wave type to the equation (1.147) in the form of an expansion

$$u = V_0(\tau, t) + \varepsilon(u_1(x,t) + V_1(\tau, x, t)) + \dots, \tag{1.148}$$

where $\tau = (x - \varphi(t))/\varepsilon$, $V_0, V_1 \in \mathcal{H}$, $\varphi, u_1 \in C^\infty$.

Substituting (1.148) into (1.147), collecting the terms according to the power of ε, and equating the coefficient of ε^0 to zero, we obtain the model equation whose solution is a function V_0 of the form (1.144) with $\gamma = 1$ and $\beta = \alpha(\varphi)/\sqrt{1-\varphi_t^2}$. Then the substitution of (1.148) into (1.147) results in

$$\varepsilon\{LV_1 + \mathcal{F}_1(\tau, x, t)\} = O(\varepsilon^2), \tag{1.149}$$

where

$$\begin{aligned}L &= (1-\varphi_t^2)\frac{\partial^2}{\partial \tau^2} - \alpha^2(x)\cos V_0, \\ \mathcal{F}_1 &= 2\varphi_t V_{0\tau t} + \varphi_{tt} V_{0\tau} - (\rho_1 + 3\rho_2 \cos V_0)\varphi_t|\varphi_t| V_{0\tau}^2 \\ &\quad - \eta\varphi_t V_{0\tau\tau\tau} - \tau\alpha_x(\varphi, t)\sin V_0 - u_1\alpha^2 \cos V_0 - \delta.\end{aligned} \tag{1.150}$$

Here we have used the fact that $V_{0\tau} < 0$.

Passing to the limit in (1.149) as $\tau \to \pm\infty$, we find the regular correction

$$u_1 = \frac{\delta(x,t)}{\alpha^2(x)}. \tag{1.151}$$

Then, applying the Taylor expansion with respect to x in the left-hand side of (1.149), collecting the terms according to the power of ε, and equating to zero the coefficient of ε^1, we obtain the equation

$$\check{L}\check{V}_1 = -\check{\mathcal{F}}_1(\tau, \varphi, t), \tag{1.152}$$

where, as before, $\check{g}$ denotes $g(\tau, x, t)|_{x=\varphi}$ for any function g.

It follows from the explicit formula (1.150) and equality (1.151) that $\check{\mathcal{F}}_1$ decreases exponentially as $\tau \to \pm\infty$. Next, the only bounded solution of the equation $\check{L}v = 0$ is the function $v = V_{0\tau}$ (the second solution increases exponentially in τ). Therefore the only condition for solvability of (1.152) is

$$\int_{-\infty}^{\infty} \mathcal{F}_1(\tau, \varphi, t) V_{0\tau}\, d\tau = 0. \tag{1.153}$$

By elementary calculations (1.153) reduces to the system of equations

$$\begin{gathered} \frac{d}{dt}(p^2 + \alpha^2(\varphi)) + \frac{2}{3}\eta(\varphi)p^2(p^2 + \alpha^2(\varphi)) - 2p^2|p|\xi(\varphi) - 2p\delta(\varphi, t) = 0, \\ \left(\frac{d\varphi}{dt}\right)^2 = \frac{p^2}{\alpha^2(\varphi) + p^2}, \end{gathered} \tag{1.154}$$

where $\xi = \frac{1}{6}(16 - 3\pi)\rho_2(\varphi) - \frac{1}{2}\pi\rho_1(\varphi)$.

It remains to notice that the solution of (1.152) (which can be found explicitly with the help of formula (1.24)) decreases exponentially as $\tau \to \pm\infty$. Therefore the first correction V_1 does not depend on x, i.e., $V_1 = \check{V}_1(\tau, t)$.

The equations (1.154) are Hugoniot type conditions for the smoothed shock wave V_0. When the coefficients and the right-hand side of (1.147) do not depend on x, the first equation (1.154) splits off and becomes an Abel equation. By specifying the quantities $\alpha, \rho_i, \eta, \delta$ one can analyze in detail the dynamics of the wave front. For instance, let $\alpha = 1$, $\eta = 0$ and let ξ and δ be constants. Consider for definiteness a wave with a positive velocity, i.e., $\varphi_t > 0$. Then the solution of the first equation (1.154) has the form

$$\begin{aligned} p &= \frac{p_0\sqrt{\xi\delta} + \delta\tan(\sqrt{\xi\delta}t)}{\sqrt{\xi\delta} - \delta p_0 \tan(\sqrt{\xi\delta}t)} && \text{for} \quad \xi\delta > 0, \\ p &= \frac{p_0\sqrt{|\xi\delta|} + \delta\tanh(\sqrt{|\xi\delta|}t)}{\sqrt{|\xi\delta|} - \delta p_0 \tanh(\sqrt{|\xi\delta|}t)} && \text{for} \quad \xi\delta < 0, \end{aligned}$$

where $p_0 = p|_{t=0} = \varphi_t/\sqrt{1 - \varphi_t^2}|_{t=0}$.

Hence it is clear that when $\delta < 0$ the impulse p is bounded for any sign of ξ (for all t if $\xi > 0$ and for $t < \pi/2\sqrt{\xi\delta}$ if $\xi < 0$). When $\delta > 0$ and $\xi < 0$, there exists a threshold value of the displacement current $\delta^* = |\xi|/p_0^2$ such that the impulse remains bounded for all t if $\delta < \delta^*$, whereas $p \to \infty$ as $t \to |\xi\delta|^{-1/2} \operatorname{arctanh} \sqrt{|\xi|/\delta p_0^2}$ if $\delta > \delta^*$. In a similar way, when $\delta > 0$, $\xi > 0$ we obtain that $p \to \infty$ as $t \to |\xi\delta|^{-1/2} \arctan \sqrt{\xi/\delta p_0^2}$. The unbounded growth of the impulse entails that $|\varphi_t| \to 1$ and the solution of (1.148) collapses. Note also that if $|p| \le \text{const}$ then $1 - \varphi_t^2 = \alpha^2/(\alpha^2 + p^2) \ge \text{const} > 0$.

3. The Benjamin–Bona–Mahogany equation (regularized long waves equation)

$$u_t + u_x + 3uu_x = \varepsilon^2 u_{xxt} \tag{1.155}$$

is an alternative model to the KdV equation for the description of waves in a rectangular channel. A detailed study of this equation was carried through in [**17**]. The first numerical solutions of equation (1.155) were given by Peregrine [**209**]; hence it is often called also the PBBM equation.

The asymptotic soliton-type solution is constructed in the same manner as for the KdV equation. In this case we obtain for the background u_0 a degenerate equation (1.155), i.e.,

$$u_{0t} + u_{0x} + 3u_0 u_{0x} = 0;$$

the distorted soliton has the standard form $V_0 = A\cosh^{-2}\beta(\tau + \varphi_1)$, where $\tau = (x - \varphi(t))/\varepsilon$ and $\beta = \sqrt{A/4\varphi_t}$, while the Hugoniot type conditions have the form

$$\frac{d\varphi}{dt} = A + 3u_0(\varphi, t) + 1, \tag{1.156}$$

$$\frac{d}{dt}\Big\{\frac{A^2}{\beta}\Big(1 + \frac{4}{5}\beta^2\Big)\Big\} + 3\frac{A}{\beta}\frac{du_0(\varphi, t)}{dt} = 0. \tag{1.157}$$

Using the formula for β and (1.156), the equation (1.157) can be easily rewritten in the standard form $A_t = F(A, \varphi, t)$. However, the explicit expression for F is very cumbersome.

1.6. Motion of a fluxon in a medium with strong dissipation

In this section we return to the problem on distributed Josephson junctions, but we consider here the case of strong dissipation and large energy pumping. In this case instead of (1.147) we have the equation (see [**205**])

$$\begin{aligned}\varepsilon^2(u_{tt} - u_{xx}) + \sin u + \varepsilon^2\big(\rho_1(x) + 3\varepsilon\rho_2(x)\cos u\big)|u_t|u_t \\ = \delta(x, t, \varepsilon) + \varepsilon^4\eta(x)u_{xxt}.\end{aligned} \tag{1.158}$$

Here $\rho_1 \neq 0$, ρ_2, δ, η are smooth functions, and in order to simplify the formulas we assume that the coefficient α in (1.147) equals one. The physical meaning of this assumption is that the so-called Josephson critical current does not depend on the spatial variable x.

It is not quite appropriate to regard the equation (1.158) as a version of the sine-Gordon equation, since the dissipative term $\varepsilon^2\rho_1|u_t|u_t$ has the same order of magnitude as $\sin u$ (in the class of functions with fast variation which are of interest for us). Moreover, the problem we consider here is not a standard one (construction of a fluxon, i.e., of a smoothed shock wave); it is rather a problem of control theory: to find the displacement current

$$\delta(x, t, \varepsilon) = \delta_0(x, t) + \varepsilon\delta_1(x, t) + \dots \tag{1.159}$$

for which the fluxon exists and its motion obeys the preassigned law. Nevertheless we begin with a formal construction of the asymptotic solution of smoothed shock wave type,

$$u = u_0(x, t) + V_0(\tau, x, t) + \varepsilon(u_1(x, t) + V_1(\tau, x, t)) + \dots, \tag{1.160}$$

where $\tau = (x - \varphi(t))/\varepsilon$, $V_0, V_1 \in \mathcal{H}$, and u_0, u_1, φ are smooth functions. To simplify the calculations, we restrict ourselves to the case of a solitary fluxon moving in the positive direction. This means that

$$\varphi_t > 0 \qquad \text{and} \qquad V_0 \to 0 \quad \text{as } \tau \to \infty. \tag{1.161}$$

Following the general scheme, we substitute the expansion (1.160) into the equation (1.158) and collect the terms of order ε^0. Then we obtain

$$(1 - \varphi_t^2)V_{0\tau\tau} + \rho_1(x)\varphi_t^2|V_{0\tau}|V_{0\tau} - \sin(u_0 + V_0) + \delta_0(x,t) = O(\varepsilon). \tag{1.162}$$

Taking the limit as $\tau \to \infty$ in (1.162), we find the relationship between the "background" u_0 and the leading term of the displacement current δ_0:

$$u_0(x,t) = \arcsin \delta_0(x,t). \tag{1.163}$$

Next we see that for (1.162) to hold uniformly in $\tau \in \mathbb{R}^1$ it is necessary that $V_0 \to 2\pi$ as $\tau \to -\infty$. This implies that the fluxon V_0 cannot depend on the slow variable x, i.e., $V_0 = V_0(\tau, t)$. Now, taking (1.163) into account, we expand the functions in the left-hand side of (1.162) by the Taylor formula around $x = \varphi(t)$ and equate to zero the terms of order ε^0 to obtain the equation

$$(1 - \varphi_t^2)V_{0\tau\tau} + \check{\rho}_1\varphi_t^2|V_{0\tau}|V_{0\tau} - \sqrt{1 - \check{\delta}_0^2}\sin V_0 + \check{\delta}_0(1 - \cos V_0) = 0. \tag{1.164}$$

It is not hard to see that (1.164) has a solution of smoothed shock wave type

$$V_0(\tau, t) = 4\arctan\exp\big(-\beta(\tau + \varphi_1)\big) \tag{1.165}$$

only if the displacement current δ_0 and the coefficient ρ_1 (the so-called tunnel current of quasi-particles) have the same sign,

$$\frac{\delta_0(\varphi,t)}{\rho_1(\varphi,t)} \geq \text{const} > 0, \tag{1.166}$$

and the quantities $\check{\delta}_0, \check{\rho}_1$ are related to the velocity φ_t of the fluxon by the equality

$$\check{\delta}_0^2 = \frac{4\check{\rho}_1^2\varphi_t^4}{(1 - \varphi_t^2)^2 + 4\check{\rho}_1^2\varphi_t^4}. \tag{1.167}$$

It is clear that both (1.163) and (1.167) imply the restriction $|\check{\delta}_0| < 1$. Moreover, when $\check{\rho}_1 \neq 0$, (1.164) has a nontrivial solution only if $\check{\delta}_0 \neq 0$. This implies the condition $\varphi_t \neq 0$. If (1.166) and (1.167) hold, then the function (1.165) is a solution of (1.164) satisfying our conditions ($V_0 \to 0$ as $\tau \to \infty$, $V_0 \to 2\pi$ as $\tau \to -\infty$) with

$$\beta = \sqrt{\sqrt{1 - \check{\delta}_0^2}/(1 - \varphi_t^2)}$$

and an as yet arbitrary phase shift $\varphi_1 = \varphi_1(t)$.

In principle, the equality (1.167) gives an answer to the question about the choice of the "control" δ_0. However, we have to verify whether condition (1.167) does not contradict the conditions that will inevitably appear when constructing the higher order terms of the asymptotic expansion. It is clear that such a contradiction, if it occurs, would imply nonexistence of the required solution of equation (1.158). Hence we consider the terms of order ε which are obtained on substituting the expansion (1.160) into (1.158) and taking (1.163), (1.164) into account:

$$\varepsilon\Big\{L\Big(x,t,\tau,\frac{\partial}{\partial\tau}\Big)V_1+\mathcal{F}_1(\tau,x,t)\Big\}=O(\varepsilon^2). \tag{1.168}$$

Here we use the inequality $V_{0\tau}<0$ and denote

$$\begin{aligned} L\Big(x,t,\tau,\frac{\partial}{\partial\tau}\Big)&=(1-\varphi_t^2)\frac{\partial^2}{\partial\tau^2}-2\rho_1(x)\varphi_t^2V_{0\tau}\frac{\partial}{\partial\tau}-\cos(u_0(x,t)+V_0),\\ \mathcal{F}_1&=2\varphi_tV_{0\tau t}+\varphi_{tt}V_{0\tau}-\eta(x)\varphi_tV_{0\tau\tau\tau}\\ &\quad-3\rho_2(x)\varphi_t^2\cos(u_0(x,t)+V_0)V_{0\tau}^2\\ &\quad+2\rho_1(x)\varphi_tV_{0\tau}(u_{0t}(x,t)+V_{0t})\\ &\quad-u_1(x,t)\cos(u_0(x,t)+V_0)+\delta_1(x,t)-\tau\check{\rho}_{1x}\varphi_t^2V_{0\tau}^2. \end{aligned} \tag{1.169}$$

Taking the limit as $\tau\to\pm\infty$ in (1.168), we find the relationship between the first corrections to the background and the displacement current:

$$u_1(x,t)=\frac{\delta_1(x,t)}{\sqrt{1-\delta_0^2(x,t)}}. \tag{1.170}$$

Using (1.170) and the explicit formula (1.169), we see that the function V_1 tends to zero exponentially as $\tau\to\pm\infty$. This enables us to set $V_1=V_1(\tau,t)$ and to expand the functions in the left-hand side of (1.168) by the Taylor formula around $x=\varphi(t)$. Now, equating to zero the term of order ε, we derive from (1.168) the equation

$$L_1V_1+F_1(\tau,t)=0, \tag{1.171}$$

where $L_1=L(\varphi,t,\tau,\partial/\partial\tau)$, $F_1=\mathcal{F}(\tau,\varphi,t)$.

Let us examine the solvability of equation (1.171) in the class $\mathcal{S}$ of functions decreasing exponentially as $\tau\to\pm\infty$. The operator L_1 is not symmetric; however, we can find the kernel of the operator L_1^* from the following simple considerations. First of all, observe that the equation $L_1z=0$ has the solutions $z_1=V_{0\tau}$ and z_2 which exponentially increases in τ. Next, the substitution

$$V=w\exp\Big(\frac{1}{2}\mu V_0\Big),\quad \mu=\frac{2\check{\rho}_1\varphi_t^2}{1-\varphi_t^2}$$

yields the formula

$$L_1V=\exp\Big(\frac{1}{2}\mu V_0\Big)L_2w,\quad L_2=(1-\varphi_t^2)\frac{\partial^2}{\partial\tau^2}-q(\tau,t)$$

with $q=\cos(\check{u}_0+V_0)+(1-\varphi_t^2)\mu^2V_{0\tau}^2/4-(1-\varphi_t^2)\mu V_{0\tau\tau}/2$.

It is clear that the functions $w_i = z_i \exp(-\mu V_0/2)$ satisfy the equation $L_2 w = 0$. Therefore

$$\left(L_1 V_1, z_1 e^{-\mu V_0}\right) = \left(e^{\mu V_0/2} L_2 e^{-\mu V_0/2} V_1, z_1 e^{-\mu V_0}\right) = \left(e^{-\mu V_0/2} V_1, L_2 w_1\right) = 0.$$

Since w_2 does not belong to the dual space of $\mathcal{S}$, it is necessary and sufficient for solvability of (1.171) in $\mathcal{S}$ that $F_1 \in \mathcal{S}$ and the following orthogonality condition be fulfilled:

$$\int_{-\infty}^{\infty} F_1(\tau, t) e^{-\mu V_0} V_{0\tau}\, d\tau = 0. \tag{1.172}$$

The condition $F_1 \in \mathcal{S}$ is fulfilled due to equality (1.170). Consider condition (1.172). To compute the corresponding integrals, observe that

$$\int_{-\infty}^{\infty} V_{0\tau} \cos V_0 e^{-\mu V_0}\, d\tau = -\mu\Big\{1 - e^{-2\pi\mu} + \mu \int_{-\infty}^{\infty} V_{0\tau} \cos V_0 e^{-\mu V_0}\, d\tau\Big\},$$

which can be easily established by integration by parts. Hence

$$\int_{-\infty}^{\infty} V_{0\tau} \cos V_0 e^{-\mu V_0}\, d\tau = -\frac{\mu}{1+\mu^2}\psi,$$

where $\psi = 1 - \exp(-2\pi\mu(t))$. In a similar way,

$$\int_{-\infty}^{\infty} V_{0\tau} \sin V_0 e^{-\mu V_0}\, d\tau = -\frac{\psi}{1+\mu^2},$$

$$\int_{-\infty}^{\infty} V_{0\tau} \sin 2V_0 e^{-\mu V_0}\, d\tau = -2\frac{\psi}{4+\mu^2},$$

$$\int_{-\infty}^{\infty} V_{0\tau}(1 - \cos V_0)(\cos V_0 - \mu \sin V_0) e^{-\mu V_0}\, d\tau = -2\frac{\psi}{\mu(4+\mu^2)}.$$

Moreover, it can be easily shown that

$$V_{0\tau}^2 = 2\beta^2(1 - \cos V_0), \quad V_{0\tau\tau} = \beta^2 \sin V_0,$$

$$V_{0t} = \frac{(\beta_t \tau + (\beta\varphi_1)_t) V_{0\tau}}{\beta}.$$

Therefore (1.172) can be rewritten as

$$\begin{aligned} &2\int_{-\infty}^{\infty} e^{-\mu V_0} \frac{\partial}{\partial t}\big(\varphi_t \beta^2 (1 - \cos V_0)\big)\, d\tau \\ &+ 2\varphi_t\beta^2\Big(2\check{\rho}_1 - \check{\rho}_{1x}\varphi_t\frac{\beta}{\beta_t}\Big) \int_{-\infty}^{\infty} (1 - \cos V_0) V_{0t} e^{-\mu V_0}\, d\tau \\ &+ 4\varphi_t\beta^2\check{\rho}_1\check{u}_{0t} \int_{-\infty}^{\infty} (1 - \cos V_0) e^{-\mu V_0}\, d\tau \\ &+ \check{\eta}\varphi_t\beta^4 \int_{-\infty}^{\infty} \Big(\sin^2 V_0 - \frac{4}{3}(1 - \cos V_0)^2\Big) e^{-\mu V_0}\, d\tau \\ &- \frac{\psi}{\mu}\Big\{\check{\delta}_1 + 2\beta^2\varphi_t^2\Big(\check{\rho}_{1x}\frac{(\beta\varphi_1)_t}{\beta_t(1+\mu^2)} + 6\check{\rho}_2\frac{\check{\delta}_0}{\mu(4+\mu^2)}\Big)\Big\} = 0. \end{aligned} \tag{1.173}$$

Let us introduce the functions

$$\sigma_1(\mu) = \int_{-\infty}^{\infty} \cosh^{-2} y e^{-\mu v(y)}\, dy,$$
$$\sigma_2(\mu) = \int_{-\infty}^{\infty} y \cosh^{-3} y e^{-\mu v(y)}\, dy,$$
$$\sigma_3(\mu) = \int_{-\infty}^{\infty} \cosh^{-4} y e^{-\mu v(y)}\, dy,$$

where $v(y) = 4\arctan(\exp(-y))$. For any function $f(x,t) \in C^1$ we will write $\dot{f} = df(\varphi, t)/dt$. Then rather simple but cumbersome transformations allow us to rewrite (1.173) as

$$\check{\delta}_1 = -\kappa_1\{(2\check{\rho}_1 + \mu)\varphi_{1t} + \dot{\rho}_1\varphi_1\} + \mu_t\kappa_2 - \kappa_3\check{\delta}_0 + \kappa_4. \tag{1.174}$$

Here

$$\kappa_1 = \left(\frac{\mu(2\rho_1 + \mu)}{\rho_1^2(1+\mu^2)}\right)^{1/2}\Bigg|_{x=\varphi},$$
$$\kappa_2 = \kappa_0\left\{\sigma_1 + \frac{\partial\sigma_1}{\partial\mu} + \frac{2\rho_1\mu - 1}{\mu(1+\mu^2)}(\sigma_1 + \mu\sigma_2)\right\},$$
$$\kappa_0 = \frac{1}{\psi}\left(\frac{8\mu^3}{\rho_1\sqrt{1+\mu^2}}\right)^{1/2}\Bigg|_{x=\varphi},$$
$$\kappa_3 = \frac{6\rho_2}{\rho_1(4+\mu^2)\sqrt{1+\mu^2}}\Bigg|_{x=\varphi},$$
$$\kappa_4 = \kappa_0\left\{\frac{\dot{\rho}_1}{\rho_1}\left[(2\rho_1 + \mu)\sigma_2 - \frac{1}{2}\sigma_1\right]\right.$$
$$\left. + \eta\frac{2\rho_1 + \mu}{2\rho_1\sqrt{1+\mu^2}}\left[\sigma_1 - \left(1 + \frac{4}{3}\mu^2\right)\sigma_3\right]\right\}\Bigg|_{x=\varphi}.$$

We assume that $\check{\rho}_1 \neq 0$ and $0 < \varphi_t < 1$. This implies that all coefficients in (1.174) are smooth bounded functions and, moreover, $\kappa_1 \neq 0$ and $2\check{\rho}_1 + \mu = 2\check{\rho}_1/(1-\varphi_t^2) \neq 0$.

Now (1.174) is obviously fulfilled. Hence we find from (1.171) that

$$V_1(\tau, t) = \omega_1(\tau, t) + \varphi_2 V_{0\tau},$$

where $\varphi_2 = \varphi_2(t)$ is the "constant" of integration and $\omega_1 \in \mathcal{S}$ is a partial solution of the nonhomogeneous equation, which can be obtained, e.g., by means of formula (1.24) using the above substitution.

The subsequent calculations are carried through in a similar way. At the jth step the condition of the form (1.172) implies a relationship, analogous to (1.174), between the jth correction $\check{\delta}_j$ to the displacement current and the "constant" φ_j.

Let us discuss the conditions (1.167) and (1.174). If the trajectory $x = \varphi(t) - \varepsilon\varphi_1(t)$ of the motion of the distorted fluxon is preassigned with $0 < \varphi_t < 1$, then, taking a natural extension of $\check{\delta}_0$ and $\check{\delta}_1$ outside the curve Γ_t, the equalities (1.167) and (1.174) determine the "control"

$$\delta_0(x,t) = \frac{2\rho_1(x)\varphi_t^2}{\sqrt{(1-\varphi_t^2)^2 + 4\rho_1^2(x)\varphi_t^4}},$$

as well as the correction to the displacement current $\delta_1(x,t)$, which is obtained from (1.174) with $\rho_i = \rho_i(x)$, $\eta = \eta(x)$, and $\dot{\rho}_1 = \rho_{1x}(x)\varphi_t$.

On the other hand, if the functions δ_0, δ_1 are specified in advance subject to the condition $|\delta_0| < 1$, then the trajectory is determined as the solution to the equations

$$\left(\frac{d\varphi}{dt}\right)^2 = \left(1 + 2\frac{\rho_1(\varphi)}{\delta_0(\varphi)}\sqrt{1-\delta_0^2(\varphi)}\right)^{-1}, \quad \frac{d\varphi_1}{dt} + p(t)\varphi_1 = F(t),$$

which can be easily obtained from (1.167), (1.174) by solving them for φ_t and φ_{1t}.

Note that in any case we arrive at a special Cauchy problem for equation (1.158), where the initial value of the derivative cannot be specified arbitrarily. It is also clear that when the coefficients and the right-hand side of (1.158) are constant, then φ and φ_1 are linear functions of t. This case was examined in [**205**].

CHAPTER 2

Nonlinear Waves in Multidimensional Media

As a multidimensional analog of the solutions discussed in Chapter 1 we consider functions whose fast variation at any time moment t is concentrated in a small neighborhood (of width $\sim \varepsilon^{1-\delta}$, $0 < \delta < 1$) of a smooth surface Γ_t of codimension 1. The position of the surface Γ_t, which is naturally referred to as the wave front, as well as its shape, varies with time, and the main problem is to determine the dynamics of the front.

Multidimensional analogs of the soliton-type solutions discussed above have the form of narrow (in the transversal direction to Γ_t) crests whose parameters vary slowly along Γ_t. Such solutions arise, e.g., in various problems of plasma physics and hydrodynamics. The physical interpretation of multidimensional solutions of smoothed shock wave type is obvious.

Actually, it is merely a technical task to carry over the ideas developed in Chapter 1 to the multidimensional case. The only new aspect here is the necessity of taking the geometry of the front Γ_t into account. This should be done in accordance with specific features of a concrete physical problem. In particular, in this chapter we will consider two multidimensional equations (Kadomtsev–Petviashvili and sine-Gordon) whose different physical natures determine different ways of describing their fronts. We will also consider examples of constructing asymptotic solutions for systems of multidimensional equations.

2.1. Dynamics of two-dimensional distorted solitons

The Kadomtsev–Petviashvili equation

$$\frac{\partial}{\partial x}\left(4\frac{\partial u}{\partial t} + 6u\frac{\partial u}{\partial x} + \varepsilon^2\frac{\partial^3 u}{\partial x^3}\right) + 3\sigma\frac{\partial^2 u}{\partial y^2} = 0, \tag{2.1}$$

where $\sigma = \pm 1$, was derived to describe the waves in weakly dispersive and weakly dissipative plasmas, and, in particular, to describe the phenomenon of self-focusing [**95**]. This equation, which is often referred to as the two-dimensional KdV equation, describes slow variations along the y-axis of a wave that propagates in the direction of the x-axis. Due to the importance of its physical applications and to the fact that (2.1) is the first multidimensional equation which was integrated by the inverse scattering problem method [**269**], the Kadomtsev–Petviashvili equation is widely known among physicists and mathematicians. A detailed bibliography on the Kadomtsev–Petviashvili equation can be found, e.g., in [**37**]. Here we note only that for $\sigma = 1$ the solution to equation (2.1) is unstable with respect to the transversal perturbations, while for $\sigma = -1$ it is asymptotically stable [**94, 264, 269**]. However, $\sigma = 1$ is the most interesting case from the viewpoint of physical applications, because this is the sign for which the equation (2.1) describes the self-focusing phenomenon.

As a natural analog to the Kadomtsev–Petviashvili model for the case of nonhomogeneous media one can consider the system

$$
\begin{aligned}
&\frac{\partial u}{\partial t} + \Big(H^{1/2} + \frac{3}{2}\alpha H^{-3/4}u\Big)\frac{\partial u}{\partial x} \\
&\quad + \varepsilon^2 \frac{1}{6} H^{3/2}\frac{\partial}{\partial x}\Big(H^{1/2}\frac{\partial}{\partial x}\Big(H^{1/2}\frac{\partial u}{\partial x}\Big)\Big) + \frac{\kappa}{2}H^{1/4}\frac{\partial}{\partial y}\Big(H\frac{\partial w}{\partial y}\Big) = 0, \\
&H^{3/4}\frac{\partial w}{\partial x} = u,
\end{aligned}
\tag{2.2}
$$

which was derived in [**52**] to describe of the dynamics of surface waves in a basin of variable depth $H = H(x,y) > 0$ (normalized by a characteristic value h). Here the meaning of the dimensionless function u is the elevation of the free surface, and w is the potential of velocity of the wave motion. The parameters $\alpha = a/h$, $\varepsilon = \lambda/h \ll 1$, and $\kappa = (\lambda/L)^2$ are related to the ratios of the amplitude a and the wave lengths λ and L in the x- and y-directions, to one another, and to the characteristic depth h.

Obviously, for $H = \text{const}$ the system (2.2) can be easily reduced to an equation of type (2.1). In the one-dimensional case ($H = H(x)$, $u = u(x,t,\varepsilon)$) the first equation (2.2), after the substitution $u = H^{1/4}u'$, differs from the equation (1.86), (1.87) in [**203**] only in the terms containing the derivatives of H. For steep slopes of the bottom the model (2.2) is claimed in [**52**] to give a more adequate description of the wave transformation than that given by (1.86), (1.87).

We will consider the dynamics of a distorted two-dimensional soliton (a crest) moving over a smooth variable background [**150**]. Denote by $\Gamma_0 = \{(x,y),\ S^0(x,y) = 0\}$ the initial front of the soliton. We will assume that Γ_0 is a smooth one-dimensional curve without self-intersections. Since the solutions of equations (2.1), (2.2) are not symmetrically dependent on the variables x and y, which was envisaged when deriving the Kadomtsev–Petviashvili equations, it is natural to assume that the curve Γ_0 has a single-valued projection on the y-axis. This implies that $\partial S^0/\partial x|_{\Gamma_0} \neq 0$, and the equation $S^0(x,y) = 0$ can be solved for x. Thus it is natural to specify the initial front of the soliton in the form $\Gamma_0 = \{(x,y),\ x = \varphi^0(y)\}$. It is also natural to assume that the existence of the single-valued projection of the front to the y-axis remains valid for some time $t \in [0,T]$. Consequently, for $t \le T$ the wave front Γ_t can be described by a smooth function $\varphi(y,t)$, so that $\Gamma_t = \{(x,y),\ x = \varphi(y,t)\}$. Consider a soliton-type function $f(\tau,x,y,t) \in C^\infty$, which tends to zero as $\tau \to \pm\infty$ exponentially together with all its derivatives (as before, we will denote the class of such functions by $\mathcal{S}$). Since the variable x plays the role of the coordinate tranversal to Γ_t, i.e., $x - \varphi(y,t)$ is a sort of "distance" from Γ_t, we have

$$
\begin{aligned}
f\Big(\frac{x-\varphi(y,t)}{\varepsilon}, x, y, t\Big) = \Big\{ & f(\tau,\varphi,y,t) + \varepsilon\tau \frac{\partial f(\tau,x,y,t)}{\partial x}\Big|_{x=\varphi} \\
& + \varepsilon^2 \frac{\tau^2}{2}\frac{\partial^2 f(\tau,x,y,t)}{\partial x^2}\Big|_{x=\varphi} + \cdots \Big\}\Big|_{\tau=(x-\varphi(y,t))/\varepsilon}.
\end{aligned}
$$

This simple fact enables us to draw the important conclusion that soliton-type functions do not depend on the "slow" variable x up to order $O(\varepsilon)$.

We will present a detailed derivation of the asymptotic soliton-type solution for the standard Kadomtsev–Petviashvili equation (2.1), and at the end of this section we will state the results related to equation (2.2).

Let us introduce the auxiliary functions involved in the leading term of the asymptotic solution to (2.1). For this we denote by $u_0(x, y, t)$ the solution to the degenerate Kadomtsev–Petviashvili equation

$$\frac{\partial}{\partial x}\Big(4\frac{\partial u_0}{\partial t} + 6u_0\frac{\partial u_0}{\partial x}\Big) + 3\sigma\frac{\partial^2 u_0}{\partial y^2} = 0. \tag{2.3}$$

Assuming that the solution to this equation remains smooth for $t < T$, the function u_0 can be viewed as the "background" over which a soliton-type wave propagates. Moreover, denote by $\varphi = \varphi(y, t)$ and $A = A(y, t)$ the solution of the equations

$$\frac{\partial \varphi}{\partial t} = \frac{3}{4}\sigma\Big(\frac{\partial\varphi}{\partial y}\Big)^2 + \frac{1}{2}A + \frac{3}{2}u_0(\varphi, y, t), \tag{2.4}$$

$$\frac{\partial A}{\partial t} = \sigma A\frac{\partial^2\varphi}{\partial y^2} + \frac{3}{2}\sigma\frac{\partial A}{\partial y}\frac{\partial\varphi}{\partial y} + Au_{0x}(\varphi, y, t), \tag{2.5}$$

and let the functions φ and A also be smooth for $t < T$, with $A > 0$.

Then the asymptotic soliton-type solution of equation (2.1) has the form

$$u = u_0(x, y, t) + A(y, t)\cosh^{-2}\Big(\sqrt{\frac{A}{2}}\frac{x - \varphi(y, t) + \varepsilon\varphi_1(y, t)}{\varepsilon}\Big) + O(\varepsilon). \tag{2.6}$$

Obviously, the function φ describes the position Γ_t of the distorted soliton, $\Gamma_t = \{(x, y),\ x = \varphi(y, t)\}$, with accuracy $O(\varepsilon)$, while the function φ_1 provides the correction to Γ_t of order ε. (The derivation of φ_1 will be given below.) It is also clear that the meaning of the function A is the amplitude of the soliton.

Let us discuss the leading term of the asymptotics. Note first of all that the model equations (2.4) and (2.5) imply the relation

$$\begin{aligned}\frac{\partial}{\partial t}\Big\{\frac{1}{2}\Big(\frac{\partial\varphi}{\partial y}\Big)^2 - \sigma A\Big\} &+ \frac{\partial}{\partial y}\Big\{A\frac{\partial\varphi}{\partial y} - \frac{1}{2}\sigma\Big(\frac{\partial\varphi}{\partial y}\Big)^3\Big\}\\ &= \Big\{\frac{3}{2}\Big(\frac{\partial\varphi}{\partial y}\Big)^2 - A\sigma\Big\}u_{0x}(\varphi, y, t) + \frac{3}{2}u_{0y}(\varphi, y, t)\frac{\partial\varphi}{\partial y},\end{aligned} \tag{2.7}$$

which obtains if we differentiate (2.4) with respect to y, multiply by $\varphi(y)$, and subtract the result from (2.5).

We will assume that the perturbations of the front and amplitude of the soliton are localized in a neighborhood of the plane $y = 0$. This means that

$$A = A_*, \qquad \frac{\partial\varphi}{\partial y} = k, \qquad u_0 = \text{const} \qquad \text{for} \quad |y| \geq Y, \tag{2.8}$$

where $Y \gg 1$ is a constant. Then, on integrating (2.7), we obtain the energy relationship

$$\begin{aligned}&\frac{d}{dt}\int_{-Y}^{Y}\Big\{\frac{1}{2}\Big(\frac{\partial\varphi}{\partial y}\Big)^2 - \sigma A\Big\}\,dy\\ &= \int_{-Y}^{Y}\Big\{\Big(\frac{3}{2}\Big(\frac{\partial\varphi}{\partial y}\Big)^2 - A\sigma\Big)u_{0x}(\varphi, y, t) + \frac{3}{2}u_{0y}(\varphi, y, t)\frac{\partial\varphi}{\partial y}\Big\}\,dy.\end{aligned} \tag{2.9}$$

As is easily seen, the integrand in the left-hand side of (2.9) does not have a fixed sign for $\sigma = 1$. This is indicative of instability of the soliton (2.6) for $\sigma = 1$. Next, observe that in the unstable case the equations (2.4), (2.5) are virtually the equations of one-dimensional gas dynamics for monatomic polytropic gas. Indeed, set

$$\rho = A^{3/2}, \quad v = -\frac{3}{2}\sigma\frac{\partial\varphi}{\partial y}.$$

Then the equations (2.4), (2.5) can be rewritten as

$$\begin{aligned}\frac{\partial\rho}{\partial t} + \frac{\partial}{\partial y}(\rho v) &= f_1,\\ \rho\left(\frac{\partial v}{\partial t} + v\frac{\partial v}{\partial y}\right) + \frac{\partial p}{\partial y} &= -f_2,\end{aligned} \tag{2.10}$$

where $p = \kappa\rho^\gamma$, $\kappa = \frac{3}{10}\sigma$, $\gamma = \frac{5}{3}$, and the functions f_1, f_2 are given by

$$f_1 = \frac{3}{2}\rho u_{0x}(\varphi, y, t), \quad f_2 = \frac{9}{4}\sigma\rho\frac{\partial}{\partial y}u_0(\varphi, y, t).$$

Thus the influence of the smooth background u_0 can be interpreted as appearance of the mass "source" f_1 and the "external force" f_2.

Subsequently we restrict ourselves to the case where $\sigma = 1$ and $u_0 \equiv 0$. Then (2.10) is a hyperbolic system, and hence conditions (2.8) are fulfilled for all finite t provided that they are fulfilled at $t = 0$. To analyze the model equations we can use all facts that are known for equations of gas dynamics (see, e.g., [**219, 262**]). In particular, along with (2.9) we have the conservation laws for mass and impulse,

$$\frac{d}{dt}\int_{-\infty}^{\infty}\rho\,dy = 0, \quad \frac{d}{dt}\int_{-\infty}^{\infty}\rho v\,dy = 0.$$

Further, it is not hard to compute the Riemann invariants r and s and to establish that

$$\begin{aligned} r &= v + \frac{3}{\sqrt{2}}\rho^{1/3} = \text{const} \quad \text{along the characteristics } Y_r\colon \frac{dY_r}{dt} = \frac{1}{3}(2r+s),\\ s &= v - \frac{3}{\sqrt{2}}\rho^{1/3} = \text{const} \quad \text{along the characteristics } Y_s\colon \frac{dY_s}{dt} = \frac{1}{3}(r+2s).\end{aligned}$$

As is well known, the solution of a gas dynamics equation substantially simplifies for special initial data which correspond to the so-called Riemann waves, i.e., in the case where either $r|_{t=0}$ or $s|_{t=0}$ are constant. Let, for example, $r|_{t=0} = r_0$ be independent of y. Then, obviously,

$$v = \frac{1}{2}(r_0 + s_0(y^0))\big|_{y^0=y^0(y,t)}, \quad \rho = \left(\frac{r_0 - s_0(y^0)}{3\sqrt{2}}\right)^3\Bigg|_{y^0=y^0(y,t)}. \tag{2.11}$$

Here $y^0(y,t)$ is the solution to equation $y = Y_s(t, y^0)$, where

$$Y_s = \frac{1}{3}(r_0 + 2s_0(y^0))t + y^0,$$

and $s_0(y) = (v - 3/\sqrt{2}\rho^{1/3})|_{t=0}$. It is clear that the solution (2.11) can exist only if the Jacobian does not vanish,

$$J = 1 + \frac{2}{3}\frac{\partial s_0}{\partial y^0} t \neq 0. \tag{2.12}$$

In accordance with (2.8), let

$$\rho^{1/3}\big|_{t=0} = \rho_*^{1/3} + (\rho^0(y))^{1/3}, \quad v\big|_{t=0} = -\frac{3}{2}k + v^0(y),$$

where $\rho_* = A_*^{3/2} = \text{const}$ and $\rho^0, v^0 = 0$ for $|y| \geq Y$. Then the assumption $r = \text{const}$ implies that $v^0 = c - 3/\sqrt{2}(\rho^0)^{1/3}$, where $c = \text{const}$. Hence

$$\frac{\partial s_0}{\partial y} = -\frac{\sqrt{2}}{(\rho^0)^{2/3}}\frac{\partial \rho^0}{\partial y},$$

so that on the characteristics Y_s originating at the points y^0 with $\partial\rho^0/\partial y|_{y=y^0} \leq 0$ a rarefaction wave is formed, whereas for $\partial\rho^0/\partial y|_{y=y^0} > 0$ a compression wave is formed. Accordingly, in the former case the characteristics diverge and in the rarefaction wave the gradients of the functions ρ and v decrease in absolute value with time t. On the contrary, in the compression wave the characteristics of the corresponding family converge and the gradients of ρ and v increase in absolute value with t. One can easily calculate the first time moment T for which the characteristics intersect:

$$T = \frac{3}{2\sqrt{2}}\left\{\max_y (\rho^0)^{-2/3}\frac{\partial \rho^0}{\partial y}\right\}^{-1}.$$

Denote by y_*^0 the value of y maximizing $\rho_y^0(\rho^0)^{-2/3}$. For $t < T$ the amplitude A and the function φ specifying the position of the front are given by the formulas

$$A = \rho^{2/3}(y,t), \quad \varphi = ky + \frac{1}{2}\Big(A_* + \frac{3}{2}k^2\Big)t - \frac{2}{3}\int_{-\infty}^{y}\Big(v(y',t) + \frac{3}{2}k\Big)\,dy'.$$

As $t \to T$, the so-called gradient catastrophe occurs, because for $y \to y_* = Y_s(T, y_*^0)$ the gradients of ρ and v increase unboundedly. Such a behavior of the solution of model equations (2.10) means that at the point (T, y_*) the functions A and φ lose the smoothness property: the front bends and the amplitude of the soliton gets a jump at the bend point.

Prior to constructing asymptotics, we discuss the results of numerical solution of the Kadomtsev–Petviashvili equation (2.1) (the computations have been carried out jointly with A. Titov).

In numerical modeling we replace the Cauchy problem by the initial boundary value problem in the square $[0, 10] \times [0, 10]$ with the Neumann conditions on its boundary. We compute the solution by using the finite difference scheme constructed with the help of the explicit Lax–Wendroff scheme.

For the initial value we take the distorted soliton

$$u\big|_{t=0} = A_* \cosh^{-2}\Big(\sqrt{\frac{A_*}{2}}\,\frac{x - \varphi^0(y)}{\varepsilon}\Big)$$

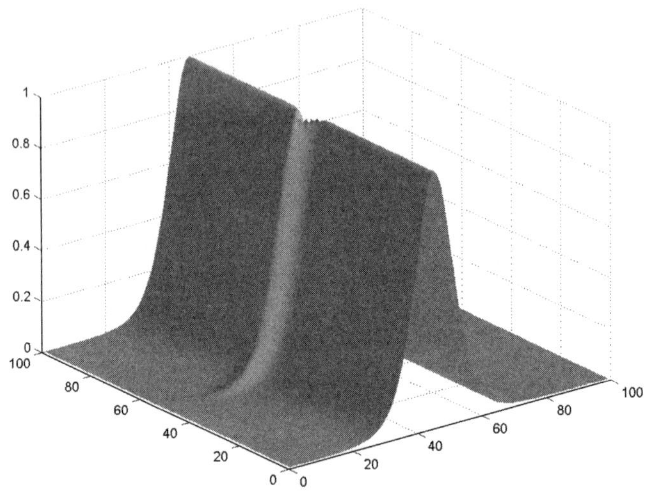

FIGURE 2.1. Spatial profile of the initial data.

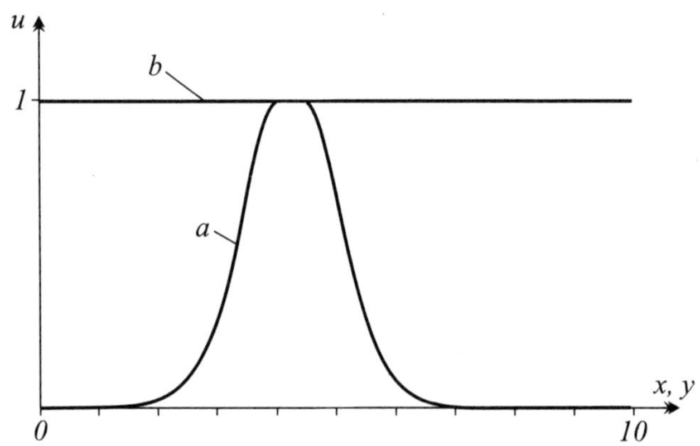

FIGURE 2.2. Profiles of the initial data. Curve a is the projection of u on the (u, x)-plane, and curve b is the amplitude of u at $t = 0$ on the line $x = 4.5$.

with unit amplitude ($A_* = 1$) concentrated near the front $\Gamma_0 = \{x = \varphi^0(y)\}$ (curve 0 in Figure 2.3), where

$$\varphi^0 = 4.5 - \frac{1}{2}\cosh^{-2}\left(\frac{y-5}{0.25\sqrt{2}}\right).$$

Figure 2.1 shows the three-dimensional plot $u\big|_{t=0}$ with $\varepsilon = 0.4$, curve a in Figure 2.2 shows the projection of the three-dimensional plot of $u\big|_{t=0}$ on the (u, x)-plane (along the y-axis), and curve b in Figure 2.2 is the amplitude of $u\big|_{t=0}$ on the front of the nonperturbed soliton, i.e., along the line $\Gamma_0' = \{x = 4.5, y \in [0, 10]\}$.

The plots in Figures 2.3–2.7 correspond to the solution of the Kadomtsev–Petviashvili equation in the stable case ($\sigma = -1$). Figure 2.3 illustrates the dynamics of the front of a distorted soliton. One can readily see that the perturbation

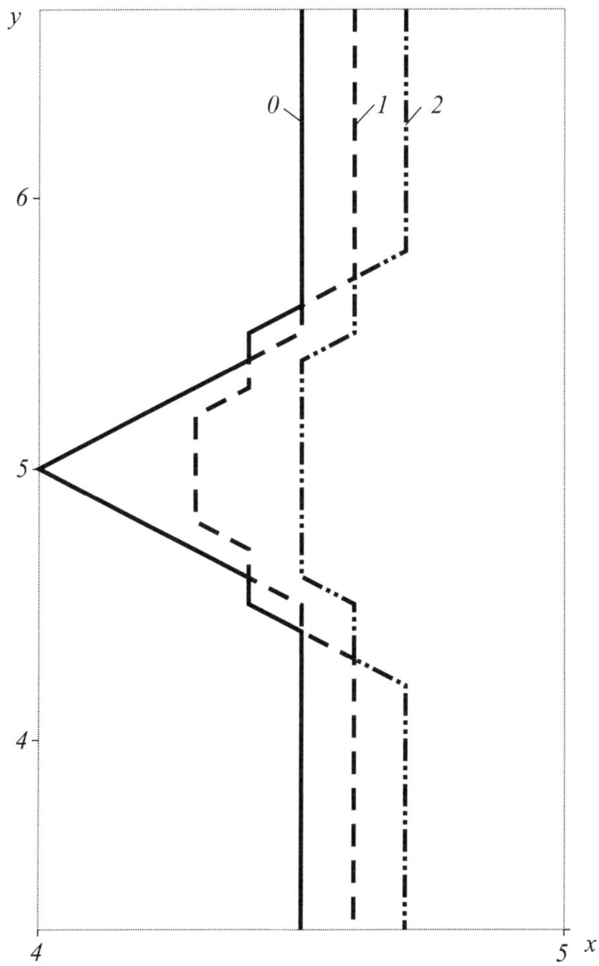

FIGURE 2.3. Dynamics of the front of the distorted soliton in the stable case ($\sigma = -1$). Curve 0 is the initial position; curves 1 and 2 are the front positions at $t = 0.1$ and $t = 0.2$.

domain of the front Γ_t is expanding with increasing t. However, in this case the perturbation amplitude decreases and the front has a clear tendency to become straight in a smooth way.

The three-dimensional plots of the solution at time moments $t = 0.1$ and $t = 0.2$ are shown in Figures 2.4 and 2.6. For better understanding, Figures 2.5 and 2.7 show the projection of the three-dimensional plot of u on the (u, x)-plane (along the y-axis) (curve a), and the amplitude of u along the front of the nonperturbed soliton (i.e., along the line $x = 4.5 + t/2$) at $t = 0.1$ and $t = 0.2$. We see that a bend in the front Γ_0 results in a soliton-type perturbation of the amplitude (curve b in Figure 2.5). This perturbation decreases with time and at the same time generates waves on the crest of the soliton, which propagate in opposite directions (curve b in Figure 2.7). One can clearly see that behind the front of the distorted soliton there appears a tail (see Figures 2.4 and 2.6 and curves a in Figures 2.5 and 2.7).

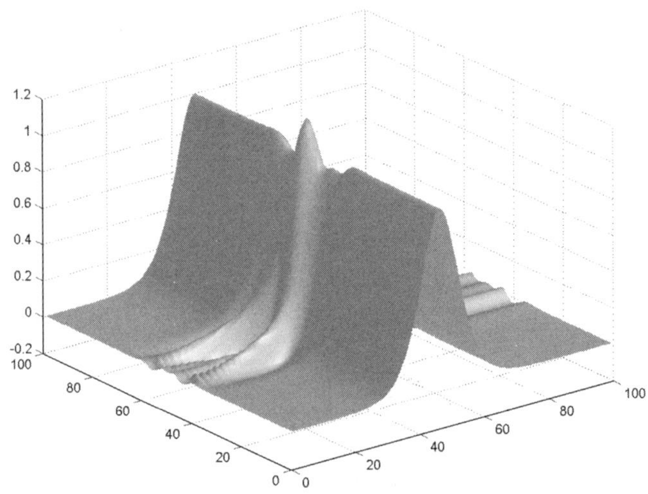

FIGURE 2.4. Stable case ($\sigma = -1$). The solution at $t = 0.1$.

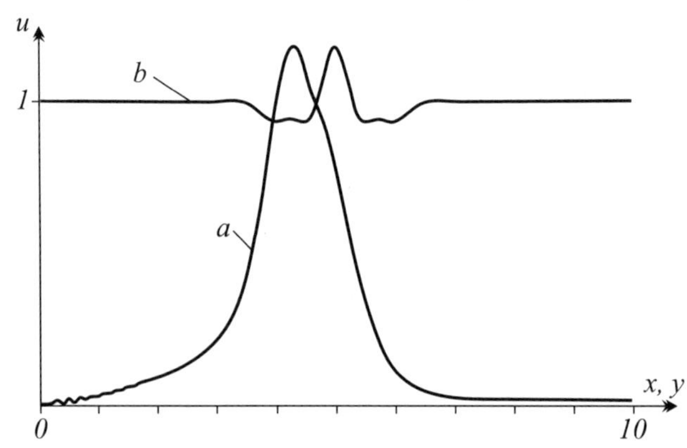

FIGURE 2.5. Profiles of the solution at $t = 0$ in the stable case. Curve a is the projection of the solution on the (u, x)-plane, and curve b is the amplitude of the solution on the front of the nonperturbed soliton.

With increasing t the tail region expands along both the y- and x-axes, which includes the direction opposite to that of the crest motion. For small x there are oscillations in the tail, which can be explained by reflection from the boundary $x = 0$ of the computation region. The amplitude of these oscillations decreases when the boundary moves to the left. A perturbation in the solution ahead of the front of the distorted soliton is related to the error of numerical modeling. This perturbation is of the same order of magnitude as the local error of the difference scheme and decreases with discretization steps.

The results of numerical computations for the Kadomtsev–Petviashvili equation with the same initial data (Figure 2.1) in the unstable case ($\sigma = 1$) are shown in Figures 2.8–2.12.

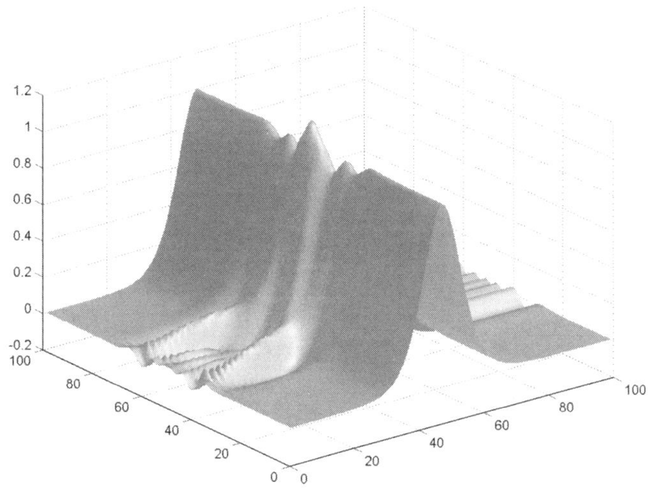

FIGURE 2.6. Stable case ($\sigma = -1$). The solution at $t = 0.2$.

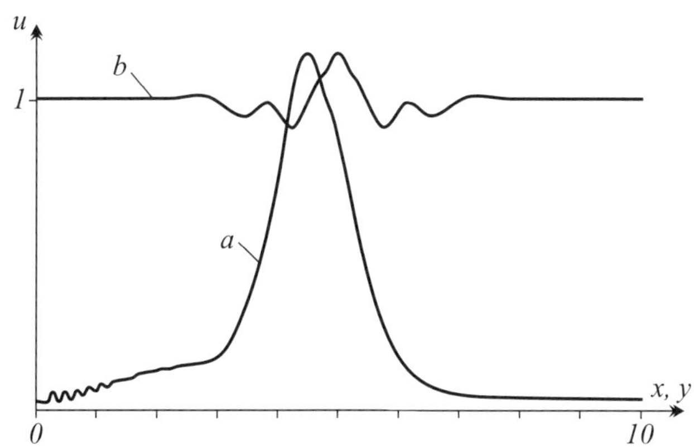

FIGURE 2.7. Profiles of the solution at $t = 0.2$ in the stable case.

Because of the change in the sign of σ, two regions on the wave front become steeper during the time $\Delta t = 0.1$, while in the stable case there was no such effect (compare curves 1 in Figures 2.3 and 2.8). Nevertheless, at $t = 0.1$ the solutions obtained for $\sigma = 1$ and $\sigma = -1$ remain close to each other qualitatively and quantitatively (compare the plots in Figures 2.4 and 2.5 and in Figures 2.9 and 2.10). However, qualitative distinctions between these solutions become clearer with increasing t. Namely, on the wave front there appear soliton-type perturbations which propagate rather fast in opposite directions (curve 2 in Figure 2.8).

As a whole, we see that the plots of the "exact" solution to equation (2.1) agree surprisingly well with the description of the solution given by the asymptotic formulas.

Now we proceed to the construction of the asymptotic soliton-type solution to equation (2.1). A natural extension of expansion (1.3) of Chapter 1 has the form

$$u = u_0(x, y, t) + V_0(\tau, y, t) + \varepsilon(u_1(x, y, t) + V_1(\tau, x, y, t)) + \dots, \tag{2.13}$$

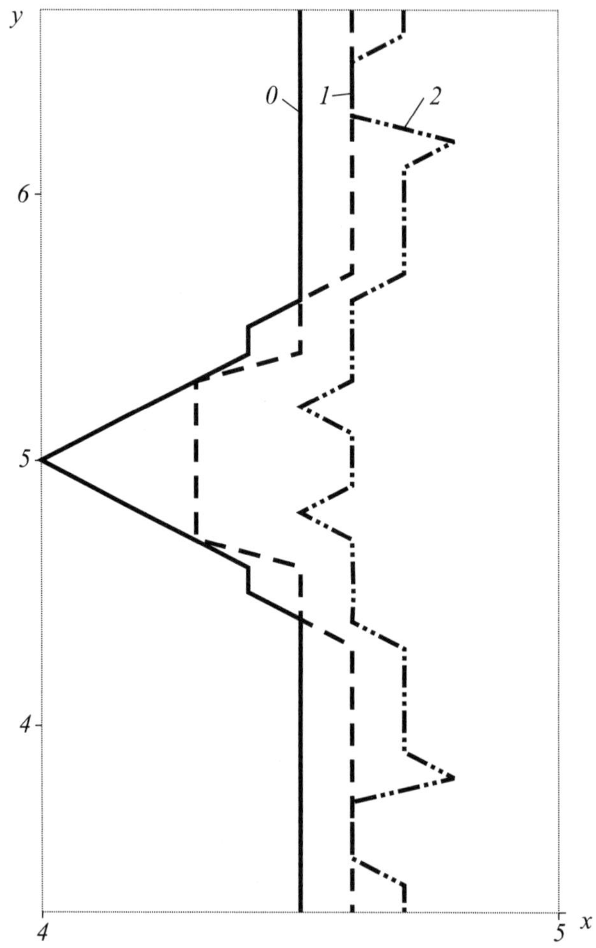

FIGURE 2.8. Dynamics of the front of the distorted soliton in the unstable case ($\sigma = 1$). Curve 0 is the initial position; curves 1 and 2 are the front positions at $t = 0.1$ and $t = 0.2$.

with $\tau = (x - \varphi(y,t))/\varepsilon$, where φ and the u_i are smooth functions, $V_0 \in \mathcal{S}$, and $V_i \in \mathcal{H}$ for $i \geq 1$. As before, we denote by $\mathcal{H}$ the class of functions stabilizing as $\tau \to \pm\infty$, i.e., $\mathcal{H} = \{f(\tau, x, y, t),\ \partial f/\partial\tau \in \mathcal{S}\}$. Moreover, we assume that $V_i \to 0$ as $\tau \to \infty$.

All constructions are carried through according to the scheme developed in the previous chapter. Namely, substituting the expansion (2.13) into (2.1), expanding the soliton-type functions by the Taylor formula around $x = \varphi(y,t)$, collecting the terms according to the order in ε, and equating to zero the coefficient of ε^{-2}, we obtain the ordinary differential equation

$$\frac{\partial^2}{\partial\tau^2}\{(3\sigma\varphi_y^2 - 4\varphi_t + 6\check{u}_0 + 3V_0)V_0 + V_{0\tau\tau}\} = 0.$$

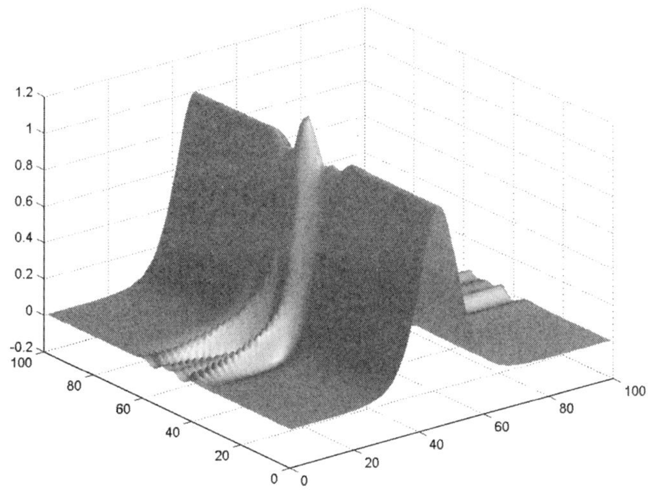

FIGURE 2.9. Unstable case ($\sigma = 1$). The solution at $t = 0.1$.

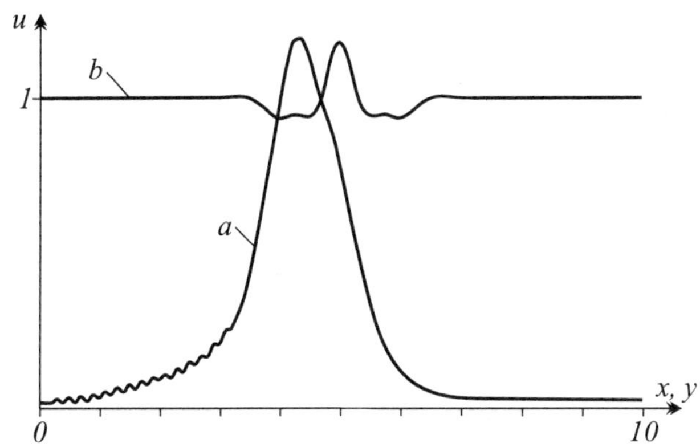

FIGURE 2.10. Profiles of the solution at $t = 0.1$ in the unstable case.

Here and subsequently, for any function f we denote $\check{f} = f(\tau, x, y, t)|_{x=\varphi(y,t)}$. Integrating this equation and assuming that

$$4\varphi_t - 3\sigma\varphi_y^2 - 6\check{u}_0 > 0,$$

we obtain the standard soliton solution of the KdV equation

$$V_0 = A\cosh^{-2}\beta(\tau + \varphi_1), \tag{2.14}$$

where $A = A(y,t) > 0$, $\beta = \sqrt{A/2}$, and $\varphi_1 = \varphi_1(y,t)$ is the "constant" of integration. We also get the first Hugoniot type condition (2.4).

Further, on repeating this procedure and equating to zero the term of order ε^{-1}, we obtain a nonhomogeneous equation in variations,

$$\frac{\partial^2}{\partial\tau^2}L\check{V}_1 + F_1(\tau, y, t) = 0, \tag{2.15}$$

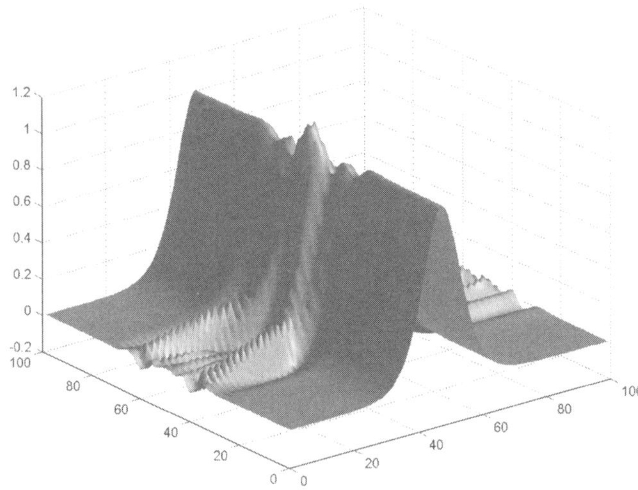

FIGURE 2.11. Unstable case ($\sigma = 1$). The solution at $t = 0.2$.

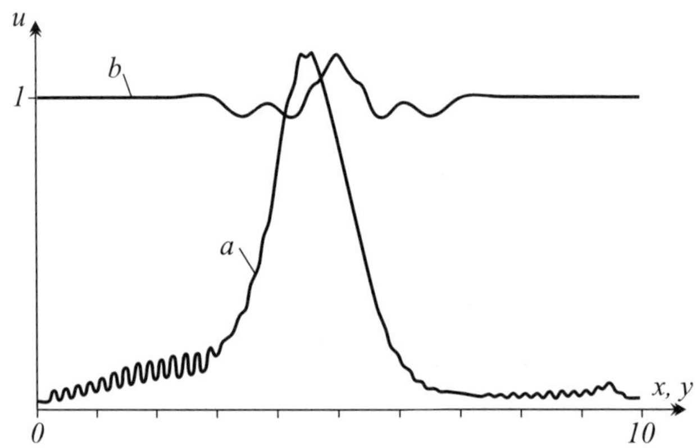

FIGURE 2.12. Profiles of the solution at $t = 0.2$ in the unstable case.

where

$$L = -2A + 6V_0 + \frac{\partial^2}{\partial \tau^2},$$
$$F_1 = \Pi V_{0\tau} + 6\check{u}_1 V_{0\tau\tau} + 6\check{u}_{0x} \tau V_{0\tau\tau},$$
$$\Pi = 4\frac{\partial}{\partial t} - 6\sigma\varphi_y \frac{\partial}{\partial y} - 3\sigma\varphi_{yy} + 12\check{u}_{0x}.$$

Integrating (2.15) with respect to τ, we obtain

$$\frac{\partial}{\partial \tau} L\check{V}_1 + \mathcal{F}_1(\tau, y, t) = 0, \tag{2.16}$$

where $\mathcal{F}_1 = \Pi V_0 + 6\check{u}_1 V_{0\tau} + 6\check{u}_{0x}(\tau V_{0\tau} - V_0)$.

Since $\mathcal{F}_1 \in \mathcal{S}$, arguing as for the KdV equation, we conclude that the following orthogonality condition is necessary and sufficient for solvability of (2.16) in $\mathcal{H}$:

$$\int_{-\infty}^{\infty} \mathcal{F}_1(\tau, y, t) V_0 \, d\tau = 0. \tag{2.17}$$

One can easily check that (2.17) yields the second Hugoniot type condition (2.5).

Now we find the solution of (2.16):

$$\check{V}_1(\tau, y, t) = \nu_1(y, t)\chi(\tau, y, t) + \omega_1(\tau, y, t),$$

and determine its extension in the form

$$V_1(\tau, x, y, t) = V_1^-(x, y, t)\chi(\tau, y, t) + \omega_1(\tau, y, t). \tag{2.18}$$

Here

$$\chi = \frac{1 - \tanh(\beta(\tau + \varphi_1))}{2}, \quad \nu_1 = \frac{\sigma\varphi_{yy} - 2\check{u}_{0x}}{\beta},$$

$\omega_1 = \varphi_2 V_{0\tau} + \omega_{11}(\tau, y, t) \in \mathcal{S}$, $\varphi_2 = \varphi_2(y, t)$ is the "constant" of integration, and the function V_1^- satisfies the boundary condition

$$V_1^-\big|_{\Gamma_t} = \nu_1(y, t). \tag{2.19}$$

Further, if we consider the term of order $O(1)$ as $\tau \to \pm\infty$ we immediately obtain the equation for the background (2.3). Next, applying the Taylor expansion and equating to zero the term of order ε^0, we obtain the equation

$$\frac{\partial^2}{\partial \tau^2} L\check{V}_2 + F_2(\tau, y, t) = 0, \tag{2.20}$$

where

$$\begin{aligned} F_2 &= (\Psi_2 + R_2)\big|_{x=\varphi}, \\ \Psi_2 &= \frac{\partial}{\partial \tau}\{4V_{1t} + 4(3u_0 - \varphi_t)V_{1x} + 3(4u_{0x} - \sigma\varphi_{yy})V_1 - 6\sigma\varphi_y V_{1y}\} \\ &\quad + 2\tau\frac{\partial^2}{\partial \tau^2}(3u_{0x}V_1 - AV_{1x}) + 3u_{0xx}(2V_0 + 4\tau V_{0\tau} + \tau^2 V_{0\tau\tau}) + 3\sigma V_{0yy}. \end{aligned}$$

Here R_2 is a function with zero average, i.e.,

$$\int_{-\infty}^{\infty} R_2 \, d\tau = 0,$$

whose explicit form is immaterial for us. Integrating (2.20) with respect to τ, we arrive at the equality

$$\frac{\partial}{\partial \tau} L\check{V}_2 + \mathcal{F}_2(\tau, y, t) = 0, \tag{2.21}$$

where

$$\mathcal{F}_2 = \int_{-\infty}^{\tau} \check{\Psi}_2 \, d\tau - \int_{-\infty}^{\infty} \check{\Psi}_2 \, d\tau + \int_{-\infty}^{\tau} \check{R}_2 \, d\tau.$$

As before, for solvability of equation (2.21) it is necessary and sufficient that $\mathcal{F}_2 \in \mathcal{S}$ and the following orthogonality condition be fulfillled:

$$\int_{-\infty}^{\infty} \mathcal{F}_2 V_0 \, d\tau = 0. \tag{2.22}$$

In contrast to the previous iteration, the condition $\mathcal{F}_2 \in \mathcal{S}$ yields the equality

$$\int_{-\infty}^{\infty} \check{\Psi}_2 \, d\tau = 0.$$

Computing the integral, we obtain the boundary condition

$$\left\{ \frac{2}{3}\frac{\partial}{\partial t} + \left(\check{u}_0 + \frac{\sigma}{2}\varphi_y^2 \right) \frac{\partial}{\partial x} - \sigma \varphi_y \frac{\partial}{\partial y} \right\} V_1^- \Big|_{\Gamma_t} = 2\sigma \beta_{yy} + \frac{1}{2}\beta \nu_1^2. \tag{2.23}$$

It is clear that condition (2.22) reduces to a differential equation for the phase shift $\varphi_1(t)$. Then from (2.21) we determine the function $\check{V}_2$ up to the "constant" $\varphi_3(y,t)$.

Now we determine the extension $V_2(\tau, x, y, t)$ in the form similar to (2.18) and consider the terms of order ε to obtain the following equations for the correction to the background u_1 and the "tail" V_1^-:

$$\frac{\partial}{\partial x}\left\{ 4\frac{\partial u_1}{\partial t} + 6\frac{\partial (u_0 u_1)}{\partial x} \right\} + 3\sigma \frac{\partial^2 u_1}{\partial y^2} = 0, \quad t > 0, \ (x,y) \in \mathbb{R}^2, \tag{2.24}$$

$$\frac{\partial}{\partial x}\left\{ 4\frac{\partial V_1^-}{\partial t} + 6\frac{\partial (u_0 V_1^-)}{\partial x} \right\} + 3\sigma \frac{\partial^2 V_1^-}{\partial y^2} = 0, \quad t > 0, \ (x,y) \in \Omega_t^-, \tag{2.25}$$

where $\Omega_t^- = \{(x,y), \ x < \varphi(y,t)\}$.

It is clear that (2.24) and (2.25) ensure that the right-hand side of the equation

$$\frac{\partial^2}{\partial \tau^2} L\check{V}_3 = -F_3(\tau, y, t)$$

belongs to $\mathcal{S}$. This implies that we can repeat the above procedure and construct the asymptotic solution with any given accuracy.

Thus to determine the leading term of the asymptotic expansion we have the equation (2.3) for the background, formula (2.14) for the distorted soliton, and the Hugoniot type conditions (2.4), (2.5). The first correction consists of correction to the background u_1 and the rapidly varying function (2.18), where the "tail" $V_1^-(x,y,t)$ is obtained from equation (2.25) completed with boundary conditions (2.19), (2.23) and initial condition $(x,y) \in \Omega_0^-$ for $t = 0$. Note that (2.23) involves the derivative along the vector $\zeta = (2/3, \check{u}_0 + \sigma\varphi_y^2/2, -\sigma\varphi_y)$ in the space $\mathbb{R}^3_{txy}$. One can readily find a normal vector ν to the surface $\mathcal{T} = \bigcup_t \Gamma_t$, namely, $\nu = (\varphi_t, -1, \varphi_y) \in \mathbb{R}^3_{txy}$. Since $(\zeta, \nu) = A/3 > 0$, the vector ζ is transversal to $\mathcal{T}$.

Now we turn to equations (2.2). Construction of the asymptotic soliton-type solution for this system is carried out following the above scheme with the only distinction that instead of (2.13) we will now use the expansion

$$\begin{aligned} u &= u_0(x,y,t) + V_0(\tau,y,t) + \varepsilon(u_1(x,y,t) + V_1(\tau,x,y,t)) + \dots, \\ w &= w_0(x,y,t) + \varepsilon(w_1(x,y,t) + W_1(\tau,x,y,t)) + \dots. \end{aligned}$$

Here the w_i are smooth functions, $W_j \in \mathcal{H}$, and all other notations have the same meaning as in (2.13).

Omitting the intermediate derivations, we will present the formulas for the functions involved in the leading term of the asymptotic expansion. Namely, let u_0, w_0 be the smooth (for $t \le T$) solution of the degenerate equations (2.2), i.e., of

$$\frac{\partial u_0}{\partial t} + \Big(H^{1/2} + \frac{3}{2}\alpha H^{-3/4} u_0\Big)\frac{\partial u_0}{\partial x} + \frac{\kappa}{2} H^{1/4} \frac{\partial}{\partial y}\Big(H \frac{\partial w_0}{\partial y}\Big) = 0,$$
$$H^{3/4}\frac{\partial w_0}{\partial x} = u_0,$$

and let the functions $\varphi = \varphi(y,t)$, $A = A(y,t) > 0$ satisfy the system of equations

$$\frac{\partial \varphi}{\partial t} = \frac{\kappa}{2}\sqrt{\check{H}}\Big(\frac{\partial \varphi}{\partial y}\Big)^2 + \frac{\alpha}{2}(A + 3\check{u}_0)\check{H}^{-3/4} + \check{H}^{1/2},$$

$$(2.26)\qquad \frac{\partial A}{\partial t} = \kappa\sqrt{\check{H}}\frac{\partial A}{\partial y}\frac{\partial \varphi}{\partial y} + \frac{2}{3}\kappa\sqrt{\check{H}}A\frac{\partial^2 \varphi}{\partial y^2} - \frac{1}{2}\frac{A}{\check{H}}\Big\{\frac{13}{6}\frac{\partial \check{H}}{\partial t} - \frac{3}{2}\kappa\sqrt{\check{H}}\frac{\partial \varphi}{\partial y}\frac{\partial \check{H}}{\partial y}$$
$$+ 2\alpha \check{H}^{-1/2}\big(u_0 H^{3/4}\big)_x\Big|_{x=\varphi} + \check{H}^{-3/4}\Big(\frac{7}{5}\alpha A - \frac{2}{3}\check{H}^{5/4}\Big)H_x\Big|_{x=\varphi}\Big\},$$

and remain smooth for $t \le T$.

Then for $t \le T$

$$u = u_0(x,y,t) + A\cosh^{-2}\Big(\beta\Big(\frac{x-\varphi}{\varepsilon} + \varphi_1\Big)\Big) + O(\varepsilon),$$
$$w = w_0(x,y,t) + O(\varepsilon),$$

where $\beta = \sqrt{(3\alpha/4)A/\check{H}^{13/8}}$, and the function $\varphi_1 = \varphi_1(y,t)$ is obtained from a condition of the form (2.22).

Making the substitution

$$v = -\kappa\sqrt{\check{H}}\frac{\partial \varphi}{\partial y}, \quad \rho = \Big(\frac{A}{\check{H}^{1/4}}\Big)^{3/2},$$

we rewrite the model equations (2.26) as equations of gas dynamics:

$$(2.27)\qquad \begin{aligned} &\frac{\partial \rho}{\partial t} + \frac{\partial}{\partial y}(\rho v) = F_1, \\ &\rho\Big(\frac{\partial v}{\partial t} + v\frac{\partial v}{\partial y}\Big) + \frac{\partial p}{\partial y} = F_2, \end{aligned}$$

with "pressure"

$$p = \frac{\alpha\kappa}{5}\rho^{5/3}$$

and "external forces"

$$F_1 = -\frac{\rho}{H}\Big\{\frac{3}{2}\alpha H^{-5/2}(u_0 H^{11/4})_x + H_x\Big(\frac{41}{20}\alpha\rho^{2/3}H^{-1/2} + H^{1/2}\Big) + H_y v\Big\}\Big|_{x=\varphi},$$

$$F_2 = \frac{\rho}{H}\Big\{v\Big(H^{1/2} + \frac{3}{4}\alpha u_0 H^{-3/4}\Big)H_x - \frac{3}{2}\alpha\kappa H^{3/2}(\check{u}_0\check{H}^{-3/4})_y$$
$$+ \frac{1}{4}(v^2 - \kappa H + \alpha\kappa\rho^{2/3})H_y\Big\}\Big|_{x=\varphi}.$$

Let us point out that in spite of the differences in coefficients of equations (2.1) and (2.2), in model equations (2.10) and (2.27) we obtain the same state equation which corresponds to the monatomic gas.

In the one-dimensional case, where the depth and the solution do not depend on the longitudinal variable y, the model equations after the normalizing substitution $u_0' = u_0 H^{-1/4}$, $A' = AH^{-1/4}$, become

$$\begin{aligned} &\frac{d}{dt}\Big((A' + 2\check{u}_0')\check{H}\Big) + \frac{3}{2}\frac{3\check{u}_0' + (8/5)A'}{A' + (2/\alpha)\check{H} + 3\check{u}_0'} A' \frac{d\check{H}}{dt} = 0, \\ &\frac{d\varphi}{dt} = \frac{\alpha}{2}\check{H}^{-1/2}\Big(A' + \frac{2}{\alpha}\check{H} + 3\check{u}_0'\Big). \end{aligned} \tag{2.28}$$

Comparison of (2.28) with (1.118) shows that the second equations in these systems are the same, whereas the first equations differ, with the difference being larger when the bottom slope is steeper.

2.2. Waves in multidimensional Josephson junctions

In this section we consider the problem of dynamics of waves in a multidimensional nonlinear medium when the equation is symmetric with respect to the spatial variables (in contrast to the Kadomtsev–Petviashvili equation). The simplest physical problem leading to this setup is the problem on excitation of the magnetic flux in a multidimensional Josephson junction. Assuming that the displacement current and the dissipation effects are small, we arrive at the sine-Gordon mathematical model

$$\varepsilon^2 \frac{\partial^2 u}{\partial t^2} - \varepsilon^2 \Delta u + \frac{\alpha^2}{\gamma} \sin \gamma u + \varepsilon F\Big(x, t, u, \varepsilon\frac{\partial u}{\partial t}, \varepsilon\frac{\partial u}{\partial x}\Big) = 0 \tag{2.29}$$

for $x \in \mathbb{R}^n$, $n = 2, 3$; here $\alpha = \alpha(x,t) > 0$ and $\gamma = \gamma(x,t) > 0$ are given smooth functions. For the "right-hand side" F we will consider $F = 0$ (which was treated in [**150**]),

$$F = \big(\rho_1(x) + 3\rho_2(x)\cos u\big)\,|\varepsilon u_t|\,\varepsilon u_t - \eta(x)\,\varepsilon^3 \Delta u_t + \delta(x,t), \tag{2.30}$$

and

$$F = f(x,t)(1 - \cos\gamma u), \tag{2.31}$$

where $\rho_i(x,t)$, $\eta(x)$, $\delta(x,t)$, and $f(x,t)$ are given smooth functions (see also §1.5 and [**205**]). Note that in the rotationally symmetric case the equation (2.29), (2.31) with constant α, γ, f was considered by E. M. Maslov [**140**].

Denote by $\Gamma_0 = \{x,\, S^0(x) = 0\}$ the initial front of the multidimensional kink. We will assume that Γ_0 is a smooth nondegenerate surface of codimension 1. For example, in the two-dimensional case Γ_0 may be a simple Jordan curve or a closed curve without self-crossings. It is clear that in general the equation $S^0(x) = 0$ can be solved for one of the variables x_i only locally. Further, denote by Γ_t the position of the kink front for $t > 0$ and let Γ_t be specified by the equation $S(x,t) = 0$. It is natural to assume that the surface Γ_t remains smooth and nondegenerate for some time $t \le T$. Moreover, let us assume that the normal velocity of the kink front does not vanish. Then the surface $\mathcal{T} = \bigcup_{t\in[0,T]} \Gamma_t \subset \mathbb{R}^n_x \times [0,T]$ admits a diffeomorphic projection onto $\mathbb{R}^n_x$. This implies that the equation $S(x,t) = 0$ can be solved for t for all $t \in [0,T]$; hence the front Γ_t can be described as the level surface of some smooth function $\varphi(x)$, i.e., $\Gamma_t = \{x,\ \varphi(x) = t\}$. Next, in the multidimensional case the sine-Gordon equation also has solutions of kink and anti-kink type, which differ by the direction of the front motion. However, in the multidimensional problem the difference between these two solutions is mostly terminological, since, unlike the case $x \in \mathbb{R}^1$, there is no specific direction which could be singled out in a natural way.

To specify the problem we complete the equation (2.29) with initial conditions

$$u\big|_{t=0} = u^0(x,\varepsilon), \qquad \varepsilon u_t\big|_{t=0} = u^1(x,\varepsilon); \tag{2.32}$$

moreover, let the function u^0 vary from 0 to $2\pi/\gamma(x,0)$, and let, up to $O(\varepsilon)$,

$$\begin{aligned} &u^0\big|_{\Gamma_0} = \frac{\pi}{\gamma(x,0)}, \qquad u^1\big|_{\Gamma_0} < 0, \\ &u^0\big|_{x\in\Omega_0^+} \to 0, \qquad u^0\big|_{x\in\Omega_0^-} \to \frac{2\pi}{\gamma(x,0)}\bigg|_{x\in\Omega_0^-} \quad \text{as} \quad \varepsilon \to 0, \\ &u^1\big|_{x\in\Omega_0^\pm} \to 0 \qquad \text{as} \quad \varepsilon \to 0. \end{aligned} \tag{2.33}$$

Here $\Omega_0^\pm$ are domains in $\mathbb{R}^n$ separated by the surface Γ_0, so that $\mathbb{R}^n = \Omega_0^- \cup \Gamma_0 \cup \Omega_0^+$. Denote by ν_0 the normal to Γ_0 directed towards Ω_0^+. Then assumptions (2.33) mean that we study the dynamics of the kink whose front moves into the domain Ω_0^-, i.e., in the direction $-\nu_0$. It is clear by analogy with the one-dimensional problem that the leading terms of the functions u^0, u^1 must be of the form

$$u^0 = \frac{4}{\gamma} \arctan e^{\beta^0 \rho(x)/\varepsilon} + O(\varepsilon), \quad u^1 = -2\frac{\beta^0}{\gamma} |V_{\nu_0}| \cosh^{-1}\Big(\beta^0 \frac{\rho(x)}{\varepsilon}\Big) + O(\varepsilon), \tag{2.34}$$

where $\rho(x)$ is the signed distance from Γ_0 with $\rho < 0$ in the direction ν_0, V_{ν_0} with $0 < |V_{\nu_0}| < 1$ is the initial velocity of the front along the normal ν_0, and $\beta^0 = \alpha(x,0)/\sqrt{1-(V_{\nu_0})^2}\big|_{\Gamma_0}$.

Now we will state the main result of this section. First, we need to introduce some auxiliary functions. Denote

$$v_0(\tau, x) = 4 \arctan e^{-\beta\tau}, \tag{2.35}$$

where $\tau = (t - \varphi(x))/\varepsilon$, $\beta = \alpha(x,\varphi(x))/\sqrt{|\nabla\varphi|^2 - 1}$, and let $\varphi(x)$ be the solution to the problem

$$\begin{aligned} &\operatorname{div}\Big(\frac{\kappa(x,\varphi)}{\sqrt{|\nabla\varphi|^2 - 1}} \operatorname{grad}\varphi\Big) - \sqrt{|\nabla\varphi|^2 - 1}\,\kappa'(x,\varphi) + G(x,\varphi) = 0, \\ &\varphi\big|_{\Gamma_0} = 0, \qquad \frac{\partial\varphi}{\partial\nu_0}\bigg|_{\Gamma_0} = -\sqrt{1 + 4\Big(\frac{\alpha(x,0)}{\gamma(x,0)u^1(x)}\Big)^2}\,\Bigg|_{\Gamma_0}. \end{aligned} \tag{2.36}$$

Here

$$\kappa(x,\varphi) = \frac{\alpha(x,t)}{\gamma^2(x,t)}\bigg|_{t=\varphi(x)}, \qquad \kappa' = \frac{\partial(\alpha(x,t)/\gamma^2(x,t))}{\partial t}\bigg|_{t=\varphi(x)},$$

$$G(x,\varphi) = \frac{1}{8\gamma(x,\varphi)} \int_{-\infty}^{\infty} F\left(x, \varphi, \frac{v_0(\tau,x)}{\gamma(x,\varphi)}, \frac{v_{0\tau}(\tau,x)}{\gamma(x,\varphi)}, -\frac{v_{0\tau}(\tau,x)}{\gamma(x,\varphi)}\frac{\partial\varphi}{\partial x}\right) v_{0\tau}(\tau,x)\,d\tau,$$

and $F(x,t,u,v,w)$ is the same function as in (2.29). Let the condition $|\nabla\varphi| > 1$ hold in a neighborhood Σ of the surface Γ_0. Then at each time moment t the kink front, within $O(\varepsilon)$, is a level surface of φ, i.e., $\Gamma_t = \{x \in \Sigma,\ \varphi(x) = t\}$, and the asymptotic solution to the problem (2.29)–(2.34) has the form

$$u = \frac{1}{\gamma(x,t)} v_0\Big(\frac{t-\varphi(x)}{\varepsilon} + \varphi_1(x), x\Big) + O(\varepsilon), \tag{2.37}$$

where φ_1 is a smooth function describing a small (of order $O(\varepsilon)$) correction to the front position.

Let us discuss the result stated above. First of all, observe that $\nabla\varphi$ is a normal vector to Γ_t. Denote $\Omega_t^+ = \{x, t-\varphi(x) > 0\}$, $\Omega_t^- = \{x, t-\varphi(x) < 0\}$. Then $\nu = -\nabla\varphi/|\nabla\varphi|$ is the unit normal vector directed into Ω_t^+. It is not hard to see that the normal velocity of the motion of points lying on the front Γ_t is given by the formula

$$V_\nu \overset{\text{def}}{=} \langle \nu, \dot{x}\rangle = -\frac{1}{|\nabla\varphi|}\Big|_{\Gamma_t}.$$

Next we observe that $|\nabla\varphi|\beta|_{\Gamma_0} = \beta^0$ and that replacing $\varphi(x)/|\nabla\varphi|$ by the distance ρ (from Γ_0 along ν_0) changes v_0 and $v_{0\tau}$ by $O(\varepsilon)$. The latter assertion can be obtained by the following elementary calculation:

$$\begin{aligned} f\Big(\beta\frac{\varphi(x)}{\varepsilon}\Big) &= f\Big(\beta^0\frac{\varphi(x)/|\nabla\varphi|}{\varepsilon}\Big) = f\Big(\beta^0\frac{\rho+\rho^2\xi(x)}{\varepsilon}\Big) \\ &= f\Big(\beta^0\frac{\rho}{\varepsilon}\Big) + \varepsilon f'_\tau\Big(\beta^0\frac{\rho}{\varepsilon}\Big)\Big(\frac{\rho}{\varepsilon}\Big)^2\beta^0\xi + \cdots = f\Big(\beta^0\frac{\rho}{\varepsilon}\Big) + O(\varepsilon), \end{aligned}$$

which is valid for any function $f(\tau)$ such that f'_τ decreases exponentially as $\tau \to \pm\infty$. Therefore the leading term of the asymptotic formula (2.37) satisfies the initial conditions (2.32)–(2.34).

For the analysis of the basic model equation (2.36), observe first of all that in spite of its divergence form, it is a quasilinear hyperbolic equation of second order, in which the normal distance from Γ_t plays the role of "time", and the coordinates on Γ_t play the role of "spatial variables". Hence the problem (2.36) is well-posed and its solution does exist, at least locally.

Next we rewrite (2.36) in a form which admits a geometric interpretation. To that end, denote by

$$\mathcal{K} = -\operatorname{div}\Big(\frac{\nabla\varphi}{|\nabla\varphi|}\Big)$$

the mean curvature of the surface Γ_t, and introduce the function a,

$$a = \frac{\langle \nu, \varphi_{xx}\nu\rangle}{|\nabla\varphi|^3},$$

which may be interpreted as the acceleration of the surface Γ_t in the direction of the normal ν. Here φ_{xx} is the Hessian of the function φ, and in this interpretation we neglect the motion of points $x \in \Gamma_t$ in directions tangent to Γ_t.

Now, after some algebra, the equation (2.36) can be rewritten as

$$-\mathcal{K} + \frac{a}{1 - V_\nu^2} = V_\nu \mu_1 + \mu_2,$$

where

$$\mu_1 = \frac{\partial}{\partial t} \log \kappa(x,t)\Big|_{t=\varphi}, \quad \mu_2 = \frac{\partial}{\partial \nu} \log \kappa(x,t)\Big|_{t=\varphi} - \frac{\sqrt{1-V_\nu^2}}{\kappa(x,\varphi)} G(x,\varphi).$$

We see from this equation that the surface Γ_t moves so as to keep the balance between the curvature, the acceleration, the normal velocity, and the "external force", which is determined by the right-hand side F of equation (2.29) and by the dependence on the spatial variable coefficients $\alpha(x,t)$ and $\gamma(x,t)$.

Finally, consider equation (2.36) for some specific functions F. Let $F = 0$, $n = 2$, $\alpha = \alpha(r)$, $\gamma = \gamma(r)$, and let Γ_0 be a circle of radius r_0 with domain Ω_0^- lying inside Γ_0. Then $\varphi = \varphi(r)$, and after integration the problem (2.36) can be rewritten in the form

$$\frac{d\varphi}{dr} = -\left(1 - \left(\frac{\kappa(r)r}{c}\right)^2\right)^{-1/2}, \quad \varphi\big|_{r=r_0} = 0,$$

where the constant of integration

$$c = -r_0 \sqrt{\kappa^2 + \left(\frac{u^1}{2\gamma}\right)^2}\Bigg|_{r=r_0}$$

is determined from the initial condition for the derivative. It is seen from this formula and the equation $\Gamma_t = \{r,\ \varphi(r) = t\}$ that the front moves towards the center as t increases. Moreover, the derivative $d\varphi/dr$ is bounded for all $r \in (0, r_0]$, provided that the product $r\kappa(r)$ does not increase for decreasing r. Thus the asymptotic formula (2.37) remains valid up to the collapse time. On the contrary, if $\partial(r\kappa(r))/\partial r < 0$ for $r \in (0, r_0]$ and there exists r^* such that $r^*\kappa(r^*) = |c|$, then $|d\varphi/dr| \to \infty$ for $r \to r^*$. This means that for $r = r^*$ the turning effect appears, i.e., the wave front begins to move in the opposite direction. However the asymptotic solution discussed here does not describe this effect.

Now let F have the form (2.31). Then $G = -\pi f(x,\varphi)/4\gamma(x,\varphi)$. To simplify the problem, assume that $n = 2$, all data in the problem are rotationally symmetric, and Γ_0 is the circle of radius r_0. Then the equation (2.36) becomes

$$\frac{1}{r}\frac{d}{dr}\left(\frac{r\kappa(r,\varphi)}{\sqrt{\varphi_r^2 - 1}}\varphi_r\right) - \sqrt{\varphi_r^2 - 1}\,\kappa'(r,\varphi) = \frac{\pi f(r,\varphi)}{4\gamma(r,\varphi)}. \tag{2.38}$$

Denote by $r = \rho(t)$ the solution to the equation $t = \varphi(r)$. Then (2.38) can be rewritten as the system

$$\begin{aligned} \frac{d\rho}{dt} &= V, \\ \frac{dV}{dt} &= -\frac{1}{\rho}(1 - V^2) + \frac{\pi f(\rho,t)}{4\kappa(\rho,t)\gamma(\rho,t)}(1 - V^2)^{3/2} + \frac{d}{dt}\log \kappa(r, \rho(t)). \end{aligned} \tag{2.39}$$

Note that in the particular case when $f = \text{const}$ and $\alpha = \gamma = 1$ the equations (2.39) were obtained in [**140**], where one can also find a detailed analysis of this system.

Finally, let F have the form (2.30) and $\gamma = 1$. Then we obtain the following formula for the term G in equation (2.36):

$$
\begin{aligned}
G = &-\frac{1}{8}\rho_1 \int_{-\infty}^{\infty} v_{0\tau}^3\, d\tau + \frac{3}{4}\rho_2 \frac{|\nabla\varphi|^2 - 1}{\alpha^2} \int_{-\infty}^{\infty} v_{0\tau}(v_{0\tau\tau})^2\, d\tau \\
&+ \frac{1}{8}\eta|\nabla\varphi|^2 \int_{-\infty}^{\infty} v_{0\tau\tau}^2\, d\tau - \frac{\delta(x,\varphi)}{4\pi}.
\end{aligned}
\tag{2.40}
$$

Computing the integrals, we obtain

$$
G = \frac{\pi}{2}\frac{\beta^2}{\gamma^3}\Big(\rho_1 - \frac{3}{2}\rho_2\Big) + \frac{\eta|\nabla\varphi|^2\beta^3}{3\gamma^2} - \frac{\delta(x,\varphi)}{4\pi}.
$$

Now we turn to the construction of the asymptotic solution. After the substitution $u = v/\gamma$, the equation (2.29) becomes

$$
\begin{aligned}
&\varepsilon^2 \frac{\partial^2 v}{\partial t^2} - \varepsilon^2 \Delta v + \alpha^2 \sin v \\
&\quad + \varepsilon\gamma\Big\{F\Big(x, t, \frac{v}{\gamma}, \varepsilon\frac{\partial}{\partial t}\Big(\frac{v}{\gamma}\Big), \varepsilon\frac{\partial}{\partial x}\Big(\frac{v}{\gamma}\Big)\Big) + 2\Big(\frac{\partial v}{\partial t}\frac{\partial}{\partial t}\frac{1}{\gamma} - \Big\langle\nabla v, \nabla\frac{1}{\gamma}\Big\rangle\Big)\Big\} \\
&\quad + \varepsilon^2 v\gamma\Big(\frac{\partial^2}{\partial t^2} - \Delta\Big)\frac{1}{\gamma} = 0.
\end{aligned}
\tag{2.41}
$$

Since the kink-type solution to the sine-Gordon equation stabilizes at a constant value (outside a small neighborhood of the front) exponentially fast, the above substitution allows us to avoid the (rather obvious) procedure of matching the solution inside and outside a neighborhood of Γ_t. Further, the above analysis of the front geometry shows that under our assumptions the phase of the asymptotic solution may be written in the form

$$
S = t - \varphi(x).
$$

Therefore, as an analog of the expansions (1.3) and (2.13), we have

$$
v = v_0(\tau, x) + \varepsilon(u_1(x,t) + v_1(\tau, x, t)) + \dots, \tag{2.42}
$$

where $\tau = (t - \varphi(x))/\varepsilon$, $v_i \in \mathcal{H}$, and the u_i are smooth functions. Substituting (2.42) into (2.41), collecting terms according to the power of ε and equating to zero the coefficient at ε^0, we arrive at the equation

$$
(1 - |\nabla\varphi|^2) v_{0\tau\tau} + \alpha^2(x,\varphi)\sin v_0 = 0.
$$

Hence we immediately find the function v_0 in the form (2.35). Further, equating to zero the coefficient at ε, we obtain the relation

$$
Lv_1 + \alpha^2(x,t)\cos v_0 u_1 + \mathcal{F}_1(\tau, x, t) = O(\varepsilon), \tag{2.43}
$$

where

$$\begin{aligned}
L &= (1-|\nabla\varphi|^2)\frac{\partial^2}{\partial\tau^2} + \alpha^2(x,t)\cos v_0,\\
\mathcal{F}_1 &= \mathcal{R}_1 + \gamma(x,t)F\Big(x,t,\frac{v_0}{\gamma},\frac{v_{0\tau}}{\gamma},-\frac{\varphi_x}{\gamma}v_{0\tau}\Big),\\
\mathcal{R}_1 &= (2\langle\nabla\varphi,\nabla\rangle + \Delta\varphi)v_{0\tau} + 2\alpha\alpha_t\big|_{t=\varphi}\tau\sin v_0\\
&\quad + 2\gamma v_{0\tau}\Big(\frac{\partial}{\partial t} + \langle\nabla\varphi,\nabla\rangle\Big)\frac{1}{\gamma}\Big|_{t=\varphi}.
\end{aligned}$$

Letting τ tend to ∞ in (2.43) and using the fact that in this case v_1 tends to zero, we obtain the equation for the regular correction u_1:

$$\alpha^2(x,t)u_1 = -\gamma(x,t)F^+,$$

where $F^+ = F(x,t,0,0,0)$. Now, taking the limit as $\tau\to-\infty$, we arrive at the equation for the limiting value $v_1^-(x,t)$:

$$\alpha^2(x,t)v_1^- = -\gamma(x,t)(F^- - F^+),$$

where $F^- = F(x,t,2\pi/\gamma,0,0)$.

Let us write the function v_1 in the form

$$v_1 = v_1^-(x,t)\chi(\tau,x) + \omega_1(\tau,x),$$

where $\chi = v_0/2\pi$ and ω_1 tends to zero as $\tau\to\pm\infty$. Then (2.43) can be rewritten as

$$\begin{aligned}
L\omega_1 &+ \mathcal{R}_1 + (1-|\nabla\varphi|^2)v_1^-\chi_{\tau\tau}\\
&+ \gamma(x,t)\big\{F + (F^+ - F^-)\chi\cos v_0 - F^+\cos v_0\big\} = O(\varepsilon).
\end{aligned} \tag{2.44}$$

It is easily seen that the expression in braces tends to zero exponentially fast as $\tau\to\pm\infty$. Thus all the terms in (2.44) decrease exponentially as $\tau\to\pm\infty$. This enables us to expand the functions in the left-hand side of (2.44) by the Taylor formula with respect to t around the point $t=\varphi$. Neglecting the terms of order $O(\varepsilon)$ (note that this bound is uniform in $\tau\in\mathbb{R}^1$), we derive from (2.44) the equation

$$(1-|\nabla\varphi|^2)\frac{\partial^2\omega_1}{\partial\tau^2} + \alpha^2(x,\varphi)\omega_1\cos v_0 + \check{\mathcal{F}}_1(\tau,x) = 0, \tag{2.45}$$

where

$$\begin{aligned}
\check{\mathcal{F}}_1 = \mathcal{R}_1 &+ (1-|\nabla\varphi|^2)v_1^-(x,\varphi)\chi_{\tau\tau}\\
&+ \gamma(x,t)\{F + (F^+ - F^-)\chi\cos v_0 - F^+\cos v_0\}\big|_{t=\varphi}.
\end{aligned}$$

Next, the condition

$$\int_{-\infty}^{\infty}\check{\mathcal{F}}_1(\tau,x)v_{0\tau}\,d\tau = 0 \tag{2.46}$$

of solvability of (2.45) in the class of functions at hand can be easily reduced to the form

$$\int_{-\infty}^{\infty} \mathcal{R}_1(\tau, x) v_{0\tau}\, d\tau + \gamma(x,\varphi) \int_{-\infty}^{\infty} F|_{t=\varphi} v_{0\tau}\, d\tau = 0.$$

Taking into account the explicit form of $\mathcal{R}_1$, we can rewrite the last equality as

$$(2.47)\qquad \begin{aligned} &\Big\langle \nabla, \Big(\int_{-\infty}^{\infty} v_{0\tau}^2\, d\tau \Big) \nabla\varphi \Big\rangle - \frac{|\nabla\varphi|^2 - 1}{\alpha} \frac{\partial \alpha}{\partial t}\Big|_{t=\varphi} \int_{-\infty}^{\infty} v_{0\tau}^2\, d\tau \\ &\quad + \gamma^2 \Big(\Big\{ \frac{\partial}{\partial t} + \langle \nabla\varphi, \nabla \rangle \Big\} \Big(\frac{1}{\gamma} \Big)^2 \Big) \Big|_{t=\varphi} \int_{-\infty}^{\infty} v_{0\tau}^2\, d\tau \\ &\quad + \gamma(x,\varphi) \int_{-\infty}^{\infty} F\Big|_{t=\varphi} v_{0\tau}\, d\tau = 0. \end{aligned}$$

It remains to observe that

$$\langle \nabla\varphi, \nabla \rangle \Big(\frac{1}{\gamma(x,\varphi)} \Big)^2 = \Big\{ \langle \nabla\varphi, \nabla \rangle \Big(\frac{1}{\gamma(x,t)} \Big)^2 + |\nabla\varphi|^2 \frac{\partial}{\partial t} \Big(\frac{1}{\gamma(x,t)} \Big)^2 \Big\} \Big|_{t=\varphi}.$$

Hence, taking into account the equality

$$\int_{-\infty}^{\infty} v_{0\tau}^2\, d\tau = 8 \frac{\alpha(x,\varphi)}{\sqrt{|\nabla\varphi|^2 - 1}},$$

it is not hard to transform condition (2.47) to the form (2.36).

2.3. Soliton-type waves in a basin of variable depth

In this section we demonstrate an application of the method expounded above to the construction of asymptotic solutions of systems of nonlinear equations. As an example we consider the physically meaningful problem on the motion of distorted soliton-type waves in a basin of variable depth. One possible application of the results discussed here is the computation and prediction of tsunami waves.

As a mathematical model we consider the system of Boussinesq equations

$$(2.48)\qquad \begin{aligned} &\frac{\partial u}{\partial t} + \nabla \Big\{ \frac{1}{2}|u|^2 + g\rho \Big\} = 0, \\ &\frac{\partial \rho}{\partial t} + \langle \nabla, (h+\rho)u + \varepsilon^2 h^3 \Delta u \rangle = 0. \end{aligned}$$

Here $x \in \mathbb{R}^2$, $u = (u^1(x,t), u^2(x,t))$ is the vector of the horizontal velocity of the fluid flow, $\rho = \rho(x,t)$ is the elevation of the free surface, $h = h(x)$ is the depth of the nonperturbed fluid, g is the acceleration of gravity, and $\varepsilon > 0$ is a small parameter characterizing the magnitude of dispersion. We assume that the function h is infinitely differentiable and positive, $h(x) \geq h_0 = \text{const} > 0$.

The system (2.48) in the KdV approximation (one-wave approximation, small nonlinearity, and small dispersion) has been studied in [**90, 203, 206, 218, 233**]. The main results of these papers were obtained for the axially symmetric case. Asymptotic soliton-type solutions of (2.48) in a general setup were first constructed in [**161, 162**].

First, we discuss the structure of the initial data. We assume that at the initial time moment we are given a distorted soliton-type wave with amplitude $R^0(S^0(x)/\varepsilon, x, \varepsilon)$ and velocity $v^0(S^0(x)/\varepsilon, x, \varepsilon)$ placed over background perturbations $\rho^0(x,\varepsilon)$ and $u^0(x,\varepsilon)$ respectively. Regarding the functions ρ^0 and $u^0 = (u^{01}, u^{02})$, we will assume that they are infinitely differentiable uniformly in the parameter $\varepsilon \geq 0$ and that

$$h + \rho^0 > 0, \tag{2.49}$$

which means that the surface of the fluid lies above the bottom. Moreover, we assume that the functions $R^0(\tau, x, \varepsilon)$ and $v^0(\tau, x, \varepsilon) = (v^{01}, v^{02})$ are smooth with respect to $(\tau, x) \in \mathbb{R}^3$ uniformly in $\varepsilon \geq 0$ and have some fixed profile. As usual, the last assumption can be stated more precisely only after the asymptotic solution is constructed. *A priori* we can only assume that the leading term of the initial wave $R^0(\tau, x, 0)$, $v^0(\tau, x, 0)$ tends to zero exponentially fast as $\tau \to \pm\infty$, i.e., it has a soliton-type form. Regarding the higher order terms, i.e., the derivatives of $R^0(\tau, x, \varepsilon)$, $v^0(\tau, x, \varepsilon)$ with respect to ε taken at the point $\varepsilon = 0$, we will assume that they stabilize as $\tau \to \pm\infty$. As is usual for soliton-type functions, it follows from these assumptions that the behavior of the phase $S^0(x) \in C^\infty(\mathbb{R}^2)$ is essential only in a small neighborhood of the set of its zeroes, $\Gamma_0 = \{x \in \mathbb{R}^2,\ S^0(x) = 0\}$. It is clear that smoothness and nondegeneracy of the phase S^0 imply that Γ_0 is a smooth one-dimensional curve without self-crossings.

Thus the basic geometric object, which is specified at $t = 0$, is the initial front Γ_0 of the soliton-type perturbation of elevation and velocity. Later we will show that the amplitude $A^0(x) = R^0(0, x, 0)|_{\Gamma_0}$ of the soliton-type elevation of the fluid is another parameter which can be specified arbitrarily for $t = 0$. The only conditions on A^0 are that it is positive and smooth.

When constructing the asymptotic solution to the system (2.48) we will assume that it is self-similar, i.e., that for $t > 0$ the wave front Γ_t remains a smooth curve without self-crossings and that the wave profile has the same structure as at the initial time moment.

Now we state the main result of this section. First we introduce some auxiliary functions. Denote by $\rho_0 = \rho_0(x,t)$, $u_0 = u_0(x,t)$ the background perturbation at time t, which satisfies the degenerate Boussinesq system

$$\begin{aligned} &\frac{\partial \rho_0}{\partial t} + \langle \nabla, (h + \rho_0) u_0 \rangle = 0, \\ &\frac{\partial u_0}{\partial t} + \nabla \Big\{ \frac{1}{2} |u_0|^2 + g\rho_0 \Big\} = 0. \end{aligned} \tag{2.50}$$

It is assumed that during some time $T > 0$ the solution of (2.50) remains smooth and continues to satisfy the natural condition

$$h(x) + \rho_0(x,t) > 0, \tag{2.51}$$

which is similar to (2.49).

Next, denote by $\varphi(x)$ the function describing the front dynamics, i.e., $\Gamma_t = \{x \in \mathbb{R}^2,\ \varphi(x) = t\}$. Then the unit normal to Γ_t is given by the formula $\nu = \nabla\varphi / |\nabla\varphi|\big|_{\Gamma_t}$, and we have the following expression for the normal velocity V_ν of the front motion:

$$V_\nu = \frac{1}{|\nabla \varphi|}. \tag{2.52}$$

Further, denote by H the total depth at the front position Γ_t, i.e.,

$$H = h(x) + \rho_0(x,t)\big|_{\Gamma_t}; \tag{2.53}$$

and let $c = \sqrt{gH}$ denote the "velocity of sound" related to the background perturbation of the fluid surface at the front position Γ_t.

Moreover, denote by κ the normal velocity of the front motion in the coordinate system moving with the normal to Γ_t component of the background velocity $u_0|_{\Gamma_t}$, i.e.,

$$\kappa = V_\nu - u_{0\nu}, \quad u_{0\nu} = \langle u_0, \nu\rangle\big|_{\Gamma_t}. \tag{2.54}$$

Define also the parameter

$$M = \kappa/c, \tag{2.55}$$

whose meaning is the Mach number for the soliton-type component of the wave, and auxiliary functions

$$\alpha = 2(\kappa - c), \quad \lambda = 2(\kappa + c). \tag{2.56}$$

Now we can write down the basic model equation, whose solution describes the dynamics of the front:

$$h^3\langle\nabla, I_0\nu\rangle - \langle\nabla, V_\nu I_1\zeta\rangle + \frac{\partial\rho_0}{\partial t}\bigg|_{t=\varphi} I_2 + \left\langle \nu, \frac{\partial u_0}{\partial t}\bigg|_{t=\varphi}\right\rangle I_1 = 0. \tag{2.57}$$

Here

$$\begin{aligned} I_0 &= \sqrt{g}\Big(\frac{H}{h}\Big)^{3/2}\Big\{\frac{1}{3}(2+M^2)\sqrt{M^2-1} - \log m\Big\}, \\ I_1 &= \frac{Hh^{3/2}}{V_\nu}\{M\sqrt{M^2-1} - \log m\}, \\ I_2 &= \frac{\sqrt{H}h^{3/2}}{V_\nu}\{-\sqrt{M^2-1} + M\log m\}, \end{aligned} \tag{2.58}$$

$$\zeta = V_\nu \nu + \check{u}_{0\xi}\xi, \tag{2.59}$$

where $m = M + \sqrt{M^2-1}$, $\check{u}_{0\xi} = \langle u_0(x,\varphi), \xi\rangle$, and ξ is the unit vector orthogonal to ν.

We complete the equation (2.57) with initial conditions

$$\varphi|_{\Gamma_0} = 0, \quad \frac{\partial\varphi}{\partial\nu}\bigg|_{\Gamma_0} = p^0|_{\Gamma_0}, \tag{2.60}$$

where

$$p^0|_{\Gamma_0} = \pm\left\{\frac{gA^0_{\rho,\mathrm{sol}}}{2c} + u^0_\nu + c\right\}^{-1}\Bigg|_{\Gamma_0},$$

$u_\nu^0 = \langle \nu, u^0 \rangle$ is the normal component of the initial background velocity $u_0|_{t=0}$, $A^0_{\rho,\text{sol}}$ is the amplitude of the elevation R^0 on the initial front Γ_0, and the sign $\pm$ means that we consider the wave whose front moves either in direction ν (sign $+$), or in direction $-\nu$ (sign $-$) with time t. We assume that $A^0_{\rho,\text{sol}} > 0$ and, moreover, that $A^0_{\rho,\text{sol}} \neq -(2c/g)(c + u^0_\nu)$. It is clear that the latter condition is surely fulfilled for $u^0_\nu \geq -c$.

It remains to introduce the auxiliary function

$$\omega_0(\tau, x) = \frac{2\alpha\lambda}{(\lambda - \alpha)\cosh \beta\tau + 4\kappa}, \qquad \beta = \frac{1}{2}\frac{\sqrt{\alpha\lambda}}{h}\frac{V_\nu}{\sqrt{gh}}, \tag{2.61}$$

which determines the profile of soliton-type components of the solution.

Now, the main result of this section is as follows: the dynamics of the front of the soliton-type solution of the system (2.48) is described by the solution of the problem (2.57), (2.60), while the profile of elevation ρ and horizontal velocity u are given by the leading term of asymptotics,

$$\begin{aligned} \rho &= \rho_0(x,t) + \frac{1}{g}\omega_0(\tau, x)\Big(\kappa - \frac{1}{2}\omega_0(\tau, x)\Big)\Big|_{\tau=(t-\varphi)/\varepsilon+\varphi_1} + O(\varepsilon), \\ u &= u_0(x,t) + \omega_0(\tau,x)\nu\big|_{\tau=(t-\varphi)/\varepsilon+\varphi_1} + O(\varepsilon), \end{aligned} \tag{2.62}$$

where $\varphi_1 = \varphi_1(x)$ is a smooth function describing the correction of order ε to the front position. The method of its computation will be given below.

Now we discuss the result stated above. Fist of all, observe that the soliton component of the velocity $\omega_0\nu$ is directed along the normal ν to the wave front Γ_t, and its maximal amplitude is

$$A_{u,\,\text{sol}} = \omega_0(0,x)\big|_{\Gamma_t} = \alpha(x)\big|_{\Gamma_t}.$$

The soliton component of the elevation has maximal amplitude

$$A_{\rho,\,\text{sol}} \stackrel{\text{def}}{=} \frac{\omega_0(0,x)}{g}\Big(\kappa - \frac{1}{2}\omega_0(0,x)\Big)\Big|_{\Gamma_t} = \alpha(x)\sqrt{\frac{H}{g}}\bigg|_{\Gamma_t}. \tag{2.63}$$

Further, the condition that the coefficients (2.58) must be real-valued implies the assumption

$$M > 1. \tag{2.64}$$

Since $M = \kappa/c$ and $\alpha = 2(\kappa - c)$, this implies that the maximal values of the velocity and amplitude are positive.

Condition (2.64) has a simple physical meaning: the relative velocity of the front motion of the distorted soliton is greater than the "velocity of sound". This can be seen by rewriting (2.64) in the form

$$V_\nu > u_{0\nu} + c.$$

This implies also that the normal velocity of the front is greater than the normal component $u_{0\nu}$ of the background velocity on the curve Γ_t. The property of "supersonic" velocity of the front motion is common for all soliton-type solutions. In particular, we have encountered this phenomenon when constructing asymptotic solutions to the Korteweg–de Vries and Kadomtsev–Petviashvili equations. Note also that the maximal amplitudes $A_{u,\mathrm{sol}}$ and $A_{\rho,\mathrm{sol}}$ are uniquely determined by the normal velocity of the front motion and the regular parameters of the problem, i.e., the background ρ_0, u_0 and the depth h, since $\alpha = 2(V_\nu - u_{0\nu} - c)$.

Further, it can be shown that under condition (2.64) the equation (2.57) is a quasilinear hyperbolic equation of second order with the distance along the normal ν playing the role of "time" and the coordinate on Γ_t playing the role of the "spatial variable".

Thus the problem (2.57), (2.60) is well posed, and its solution exists at least in a small neighborhood of the curve Γ_0. This implies, in particular, that $p^0|_{\Gamma_0}$, i.e., the normal derivative of φ taken on Γ_0, can be any function satisfying the condition

$$0 < \left|p^0|_{\Gamma_0}\right| < (c + u_\nu^0)^{-1}\big|_{\Gamma_0},$$

which is a consequence of (2.64). However, it is more natural for physical applications to specify for $t = 0$ an arbitrary amplitude $A^0_{\rho,\mathrm{sol}} > 0$ of the soliton component of elevation and then to compute p^0 by (2.60).

It is obvious that the initial background functions $u^0 = u_0(x, 0)$ and $\rho^0 = \rho_0(x, 0)$ may be specified arbitrarily (provided, naturally, that they are smooth and satisfy (2.49)). Hence the only quantities involved in the leading term of the asymptotic expansion which can be specified arbitrarily at the initial time moment are the initial front position Γ_0, the amplitude $A^0_{\rho,\mathrm{sol}}$ of elevation on Γ_0, and the initial values of the background u^0, ρ^0. Then it is natural to specify the initial phase $S^0(x)$ by the distance along the normal to Γ_0.

The equation (2.57) is in general very complicated. Therefore it will be expedient to consider particular cases where it takes a simpler form.

2.3.1. Motion on zero background. Let $\rho^0 = u^0 = 0$. Then equation (2.57) becomes

$$\langle \nabla, (M^2 - 1)^{3/2}\nu \rangle + \frac{9}{2} M I_3 \frac{\partial}{\partial \nu} \log h = 0, \tag{2.65}$$

where the Mach number is given by the formula

$$M = \frac{V_\nu}{\sqrt{gh}},$$

$I_3 = M\sqrt{M^2 - 1} - \log m$ with $m = M + \sqrt{M^2 - 1}$, and $\partial/\partial\nu = \langle \nu, \nabla \rangle$ is the derivative along the normal ν.

This equation can be easily rewritten in a geometric form:

$$\mu_1 \mathcal{K} + \mu_2 \frac{V_\nu}{c} = \frac{a}{g},$$

which shows that the front moves so as to keep the balance between the mean curvature $\mathcal{K} = \operatorname{div} \nu$, the normal velocity V_ν, and the acceleration $a = \langle \nu, \varphi_{xx}\nu \rangle / |\nabla\varphi|^3$. Here

$$\mu_1 = \frac{1}{3} h(M^2 - 1), \quad \mu_2 = \left(M - \frac{3}{2} \frac{\log m}{\sqrt{M^2 - 1}} \right) \frac{\partial h}{\partial \nu}.$$

2.3.2. Axially symmetric wave on zero background in the case of constant depth. Let $\rho^0 = u^0 = 0$, $h = \text{const}$, and let Γ_0 be the circle of radius $r_0 > 0$. Moreover, let the initial amplitude $A^0 = A^0_{\rho,\text{sol}}$ be constant on Γ_0. Let us consider, for definiteness, the wave moving outwards, i.e., take the sign + in (2.60). Then integrating (2.65) we get

$$\varphi(r) = \varphi_0 + \frac{1}{c}\{(E + r^{2/3})^{3/2} - 3E(E - r^{2/3})^{1/2}\}, \tag{2.66}$$

where the constant $E = E(A^0, r^0)$ is found from the second condition (2.60) and the constant $\varphi_0 = \varphi_0(r_0, A^0)$ is determined from the condition $\varphi(r_0) = 0$.

Formula (2.66) enables us, in particular, to compute the amplitude of elevation on the front Γ_t:

$$A_{\rho,\text{sol}} = 2h\{\sqrt{1 + Er^{-2/3}} - 1\}. \tag{2.67}$$

It is seen from this formula that for large r the amplitude of the wave decreases as $r^{-2/3}$, which agrees with the well-known result [**203, 206, 233**] obtained under the assumption $A/h \ll 1$ and the assumptions adopted in this section.

2.3.3. Dynamics of waves of small amplitude. Consider the case where the amplitude of elevation is small compared to the total depth, i.e.,

$$\left.\frac{A_{\rho,\text{sol}}}{H}\right|_{\Gamma_t} \ll 1. \tag{2.68}$$

Denote by $\mu \ll 1$ the small parameter characterizing the magnitude of this ratio. Then the function φ can be represented by the expansion

$$\varphi = \Phi(x) - \mu\psi(x) + O(\mu^2). \tag{2.69}$$

Put (2.69) into (2.57) and make the corresponding expansions taking (2.68) into account. Then for determination of the leading term Φ we obtain the Hamilton–Jacobi equation (eikonal)

$$\sqrt{gH}|\nabla\Phi| + \langle\nabla\Phi, u_0\rangle\big|_{t=\Phi} = 1.$$

For zero background this equation simplifies further and becomes

$$|\nabla\Phi| = (gh)^{-1/2}. \tag{2.70}$$

In other words, the leading term of the normal velocity of the front $V_\nu^{(0)}$ equals the velocity of sound c. In this case the equation for the correction ψ also simplifies considerably. Namely, define the auxiliary function χ related to ψ by

$$\chi = c\frac{\partial\psi}{\partial\nu_0}, \tag{2.71}$$

where $\nu_0 = \nabla\Phi/|\nabla\Phi|$. Then using (2.68)–(2.70) we obtain from (2.65) the equation

$$\frac{\partial\chi}{\partial\nu_0} + \frac{2}{3}\frac{\chi}{h^2}\langle\nabla, h^2\nu_0\rangle = 0. \tag{2.72}$$

In the axially symmetric case the equation (2.72) can be easily integrated, so that we obtain

$$\chi = \chi_0 r^{-2/3} h^{-4/3}(r),$$

where χ_0 is the constant of integration.

It is clear that equation (2.71) then becomes

$$\frac{\partial \psi}{\partial r} = \frac{\chi_0}{\sqrt{g}} r^{-2/3} h^{-11/6}. \tag{2.73}$$

In its turn, the second initial condition (2.60) implies that (taking the sign +)

$$\left.\frac{\partial \varphi}{\partial r}\right|_{r=r_0} = \frac{1}{c}\Big(1 - \frac{1}{2}\frac{A^0}{h_0} + O(\mu^2)\Big),$$

where $h_0 = h(r_0)$.

Since $\partial\Phi/\partial r = c^{-1}$ for all $r > r_0$ by (2.70), for the correction ψ we obtain the initial condition

$$\left.\frac{\partial \psi}{\partial r}\right|_{r=r_0} = \frac{1}{2c},$$

where we set $\mu = A^0/h_0$. From this relation and (2.73) we find the constant $\chi_0 = \frac{1}{2} r_0^{2/3} h_0^{4/3}$.

Now we can find the amplitude of elevation. By (2.63) and (2.70) we have

$$A_{\rho,\mathrm{sol}} = 2\mu\chi_0 r^{-2/3} h^{-1/3}(r) + O(\mu^2).$$

This implies, in particular, that for constant depth $h = h_0$ we again get the law of decay of the amplitude $A_{\rho,\mathrm{sol}}$ as $r^{-2/3}$. However, we see that this is by no means a universal law.

Now we proceed to determination of a self-similar asymptotic solution. Since (2.48) is the first example of a system of equations, we will give a detailed exposition of our constructions. As before, we represent the asymptotic solution in the form of an expansion

$$\begin{aligned} u &= u_0(x,t) + v_0(\tau,x) + \varepsilon\big(u_1(x,t) + v_1(\tau,x,t)\big) + \dots, \\ \rho &= \rho_0(x,t) + R_0(\tau,x) + \varepsilon\big(\rho_1(x,t) + R_1(\tau,x,t)\big) + \dots, \end{aligned} \tag{2.74}$$

where the u_i, v_i and ρ_i, R_i are smooth vector and scalar functions respectively to be determined. It is assumed that $v_0, R_0 \in \mathcal{S}$, i.e., that they decrease exponentially as $\tau \to \pm\infty$, while $v_j, R_j \in \mathcal{H}$ for $j \geq 1$, i.e., they tend exponentially fast to zero as $\tau \to \infty$ and to some smooth functions $v_j^-(x,t), R_j^-(x,t)$ as $\tau \to -\infty$. As we did for the sine-Gordon equation, we write the phase in the form $t - \varphi(x)$. Hence $\tau = (t - \varphi(x))/\varepsilon$.

Substitute the expansion (2.74) into equations (2.48) and consider them for $\tau \to \infty$. Equating to zero the coefficients of corresponding powers of ε, we obtain the system (2.50) for determination of u_0, ρ_0, while for determination of u_k, ρ_k for $k \geq 1$ we obtain the linearized system

$$\begin{aligned} &\frac{\partial \rho_k}{\partial t} + \langle \nabla, (h + \rho_0)u_k + \rho_k u_0 \rangle = f_{k,\rho}, \\ &\frac{\partial u_k}{\partial t} + \nabla\{\langle u_0, u_k\rangle + g\rho_k\} = f_{k,u}, \end{aligned} \tag{2.75}$$

where the $f_{k,\rho}, f_{k,u}$ are polynomials in $\rho_0, u_0, \dots, \rho_{k-1}, u_{k-1}$ and their derivatives. In particular, $f_{1,\rho} = 0$, $f_{1,u} = 0$. It is clear that equations (2.75) are solvable on the same time interval $[0, T]$ as the system (2.50).

Now, without letting τ tend to ∞, consider the terms of order ε^{-1} which arise when putting (2.74) into (2.48). Expanding the functions of t involved in these terms into Taylor series about the point $t = \varphi$, we obtain the equations

$$\begin{aligned}\frac{\partial}{\partial\tau}\Big\{R_0 - \Big\langle\nabla\varphi, \Big((h+\rho_0(x,\varphi)+R_0)v_0 + u_0(x,\varphi)R_0 \\ + h^3|\nabla\varphi|^2\frac{\partial^2 v_0}{\partial\tau^2}\Big)\Big\rangle\Big\} = 0, \\ \frac{\partial}{\partial\tau}\Big\{v_0 - \nabla\varphi\Big(\Big\langle u_0(x,\varphi) + \frac{1}{2}v_0, v_0\Big\rangle + gR_0\Big)\Big\} = 0.\end{aligned} \tag{2.76}$$

On integrating the second equation in (2.76) we immediately obtain that the vector v_0 is directed along $\nabla\varphi$, so that we have the representation

$$v_0(\tau, x) = \omega_0(\tau, x)\nu, \tag{2.77}$$

where ω_0 is a scalar soliton-type function which is related to R_0 by

$$R_0 = \frac{\omega_0}{g}\Big(\kappa - \frac{1}{2}\omega_0\Big). \tag{2.78}$$

Here and below we again use the notation (2.52), (2.54).

Now, integrating the first equation (2.76) and using (2.77), (2.78), we arrive at the equation

$$gh^3|\nabla\varphi|^2\frac{\partial^2\omega_0}{\partial\tau^2} = \omega\Big\{\Big(\kappa - \frac{1}{2}\omega\Big)(\kappa - \omega) - gH\Big\}, \tag{2.79}$$

where we made use of notation (2.53).

It is not hard to see that (2.79) is solvable in the class of functions decreasing as $\tau \to \pm\infty$ if and only if

$$\kappa > \sqrt{gH},$$

which is equivalent to (2.64). Under this condition, by elementary integration we find the solution to (2.79) in the form (2.61), where the auxiliary functions α and λ are defined by (2.56), $\tau = (t-\varphi)/\varepsilon + \varphi_1$, and φ_1 is the "constant" of integration.

Now we proceed to the construction of higher order terms of the asymptotic expansion. Equating to zero the terms of order ε^{k-1} which result from substituting the expansion (2.74) into (2.48) and taking the previous Taylor expansions into account, we arrive at the relations

$$\begin{aligned}\frac{\partial}{\partial\tau}\Big\{R_k - \Big\langle\nabla\varphi, \Big((h+\rho_0(x,t)+R_0)v_k + (u_0(x,t)+v_0)R_k \\ + h^3|\nabla\varphi|^2\frac{\partial^2 v_k}{\partial\tau^2}\Big)\Big\rangle\Big\} + \mathcal{F}_{k,R}(x,t,\tau) = 0 \pmod{O(\varepsilon)}, \\ \frac{\partial}{\partial\tau}\{v_k - \nabla\varphi(\langle u_0(x,t)+v_0, v_k\rangle + gR_k)\} + \mathcal{F}_{k,v}(x,t,\tau) = 0 \pmod{O(\varepsilon)},\end{aligned} \tag{2.80}$$

where $k = 1, 2, \ldots$, and the $\mathcal{F}_{k,R}$ and $\mathcal{F}_{k,v}$ are polynomials in $\tau, \rho_0, u_0, R_0, v_0, \ldots$, R_{k-1}, V_{k-1}, ρ_k, u_k and their derivatives. In particular, for $k = 1$ we have

$$
\begin{aligned}
\mathcal{F}_{1,R} = &-\frac{\partial}{\partial\tau}\Big\langle \nabla\varphi, R_0 u_1(x,t) + \rho_1(x,t)v_0 - h^3\widehat{\Pi}\frac{\partial v_0}{\partial\tau}\Big\rangle \\
&+ \Big\langle \nabla, R_0 u_0(x,t) + (h(x) + \rho_0(x,t) + R_0)v_0 + h^3|\nabla\varphi|^2\frac{\partial^2 v_0}{\partial\tau^2}\Big\rangle \\
&- \tau\frac{\partial}{\partial\tau}\Big\{R_0\Big\langle\nabla\varphi, \frac{\partial u_0(x,t)}{\partial t}\Big\rangle + \langle\nabla\varphi, v_0\rangle\frac{\partial\rho_0(x,t)}{\partial t}\Big\}\Big|_{t=\varphi}, \\
\mathcal{F}_{1,v} = &-\nabla\varphi\frac{\partial}{\partial\tau}\langle u_1(x,t), v_0\rangle + \nabla\Big\{\langle u_0(x,t), v_0\rangle + \frac{1}{2}|v_0|^2 + gR_0\Big\} \\
&- \nabla\varphi\tau\frac{\partial}{\partial\tau}\Big\langle v_0, \frac{\partial u_0(x,t)}{\partial t}\Big\rangle\Big|_{t=\varphi},
\end{aligned}
\tag{2.81}
$$

where $\widehat{\Pi} = 2\langle\nabla\varphi, \nabla\rangle + \Delta\varphi$.

The left-hand sides of relations (2.80) decrease exponentially as $\tau \to \pm\infty$. It is clear that the right-hand sides must possess the same property. To ensure that the $\mathcal{F}_{k,R}, \mathcal{F}_{k,v}$ decrease as $\tau \to \infty$, we determine the background functions ρ_{k-1}, v_{k-1} as a solution to equations (2.75). One should bear in mind, however, that while $R_i, v_i \to 0$ as $\tau \to \infty$ by construction, these functions stabilize as $\tau \to -\infty$, i.e., $R_i \to R_i^-(x,t)$, $v_i \to v_i^-(x,t)$. Hence the condition $\mathcal{F}_{k,R} \to 0$, $\mathcal{F}_{k,v} \to 0$ as $\tau \to -\infty$ is, in general, nontrivial (it is automatically fulfilled only for $k = 1$, since $R_0, v_0 \to 0$ as $\tau \to -\infty$). To satisfy this condition, we will use a special choice of R_{k-1}^-, v_{k-1}^-. Namely, let R_{k-1}^-, v_{k-1}^- satisfy the equations, similar to (2.75),

$$
\begin{aligned}
\frac{\partial R_{k-1}^-}{\partial t} + \big\langle\nabla, (h+\rho_0)v_{k-1}^- + R_{k-1}^- u_0\big\rangle &= f_{k-1,R}^-, \\
\frac{\partial v_{k_1}^-}{\partial t} + \nabla\{\langle u_0, v_{k-1}^-\rangle + gR_{k-1}^-\} &= f_{k-1,v}^-,
\end{aligned}
\tag{2.82}
$$

which are, however, considered in the domain $D^- = \{(t,x),\ t > 0,\ \varphi(x) > t\}$. In (2.82) $k = 2, 3, \ldots$, and the $f_{k-1,R}^-, f_{k-1,v}^-$ are polynomials in $\rho_1, u_1, R_1^-, v_1^-, \ldots$, R_{k-2}^-, v_{k-2}^- and their derivatives.

Then, as is easily seen, $\mathcal{F}_{k,R}, \mathcal{F}_{k,v} \to 0$ as $\tau \to -\infty$, and we can expand the functions involved in (2.80) by the Taylor formula with respect to t around $t = \varphi(x)$. Using the equality $t - \varphi = \varepsilon\tau$ and taking in the next approximation the terms of order $O(\varepsilon)$ into account, on integrating (2.80) we obtain the equations

$$
\begin{aligned}
&\check{R}_k - \Big\langle\nabla\varphi, (H + R_0)\check{v}_k + (\check{u}_0 + v_0)\check{R}_k + h^3|\nabla\varphi|^2\frac{\partial^2\check{v}_k}{\partial\tau^2}\Big\rangle = F_{k,R}(x,\tau), \\
&\check{v}_k - \nabla\varphi\big(\langle\check{u}_0 + v_0, \check{v}_k\rangle + g\check{R}_k\big) = F_{k,v}(x,\tau).
\end{aligned}
\tag{2.83}
$$

Here $H = h(x) + \rho_0(x,\varphi)$, the "cap" denotes the substitution of $t = \varphi$, i.e., $\check{\psi}_k = \psi_k(\tau, x, t)|_{t=\varphi}$,

$$
F_{k,R} = \int_\tau^\infty \mathcal{F}_{k,R}(x,\varphi,\tau')\,d\tau', \quad F_{k,v} = \int_\tau^\infty \mathcal{F}_{k,v}(x,\varphi,\tau')\,d\tau'.
\tag{2.84}
$$

Note that in the passage from (2.80) to (2.83) we have chosen the "constants" of integration so that to ensure the condition $\check{R}_k, \check{v}_k \to 0$ as $\tau \to \infty$.

Represent the vector function $\check{v}_k$ in the form

$$\check{v}_k(\tau, x) = \omega_k(\tau, x)\nu + w_k(\tau, x)\xi, \tag{2.85}$$

where ξ is a unit vector orthogonal to $\nu = \nabla\varphi/|\nabla\varphi|$ and ω_k, w_k are the coefficients in the decomposition of the vector $\check{v}_k$ with respect to the basis $\{\nu, \xi\}$. Then the second equation (2.83) immediately implies the equality

$$w_k(\tau, x) = \langle \xi, F_{k,v}(x, \tau)\rangle, \tag{2.86}$$

as well as the relationship between ω_k and $\check{R}_k$,

$$g\check{R}_k = (\kappa - \omega_0)\omega_k - \big\{V_\nu\langle F_{k,v}, \nu\rangle + \check{u}_{0\xi}\langle F_{k,v}, \xi\rangle\big\}, \tag{2.87}$$

where $\check{u}_{0\xi} = \langle \check{u}_0, \xi\rangle$.

Now, using (2.85)–(2.87), the first equation (2.83) can be easily rewritten as a nonhomogeneous equation in variations with respect to (2.79),

$$gh^3|\nabla\varphi|^2\frac{\partial^2\omega_k}{\partial\tau^2} - \Big(\kappa^2 - c^2 - 3\kappa\omega_0 + \frac{3}{2}\omega_0^2\Big)\omega_k = -G_k(\tau, x), \tag{2.88}$$

where

$$G_k = gV_\nu F_{k,R} + (\kappa - \omega_0)\big(V_\nu\langle F_{k,v}, \nu\rangle + \check{u}_{0\xi}\langle F_{k,v}, \xi\rangle\big).$$

As we pointed out above, the following orthogonality condition is necessary and sufficient for solvability of (2.88) in the class of functions we are interested in:

$$\int_{-\infty}^{\infty} G_k(\tau, x)\frac{\partial\omega_0}{\partial\tau}\, d\tau = 0. \tag{2.89}$$

Using formulas (2.84) and integrating by parts, one can easily transform condition (2.89) to the form

$$V_\nu\int_{-\infty}^{\infty}\omega_0\mathcal{F}_{k,R}(x, \varphi, \tau)\, d\tau + \Big\langle\zeta, \int_{-\infty}^{\infty} R_0\mathcal{F}_{k,v}(x, \varphi, \tau)\, d\tau\Big\rangle = 0, \tag{2.90}$$

where we have used the notation (2.59).

Obviously, the solution to equation (2.88) has the form

$$\omega_k = z_k(\tau, x) + \varphi_{k+1}(x)\omega_{0\tau},$$

where z_k is a particular solution of the nonhomogeneous equation which is computed by the formula (1.24) and φ_{k+1} is the "constant" of integration. It is clear that conditions (2.90) for $k = 1, 2, \ldots$ enable us to determine the phase φ, as well as the integration "constants" φ_i; namely, (2.90) with $k = 1$ allows us to find $\varphi(x)$, and (2.90) with $k > 1$ yields $\varphi_{k-1}(x)$.

Next, letting $\tau \to -\infty$ in (2.88), we obtain

$$\omega_k^- = \psi_k(x), \tag{2.91}$$

where

$$\psi_k^- = (\kappa^2 - c^2)^{-1}\Big\{gV_\nu \int_{-\infty}^{\infty} \mathcal{F}_{k,R}(x,\varphi,\tau)\,d\tau + \kappa\Big\langle \zeta, \int_{-\infty}^{\infty} \mathcal{F}_{k,v}(x,\varphi,\tau)\,d\tau\Big\rangle\Big\}.$$

In a similar way, from (2.86) and (2.87) we have

$$w_k^- = \sigma_k(x), \quad \check{R}_k^- = \gamma_k(x), \tag{2.92}$$

where

$$\sigma_k = \Big\langle \xi, \int_{-\infty}^{\infty} \mathcal{F}_{k,v}(x,\varphi,\tau)\,d\tau\Big\rangle,$$
$$\gamma_k = \frac{1}{g\kappa}\Big\{c^2\psi_k + gV_\nu \int_{-\infty}^{\infty} \mathcal{F}_{k,R}(x,\varphi,\tau)\,d\tau\Big\}.$$

Thus we have the following representation for the functions $\check{v}_k, \check{R}_k$ (which are defined only on the front $\mathcal{T} = \bigcup_{t\in[0,T]} \Gamma_t$):

$$\begin{aligned} \check{v}_k(\tau,x) &= \Phi_k(x)\chi(\tau,x) + v_{k,\mathrm{sol}}(\tau,x), \\ \check{R}_k(\tau,x) &= \gamma_k(x)\chi(\tau,x) + R_{k,\mathrm{sol}}(\tau,x), \end{aligned} \tag{2.93}$$

where $\Phi_k(x) = \psi_k(x)\nu + \sigma_k(x)\xi$, χ is a function from $\mathcal{H}$ such that $\chi \to 1$ as $\tau \to -\infty$, and the $v_{k,\mathrm{sol}}$, $R_{k,\mathrm{sol}}$ are functions decreasing as $\tau \to \pm\infty$ which are determined by the solution of (2.88) and formulas (2.85)–(2.87).

Define the extensions of the functions $\check{v}_k, \check{R}_k$ into $R^{2+1}_{x,\tau} \times [0,T]$ in the form similar to (2.93):

$$\begin{aligned} v_k(\tau,x,t) &= u_k^-(x,t)\chi(\tau,x) + v_{k,\mathrm{sol}}(\tau,x), \\ R_k(\tau,x,t) &= \rho_k^-(x,t)\chi(\tau,x) + R_{k,\mathrm{sol}}(\tau,x). \end{aligned} \tag{2.94}$$

Obviously, the equalities $v_k|_{\Gamma_t} = \check{v}_k$, $R_k|_{\Gamma_t} = \check{R}_k$ imply the condition

$$u_k^-\big|_{\Gamma_t} = \Phi_k\big|_{\Gamma_t}, \quad \rho_k^-\big|_{\Gamma_t} = \gamma_k\big|_{\Gamma_t}.$$

We construct the functions u_k^-, ρ_k^- following the same scheme as before. Namely, we determine the functions R_k^-, v_k^- from equations of the form (2.82) completed with boundary conditions on the front

$$v_k^-\big|_{\Gamma_t} = \Phi_k\big|_{\Gamma_t}, \quad R_k^-\big|_{\Gamma_t} = \gamma_k\big|_{\Gamma_t}, \tag{2.95}$$

as well as with natural initial conditions for $t = 0$ and x: $\varphi(x) < 0$. Then we take the u_k^-, ρ_k^- to be infinitely differentiable extensions of the v_k^-, R_k^- into the domain $D_\varepsilon^- = \{(x,t),\ t > 0,\ \varphi(x) > t - \sqrt{\varepsilon}\}$. It is clear that for sufficiently small ε such extensions certainly exist.

We should emphasize once more that the distorted soliton moves at a "supersonic" velocity. The same condition (2.64), which guaranteed solvability of equation (2.79) in the class of soliton-type functions, ensures that the basic model equation (2.57) is hyperbolic. Now we see that this condition (more precisely, the property of a solution) is of fundamental importance for the construction of higher order terms of the asymptotic expansion as well, since only under condition (2.64) is the problem (2.82), (2.95) well posed.

It remains to derive the equation (2.57). Note first of all that, using the inequalities (2.77)–(2.79), one can prove the identities

$$\begin{aligned}
&\Big\langle \nabla, \nu\Big\{\omega_0(h+\rho_0(x,t)+R_0)+h^3|\nabla\varphi|^2\frac{\partial^2\omega_0}{\partial\tau^2}\Big\}\Big\rangle\Big|_{t=\varphi} \\
&\qquad = \langle\nabla, R_0\kappa\nu\rangle - \omega_0|\nabla\varphi|\frac{\partial\rho_0}{\partial t}\Big|_{t=\varphi}, \\
&\nabla\Big\{\langle u_0(x,t), v_0\rangle + \frac{1}{2}v_0^2 + gR_0\Big\}\Big|_{t=\varphi} = \nabla(\omega V_\nu) - \Big\langle \nu, \frac{\partial u_0}{\partial t}\Big|_{t=\varphi}\Big\rangle\nabla\varphi.
\end{aligned} \tag{2.96}$$

Using (2.96) and the explicit form (2.81) of the functions $\mathcal{F}_{1,R}$, $\mathcal{F}_{1,v}$, we can easily transform the orthogonality condition (2.90) for $k=1$ into the form

$$h^3\langle\nabla, \nu I_0\rangle - \langle\nabla, V_\nu I_1\zeta\rangle + \frac{\partial\rho_0}{\partial t}\Big|_{t=\varphi} I_2 + \frac{\partial u_{0\nu}}{\partial t}\Big|_{t=\varphi} I_1 = 0,$$

where

$$I_0 = |\nabla\varphi|\int_{-\infty}^{\infty}\Big(\frac{\partial\omega_0}{\partial\tau}\Big)^2 d\tau, \quad I_1 = \int_{-\infty}^{\infty} R_0\omega_0\,d\tau, \quad I_2 = \frac{1}{2}\int_{-\infty}^{\infty}\omega_0^2\,d\tau.$$

On computing these integrals we obtain the equation (2.57).

2.4. A modified Boussinesq system

As is well known, there are several models describing wave processes on a fluid surface in a shallow water approximation. We considered one of them in the previous section; another one, also called a Boussinesq system, has the same order of accuracy as (2.48) for $h=\text{const}$ and is given by the equations

$$\begin{aligned}
&\frac{\partial u}{\partial t} + \nabla\Big\{\frac{\alpha}{2}|u^2| + \rho + \varepsilon^2\Delta\rho\Big\} = 0, \\
&\frac{\partial\rho}{\partial t} + \langle\nabla, (1+\alpha\rho)u\rangle = 0
\end{aligned} \tag{2.97}$$

(see, e.g., [**262**]). Here ρ is a scalar function describing the elevation of the free surface normalized by the coefficient $\alpha = \rho^0/h$, $h>0$ is the depth of the nonperturbed fluid, which is assumed to be constant, $\rho^0>0$ is the characteristic magnitude of deviation of the free surface from the equilibrium state, $u=(u^1,u^2)$ is the vector function describing the horizontal velocity, $x\in\mathbb{R}^2$, and $\varepsilon>0$ is a small parameter.

Construction of the asymptotic solution to the system (2.97) is presented in detail in [**162**] and is carried out similarly to §2.3. Hence we only formulate and analyze the main result and discuss a technical aspect which arises in the treatment of (2.97).

The Boussinesq system (2.97) has a soliton-type asymptotic solution similar to (2.62). For its description we will need auxiliary functions. Let the background perturbation $\rho_0(x,t), u_0(x,t)$ be a solution to the degenerate system (2.97), i.e.,

$$\begin{aligned}&\frac{\partial u_0}{\partial t}+\nabla\Big\{\frac{\alpha}{2}|u_0^2|+\rho_0\Big\}=0,\\&\frac{\partial \rho_0}{\partial t}+\langle\nabla,(1+\alpha\rho_0)u_0\rangle=0.\end{aligned}\tag{2.98}$$

We will assume that for $t\in[0,T]$ the functions ρ_0, u_0 are smooth and satisfy the condition

$$1+\alpha\rho_0(x,t)>0,$$

which has the same meaning as (2.51), i.e., that the free surface stays above the bottom.

As before, denote by $\varphi(x)$ the function describing the front position Γ_t of the soliton-type wave at time moment $t\in[0,T]$, i.e., $\Gamma_t=\{x,\ \varphi(x)=t\}$. Moreover, we will introduce the notation

$$c=\sqrt{1+\alpha\rho_0(x,\varphi)},\qquad \kappa=V_\nu-\alpha\langle u_0(x,\varphi),\nu\rangle,\qquad A=\kappa^2-c^2,$$

where $\nu=\nabla\varphi/|\nabla\varphi|$, $V_\nu=1/|\nabla\varphi|$, and c,κ,V_ν have the same meaning as in §2.3.

The main result of this section is the model equation

$$V_\nu\big\langle\nabla,u_0(x,\varphi)J_1+\nu J_2\big\rangle=J_1\frac{\partial u_0}{\partial t}\Big|_{t=\varphi}-J_3\frac{\partial\rho_0}{\partial t}\Big|_{t=0},\tag{2.99}$$

which describes the dynamics of the front Γ_t. Here

$$\begin{aligned}J_1&=2\alpha\kappa\Big\{\sqrt{A}c+(A-c^2)\arctan\frac{\sqrt{A}}{c}\Big\},\\J_2&=\frac{1}{2}(3A-c^2)\Big\{\sqrt{A}c\frac{3A+c^2}{3A-c^2}+\kappa^2\arctan\frac{\sqrt{A}}{c}\Big\},\\J_3&=2\alpha\kappa^2\Big\{\frac{\sqrt{A}c}{A+c^2}-\arctan\frac{\sqrt{A}}{c}\Big\}.\end{aligned}$$

We complete the equation (2.99) with initial conditions

$$\varphi\big|_{\Gamma_0}=0,\qquad \frac{\partial\varphi}{\partial\nu}\Big|_{\Gamma_0}=p^0.\tag{2.100}$$

Here $p^0=p^0(x)|_{\Gamma_0}$ is a smooth function related to the amplitude $A^0_{\rho,\mathrm{sol}}$ of elevation on the initial front Γ_0 by equality

$$p^0=\Big\{\alpha\check{u}^0_\nu\pm\sqrt{\alpha A^0_{\rho,\mathrm{sol}}+c^2}\Big\}^{-1},$$

where $\check{u}^0_\nu=\langle\nu,u(x,0)\rangle|_{\Gamma_0}$ and the sign $\pm$ refers to the choice of the direction of the wave motion, either along ν or along $-\nu$. It is assumed that $A^0_{\rho,\mathrm{sol}}>0$ and, moreover, that $\alpha A^0_{\rho,\mathrm{sol}}+c^2\neq(\alpha\check{u}^0_\nu)^2$.

It can be proved that under the condition

$$\frac{\kappa}{c} > 1 \tag{2.101}$$

the equation (2.99) is hyperbolic; hence the problem (2.99), (2.100) is well-posed. It is clear that condition (2.101) is similar to (2.64) and implies that the function $A = \alpha A^0_{\rho,\mathrm{sol}}$ is positive.

The leading term of the asymptotic solution to the system (2.97) is given by the formulas

$$\begin{aligned} \rho &= \rho_0(x,t) + R_0(\tau,x) + O(\varepsilon), \\ u &= u_0(x,t) + \kappa\frac{R_0(\tau,x)}{c^2+\alpha R_0(\tau,x)}\nu + O(\varepsilon), \end{aligned} \tag{2.102}$$

where $\tau = (t-\varphi(x))/\varepsilon+\varphi_1(x)$, φ_1 is a smooth function which determines a small (of order ε) correction to the front position Γ_t, and the function $R_0(\tau,x)$ is determined by the equation

$$\log\left\{\frac{\sqrt{A}\Psi+c}{\sqrt{A}\Psi-c}\right\} + 2\frac{\sqrt{A}}{c}\left\{\frac{\pi}{2} - \arctan\Psi\right\} = \frac{\sqrt{A}}{c|\nabla\psi|}|\tau|,$$

where $\Psi = \sqrt{(c^2+\alpha R_0)/(A-\alpha R_0)}$.

It is not hard to see that R_0 is an infinitely differentiable function which tends to zero together with all its derivatives exponentially fast as $\tau\to\pm\infty$. Moreover, $R_0>0$, and the maximum of R_0 is attained at the point $\tau=0$ and equals A/α.

Thus the amplitude $A_{\rho,\mathrm{sol}}$ of the soliton component of the elevation on the front Γ_t equals A/α, and the amplitude $A_{u,\mathrm{sol}}$ equals $A/\alpha\kappa$.

We consider particular cases where the equation (2.99) takes a simpler form.

2.4.1. Axially symmetric case with zero background. Let $u_0=\rho_0=0$, let Γ_0 be the circle of radius $r_0>0$, and let $A|_{\Gamma_0}=\mathrm{const}$. Put $\varphi=\varphi(r)$. Then the equation (2.99) can be integrated to become

$$r(3A-1)\left\{\sqrt{A}\frac{3A+1}{3A-1} + (A+1)\arctan\sqrt{A}\right\} = E. \tag{2.103}$$

This relation implicitly determines A as a function of r. Here $E>0$ is the constant of integration, which is specified by the initial condition $A|_{r=r_0} = \alpha A^0_{\rho,\mathrm{sol}}$. In its turn, the phase φ is determined by A in view of the equation

$$\frac{d\varphi}{dr} = \pm(1+A)^{-1/2}. \tag{2.104}$$

For large r we find from (2.103), (2.104) that

$$V_r = 1 + E_1 r^{-2/3} + O(r^{-4/3}), \quad A = E_2 r^{-2/3} + O(r^{-4/3}),$$

where $E_1 = \frac{1}{8}(\frac{3}{2}E)^{2/3}$, $E_2 = 2E_1/\alpha$.

Thus under the above assumptions we again arrive at the law that the wave amplitude decays as $r^{-2/3}$.

2.4.2. Small nonlinearity approximation. Let the amplitude of the distorted soliton be small compared to the total depth, i.e., $A/c^2 \ll 1$. Denote by μ the small parameter characterizing the magnitude of this ratio. Representing the functions R_0 and φ in the form of power series in μ, we can see that

$$R_0 = \frac{A}{\alpha} \cosh^{-2}(\beta\tau) + O(\mu^2),$$
$$\varphi = \Phi(x) - \mu\psi(x) + O(\mu^2),$$

where $\beta = \frac{1}{2}\sqrt{A}/(c|\nabla\varphi|)$ and Φ is the solution to the Hamilton–Jacobi equation

$$|\nabla\Phi|\sqrt{1 + \alpha\rho_0(x, \Phi)} + \alpha\langle u_0(x, \Phi), \nabla\Phi\rangle = 1. \tag{2.105}$$

In the case of zero background the equation for the correction ψ substantially simplifies, and we get

$$\frac{\partial\psi}{\partial\nu} = \frac{1}{2}\chi, \quad \frac{\partial\chi}{\partial\nu} + \frac{2}{3}\Delta\Phi\chi = 0.$$

Here $\nu = \nabla\Phi$, and we have used the fact that when $\rho_0 = u_0 = 0$ equation (2.105) becomes $|\nabla\Phi| = 1$.

The analysis of the solution thus obtained in the axially symmetric case yields the same results as in §2.3.

Let us point out a technical aspect related to the fact that the function R_0 is given in an implicit form. Namely, when deriving the model equation (2.99) the corresponding orthogonality condition reduces to

$$\langle\nabla, u_0(x, \varphi)J_1 + \nu J_2\rangle = \langle u_{0t}(x, \varphi), \nabla\varphi\rangle J_1 - |\nabla\varphi|\rho_{0t}(x, \varphi)J_3, \tag{2.106}$$

where

$$J_1 = \alpha\kappa|\nabla\varphi|^{-1} \int_{-\infty}^{\infty} (\beta + \alpha R_0)^{-1} R_0 \, d\tau,$$
$$J_2 = |\nabla\varphi|^{-1} \int_{-\infty}^{\infty} R_0^2 \, d\tau,$$
$$J_3 = \frac{1}{2}\alpha\kappa^2|\nabla\varphi|^{-1} \int_{-\infty}^{\infty} (\beta + \alpha R_0)^{-2} R_0^2 \, d\tau.$$

To compute the corresponding integrals, observe that the following equality holds:

$$\frac{|\nabla\varphi|}{R_0}\sqrt{\frac{\beta + \alpha R_0}{A - \alpha R_0}} dR_0 = \pm d\tau,$$

where the signs plus and minus are taken for $\tau < 0$ and $\tau > 0$ respectively. Passing from τ to the new variable $y = R_0$, we find, in particular, that

$$J_1 = 2\frac{\kappa}{\alpha} \int_0^A \frac{y \, dy}{\sqrt{(A - y)(\beta + y)}}.$$

Making similar transformations of J_2 and J_3 and computing the resulting integrals, we can rewrite equation (2.106) in the form (2.99).

2.5. Dynamics of the smoothed shock wave in a gas with small viscosity: Large Prandtl numbers

In Chapter 1 we set out in detail the method of constructing the asymptotic solution of smoothed shock wave type as applied to the model example of the Korteweg–de Vries–Burgers equation (see §1.2). As was to be expected, in the example treated in Chapter 1 the profile of the wave was determined easily, but the description of the front dynamics reduced to computation of parameters of the limiting solution as $\varepsilon \to 0$, i.e., of the shock wave. As is well known, the latter problem is rather difficult even for the model example of §1.2, since it consists in solving a Hopf equation (i.e., a degenerate as $\varepsilon \to 0$ KdV–Burgers equation) to the right and to the left of the wave front, considered jointly with an ordinary equation on the front (the Hugoniot condition). In this section (and in §§2.6 and 5.4) we will obtain a similar result for equations of gas dynamics with small viscosity and heat conductivity. Namely, to describe of the wave profile we will derive an ordinary differential equation, while to determine the front of the smoothed wave we will obtain a classical problem on the motion of a shock wave in a gas, i.e., the equations of ideal gas dynamics to the right and to the left of the jump, as well as the Rankine–Hugoniot conditions on the jump front. The fact that the front motion is described by the solution of a global problem, i.e., the front dynamics cannot be described by a finite system of differential equations considered only on the front, is common for solutions of shock wave type. This is what distinguishes solutions of shock wave type from soliton-type solutions, since in the latter case the problem of determining the front virtually splits off and becomes a local one. Note that the kink-type solution of the sine-Gordon equation has a smoothed shock wave profile, but the front dynamics is described by the solution to a model equation considered on the front. This fact is related to the property that the kink of the sine-Gordon equation has fixed values outside a small neighborhood of the front (0 and 2π or $2\pi/\gamma$, where γ is a known function). In this sense the kink-type solutions are closer to soliton-type solutions than to solutions of smoothed shock wave type. The fact that the sine-Gordon equation is hyperbolic is not very important in this respect; in what follows we will show that parabolic systems also have soliton-type solutions.

The necessity of solving the equations of ideal gas dynamics ahead of and behind the shock wave front makes the computations much more complicated. Therefore it is very important to develop a method avoiding these difficulties. The first attempt in this direction was apparently made by Whitham [**255–258**] in 1957–1959; he proposed, based on physical considerations, a system of model equations describing the front dynamics and considered only on the front. However, as was shown later [**217, 236**], Whitham's equations adequately describe the dynamics of the shock wave front only when the amplitude of the density jump is sufficiently small. A rigorous mathematical theory giving a solution to this problem for an arbitrary amplitude of the density jump was developed by Maslov [**143**] in 1977. He derived the required equations of three-dimensional gas dynamics, but it turned out that they form an infinite chain (which he called the chain of Hugoniot type conditions). This theory was further developed in [**217, 236**], where a similar chain was obtained for shock waves of gas dynamics taking the effects of heat transfer into account, in [**25**] dealing with the motion of the eye of a hurricane in a shallow water approximation, and in [**45, 51**] where the dynamics of other types of singularities was studied. Lately the problem of describing the dynamics of singularities has been actively explored in the framework of the algebraic theory of generalized functions, see [**9, 18–20, 39, 43, 59, 91, 138, 145, 167, 187, 210**].

Of course, infinite chains of equations cannot be used for numerical computations. Recall that a similar problem is widely known in statistical physics as the problem of truncating the Bogolyubov–Born–Green–Kirkwood–Yvon chains of equations [**247**]. The simplest truncation procedure consists in taking the first N equations of the chain and dropping all the additional unknown functions. The numerical computations where the solution obtained by chain truncation was compared with the exact solution [**217**] and with actual hurricane observation data [**25**] demonstrated a good fit within a reasonably large time interval. These results, combined with the first results on justification of the truncation procedure [**47, 48**], allow for the use of the chains of Hugoniot type conditions when dealing with concrete physical problems.

To conclude the introductory part of this section, note that the chains of Hugoniot type conditions can be derived both for quasilinear hyperbolic systems (in particular, the equations of ideal gas dynamics) and for asymptotic solutions of equations with small viscosity, in particular, when constructing asymptotic solutions of smoothed shock wave type for equations of gas dynamics. The corresponding theory was developed in [**166**].

In order to explain the method of constructing asymptotic solutions of shock wave type for systems of multidimensional equations we consider a physically meaningful problem on the motion of a smoothed shock wave in a polytropic gas with small viscosity. A similar problem for isentropic gas was considered in [**166**]. Of course, there are a lot of papers, both mathematical and physical, where the problems of gas dynamics are treated (see, e.g., [**40, 118, 121, 125, 251, 262, 271**] and references therein). For this reason we will focus on the mathematical aspects of the problem of constructing the asymptotic solution, making only a few comments on the physical meaning of the results.

Now, consider the dimensionless equations of three-dimensional gas dynamics in the divergence form

$$\begin{aligned}
&\frac{\partial \rho}{\partial t} + \langle \nabla, \rho u\rangle = 0,\\
&\frac{\partial}{\partial t}(\rho u_{(i)}) + \langle \nabla, \rho u u_{(i)}\rangle + \frac{\partial p}{\partial x_i} = \varepsilon \frac{\partial}{\partial x_j}\sigma_{ij}, \quad i = 1,2,3, \qquad (2.107)\\
&\frac{\partial}{\partial t}\Big(\frac{1}{2}\rho u^2 + \rho\mathcal{E}\Big) + \Big\langle \nabla, \rho\Big(\frac{1}{2}u^2 + w\Big)u - \varepsilon\sigma\cdot u - \varepsilon\frac{\kappa}{\mathrm{Pr}}\nabla T\Big\rangle = 0.
\end{aligned}$$

Here $\rho = \rho(x,t,\varepsilon)$ is the gas density, $u = u(x,t,\varepsilon) = (u_{(1)}, u_{(2)}, u_{(3)})$ is the gas velocity, $T = T(x,t,\varepsilon)$ is the gas temperature and $p = p(x,t,\varepsilon)$ is the gas pressure, Pr is the Prandtl number, $x \in \mathbb{R}^3$, $t > 0$. Assuming that the gas is polytropic with adiabatic coefficient $\gamma > 1$, we will use the standard formulas for internal energy $\mathcal{E}$ and enthalpy w of the gas:

$$\mathcal{E} = \frac{p}{(\gamma-1)\rho}, \qquad w = \frac{\gamma p}{(\gamma-1)\rho}, \qquad (2.108)$$

which correspond to the state equation

$$p = \rho T. \qquad (2.109)$$

In (2.107) we denoted by σ the tensor of viscosity stress

$$\sigma_{ij} = \eta\left(\frac{\partial u_{(i)}}{\partial x_j} + \frac{\partial u_{(j)}}{\partial x_i} - \frac{2}{3}\delta_{ij}\frac{\partial u_{(k)}}{\partial x_k}\right) + \zeta\delta_{ij}\frac{\partial u_{(k)}}{\partial x_k},$$

where δ_{ij} is the Kronecker delta and summation over repeated indices is assumed.

Moreover, in (2.107) we denote by ε the parameter inversely proportional to the Reynolds number, and we consider (2.107) for large Reynolds numbers, i.e., for $\varepsilon \ll 1$. To simplify the subsequent analysis, we will assume that the viscosity coefficients $\eta > 0$, $\zeta > 0$ and the heat conductivity coefficient κ are slowly varying functions,

$$\eta = \eta_0 + \varepsilon\eta_1(T), \quad \zeta = \zeta_0 + \varepsilon\zeta_1(T), \quad \kappa = \kappa_0 + \varepsilon\kappa_1(T), \tag{2.110}$$

where η_0, ζ_0, and κ_0 are constants. In this section we will consider a simplified case where the effects of heat conductivity in the medium are small as compared to the effects of viscosity. Accordingly, we set $\kappa/\Pr = \varepsilon\kappa'$, so that the heat conductivity coefficient $\varepsilon^2\kappa'$ will be small relative to the viscosity coefficients $\varepsilon\eta$ and $\varepsilon\zeta$. This assumption, which means that the Prandtl number is large, is nonstandard for usual gases; however, it holds, e.g., in a strongly magnetized plasma [**251**].

The settings with finite or small Prandtl number, which are more meaningful physically, will be considered in the next section and in §5.4.

We will seek the asymptotic solution to the system (2.107) in the form of an asymptotic expansion

$$\begin{aligned}
u &= u_0(x,t) + U_0(\tau,x,t) + \varepsilon\big(u_1(x,t) + U_1(\tau,x,t)\big) + \dots,\\
\rho &= \rho_0(x,t) + R_0(\tau,x,t) + \varepsilon\big(\rho_1(x,t) + R_1(\tau,x,t)\big) + \dots,\\
p &= p_0(x,t) + P_0(\tau,x,t) + \varepsilon\big(p_1(x,t) + P_1(\tau,x,t)\big) + \dots,\\
T &= T_0(x,t) + \theta_0(\tau,x,t) + \varepsilon\big(T_1(x,t) + \theta_1(\tau,x,t)\big) + \dots,
\end{aligned} \tag{2.111}$$

where the u_i, ρ_i, p_i, T_i are smooth functions which form the background, $\tau = S(x,t)/\varepsilon$, the U_i, R_i, P_i, θ_i are functions with fast variation from $\mathcal{H}$, i.e., they converge exponentially fast to zero as $\tau \to -\infty$ and to smooth functions $U_i^+(x,t)$, $R_i^+(x,t)$, $P_i^+(x,t)$, $\theta_i^+(x,t)$ as $\tau \to +\infty$. We will assume that for each fixed $t \geq 0$ the front $\Gamma_t = \{x,\ S(x,t) = 0\}$ of the smoothed shock wave is a smooth surface in $\mathbb{R}^3$ of codimension 1. Assuming that $\partial S/\partial t|_{\Gamma_t} \neq 0$, we will seek the phase S in the form

$$S = t - \varphi(x).$$

Note that the state equation (2.109) implies relationships between the coefficients in the expansions for p, ρ, and T. In particular,

$$\begin{aligned}
p_0 &= \rho_0 T_0, & P_0 &= \rho_0\theta_0 + R_0(T_0 + \theta_0),\\
p_1 &= \rho_0 T_1 + \rho_1 T_0, & P_1 &= \rho_0\theta_1 + \rho_1\theta_0 + R_0(T_1 + \theta_1) + R_1(T_0 + \theta_0).
\end{aligned} \tag{2.112}$$

Substitute the expansions (2.111) into (2.107) and, using the equalities (2.108), (2.109), (2.112), consider the terms of order ε^{-1}. Multiplying by ε, we obtain the relations

$$
\begin{aligned}
&\frac{\partial}{\partial\tau}\{R_0 - \langle\nabla\varphi, \rho_0 U_0 + R_0(u_0+U_0)\rangle\} = O(\varepsilon),\\
&\frac{\partial}{\partial\tau}\Big\{\rho_0 U_0 + R_0(u_0+U_0) - \langle\nabla\varphi,(u_0+U_0)\rangle(\rho_0 U_0 + R_0(u_0+U_0))\\
&\qquad - \langle\nabla\varphi, U_0\rangle\rho_0 u_0 - \nabla\varphi P_0\\
&\qquad - \frac{\partial}{\partial\tau}\Big(|\nabla\varphi|^2\eta_0 U_0 + \Big(\zeta_0+\frac{\eta_0}{3}\Big)\langle\nabla\varphi, U_0\rangle\nabla\varphi\Big)\Big\} = O(\varepsilon),\\
&\frac{\partial}{\partial\tau}\Big\{\frac{1}{2}(\rho_0+R_0)(u_0+U_0)^2 + \frac{1}{\gamma-1}P_0\\
&\qquad - \Big\langle\nabla\varphi, \Big(\frac{1}{2}(\rho_0+R_0)(u_0+U_0)^2 + \frac{\gamma}{\gamma-1}(p_0+P_0)\Big)(u_0+U_0)\Big\rangle\\
&\qquad - \frac{1}{2}\frac{\partial}{\partial\tau}\Big(\eta_0|\nabla\varphi|^2(u_0+U_0)^2 + \Big(\zeta_0+\frac{\eta_0}{3}\Big)\big(\langle\nabla\varphi, u_0+U_0\rangle\big)^2\Big)\Big\}\\
&\qquad = O(\varepsilon).
\end{aligned}
\tag{2.113}
$$

Note that all expressions in the left-hand sides of (2.113) tend to zero exponentially fast as $\tau\to\pm\infty$. This allows us to expand all the functions in (2.113) by the Taylor formula around the point $t=\varphi(x)$. We employ the equality $t-\varphi(x)=\varepsilon\tau$ and neglect the terms of order $O(\varepsilon)$ in the leading term (i.e., we take them into account in the higher order terms). Then on integrating we obtain

$$
\check{R}_0 - \langle\nabla\varphi, \check{\rho}_0\check{U}_0 + \check{R}_0(\check{u}_0+\check{U}_0)\rangle = 0, \tag{2.114}
$$

$$
\begin{aligned}
&\check{\rho}_0\check{U}_0 + \check{R}_0(\check{u}_0+\check{U}_0) - \langle\nabla\varphi,(\check{u}_0+\check{U}_0)\rangle\\
&\qquad\times(\check{\rho}_0\check{U}_0 + \check{R}_0(\check{u}_0+\check{U}_0)) - \langle\nabla\varphi,\check{U}_0\rangle\check{\rho}_0\check{u}_0 - \nabla\varphi\check{P}_0\\
&\qquad = \frac{\partial}{\partial\tau}\Big(|\nabla\varphi|^2\eta_0\check{U}_0 + \Big(\zeta_0+\frac{\eta_0}{3}\Big)\langle\nabla\varphi,\check{U}_0\rangle\nabla\varphi\Big),
\end{aligned}
\tag{2.115}
$$

$$
\begin{aligned}
&\frac{1}{2}(\check{\rho}_0+\check{R}_0)(\check{u}_0+\check{U}_0)^2 + \frac{1}{\gamma-1}\check{P}_0 - \Big\langle\nabla\varphi, \Big(\frac{1}{2}(\check{\rho}_0+\check{R}_0)(\check{u}_0+\check{U}_0)^2\\
&\qquad + \frac{\gamma}{\gamma-1}(\check{p}_0+\check{P}_0)\Big)(\check{u}_0+\check{U}_0)\Big\rangle\\
&\qquad = \frac{1}{2}\frac{\partial}{\partial\tau}\Big(\eta_0|\nabla\varphi|^2(\check{u}_0+\check{U}_0)^2 + \Big(\zeta_0+\frac{\eta_0}{3}\Big)\big(\langle\nabla\varphi,\check{u}_0+\check{U}_0\rangle\big)^2\Big)\\
&\qquad + \frac{1}{2}\check{\rho}_0\check{u}_0^2 - \Big\langle\nabla\varphi, \Big(\frac{1}{2}\check{\rho}_0\check{u}_0^2 + \frac{\gamma}{\gamma-1}\check{p}_0\Big)\check{u}_0\Big\rangle.
\end{aligned}
\tag{2.116}
$$

In integrating the equations (2.113), we have chosen the "constants" of integration so as to ensure the fulfillment of the conditions $\check{R}_0, \check{U}_0, \check{P}_0\to 0$ as $\tau\to-\infty$. Moreover, we use the same notation $\check{f}$ for functions considered on $\mathcal{T}\times\mathbb{R}^1_\tau$, where $\mathcal{T}=\bigcup_{t\geq 0}\Gamma_t$, i.e., $\check{f}=f(\tau,x,t)|_{t=\varphi(x)}$.

Denote by $\nu=\nabla\varphi/|\nabla\varphi|\big|_{\Gamma_t}$ the unit normal to Γ_t in the direction where the function φ grows. Let $V_\nu=1/|\nabla\varphi|$ denote the velocity of motion of the surface Γ_t in the direction of ν, and let $u_{0,\nu}$, $U_{0,\nu}$ be the projections of vectors u_0, U_0 onto ν.

Consider the equations (2.114)–(2.116) as $\tau \to +\infty$. It is clear that we obtain the algebraic equalities

$$V_\nu \check{R}_0^+ = \check{\rho}_0 \check{U}_{0,\nu}^+ + \check{R}_0^+ (\check{u}_{0,\nu} + \check{U}_{0,\nu}^+), \tag{2.117}$$

$$\begin{aligned} &V_\nu \{\check{\rho}_0 \check{U}_0^+ + \check{R}_0^+ (\check{u}_0 + \check{U}_0^+)\} \\ &\quad = (\check{u}_{0,\nu} + \check{U}_{0,\nu}^+)(\check{\rho}_0 \check{U}_0^+ + \check{R}_0^+ (\check{u}_0 + \check{U}_0^+)) + \check{U}_{0,\nu}^+ \check{\rho}_0 \check{u}_0 + \nu \check{P}_0^+, \end{aligned} \tag{2.118}$$

$$\begin{aligned} &V_\nu \Big\{ \frac{1}{2}(\check{\rho}_0 + \check{R}_0^+)(\check{u}_0 + \check{U}_0^+)^2 + \frac{1}{\gamma - 1} \check{P}_0^+ \Big\} \\ &\quad = \Big(\frac{1}{2}(\check{\rho}_0 + \check{R}_0^+)(\check{u}_0 + \check{U}_0^+)^2 + \frac{\gamma}{\gamma - 1}(\check{p}_0 + \check{P}_0^+) \Big)(\check{u}_{0,\nu} + \check{U}_{0,\nu}^+) \\ &\quad + \frac{1}{2} V_\nu \check{\rho}_0 \check{u}_0^2 - \Big(\frac{1}{2}\check{\rho}_0 \check{u}_0^2 + \frac{\gamma}{\gamma - 1}\check{p}_0 \Big) \check{u}_{0,\nu}. \end{aligned} \tag{2.119}$$

Recall that as $\varepsilon \to 0$, the treatment of functions $f(\tau, x)$ for $\tau \to \pm\infty$ means actually that they are considered at points x lying at a distance $\sim \varepsilon^{1-\delta}$, $0 < \delta < 1$, from Γ_t. Hence the difference $f|_{\tau \to +\infty} - f|_{\tau \to -\infty}$ describes the magnitude of the limiting jump of the function of shock wave type on the front Γ_t. Moreover, observe that letting $\tau = (t - \varphi)/\varepsilon$ and choosing the normal ν in the direction of growing φ, we obtain that $f^+ = f|_{\tau \to +\infty}$ describes the limiting value of f *behind* the shock wave front, while $f^- = f|_{\tau \to -\infty}$ describes its limiting value *ahead* of the front. The front itself moves in the direction of the normal ν.

Having this remark in mind, we introduce the notation $[\cdot]$ for the amplitude of the jump on Γ_t:

$$[f] = \lim_{\tau \to -\infty} \check{f}(\tau, x) - \lim_{\tau \to +\infty} \check{f}(\tau, x).$$

Next we observe that in view of the structure (2.111) of the solution, the limiting values of density ρ ahead of and behind the front of the shock wave are $\check{\rho}_0$ and $\check{\rho}_0 + \check{R}_0^+$ respectively. Hence

$$[\rho] = \check{\rho}_0 - (\check{\rho}_0 + \check{R}_0^+) = -\check{R}_0^+. \tag{2.120}$$

By a similar argument we obtain

$$\begin{aligned} [\rho u] &= -\{\check{\rho}_0 \check{U}_0 + \check{R}_0^+ (\check{u}_0 + \check{U}_0)\}, \quad [p] = -\check{P}_0^+, \\ [\rho u_\nu u] &= \check{\rho}_0 \check{u}_{0,\nu} \check{u}_0 - (\check{\rho}_0 + \check{R}_0^+)(\check{u}_{0,\nu} + \check{U}_{0,\nu}^+)(\check{u}_0 + \check{U}_0^+), \\ [\rho u^2] &= \check{\rho}_0 \check{u}_0^2 - (\check{\rho}_0 + \check{R}_0^+)(\check{u}_0 + \check{U}_0^+)^2, \\ [u_\nu p] &= \check{u}_{0,\nu} \check{p}_0 - (\check{u}_{0,\nu} + \check{U}_{0,\nu}^+)(\check{p}_0 + \check{P}_0^+), \\ [\rho u_\nu u^2] &= \check{\rho}_0 \check{u}_{0,\nu} \check{u}_0^2 - (\check{\rho}_0 + \check{R}_0^+)(\check{u}_{0,\nu} + \check{U}_{0,\nu}^+)(\check{u}_0 + \check{U}_0^+)^2. \end{aligned} \tag{2.121}$$

Using this notation, the equalities (2.117)–(2.119) can be easily rewritten in the form of conditions on the amplitudes and velocities of the jump of the limiting solution as $\varepsilon \to 0$:

$$V_\nu [\rho] = [\rho u_\nu], \tag{2.122}$$

$$V_\nu [\rho u] = [\rho u_\nu u + \nu p], \tag{2.123}$$

$$V_\nu \Big[\frac{1}{2}\rho u^2 + \frac{1}{\gamma - 1} p \Big] = \Big[\frac{1}{2}\rho u_\nu u^2 + \frac{\gamma}{\gamma - 1} p u_\nu \Big]. \tag{2.124}$$

Next, take the scalar product of equality (2.123) by ν and by locally defined vectors ξ_i, $i = 1, 2$, orthogonal to ν. Then, using (2.122), we immediately obtain the equalities

$$V_\nu[\rho u_\nu] = [\rho u_\nu^2 + p] \tag{2.125}$$

and

$$[\rho(V_\nu - u_\nu)u_{\xi_i}] = 0, \quad i = 1, 2. \tag{2.126}$$

Thus we see that (2.117) is the condition of continuity of the substance flow through the jump surface, (2.118) is the condition of continuity of the impulse flow, and (2.119) is the condition of continuity of the energy flow. The standard form of these conditions is given by (2.122), (2.124)–(2.126).

As is well known, the conditions of continuity of the flows admit the existence of two kinds of singularities: tangential discontinuities when

$$[u_\nu] = 0, \quad [p] = 0,$$

and shock waves when

$$[u_{\xi_i}] = 0, \quad i = 1, 2. \tag{2.127}$$

Since our goal is to consider the dynamics of shock waves, we restrict ourselves to the case (2.127). It is easily seen, using once more the second equality (2.108), that for shock waves the conditions (2.122)–(2.124) take the standard form of Rankine–Hugoniot conditions:

$$\begin{aligned} [\rho(V_\nu - u_\nu)] &= 0, \\ [\rho(V_\nu - u_\nu)^2 + p] &= 0, \\ [\tfrac{1}{2}(V_\nu - u_\nu)^2 + w] &= 0. \end{aligned} \tag{2.128}$$

Finally, note that (2.128) allows us to express the velocity of the front motion V_ν in terms of the magnitudes of density and pressure jumps, as well as the values of velocity and density ahead of the jump,

$$V_\nu - \check{u}_{0,\nu} = \sqrt{\frac{1 + \mu_\rho}{\mu_\rho} \frac{|[p]|}{\check{\rho}_0}}. \tag{2.129}$$

Here we denote by μ_ρ the relative magnitude of the density jump, i.e., $\mu_\rho = |[\rho]|/\check{\rho}_0$, and we take into account the fact that the stable shock wave is a compression wave, i.e., $[p] < 0$, $[\rho] < 0$.

Note also that the problem of constructing an asymptotic solution describing a smoothed tangential discontinuity can also be solved in the framework of the method set forth here. This was done by A. I. Shafarevich [**229**].

Now we return to equations (2.114)–(2.116), which describe the structure of the solution in a neighborhood of the front. Note that conditions (2.127) mean that the vector $\check{U}_0$ is directed along the normal to the front. Thus we obtain the representation

$$\check{U}_0 = v_0(\tau, x)\nu, \tag{2.130}$$

where v_0 is a scalar function to be determined such that $v_0 \to 0$ as $\tau \to -\infty$.

Now we easily derive from (2.114) a relationship between $\check{R}_0$ and v_0:

$$(\alpha - v_0)\check{R}_0 = \check{\rho}_0 v_0. \tag{2.131}$$

Here we denote by $\alpha = V_\nu - \check{u}_{0,\nu}$ the normal velocity of the front motion in the coordinate system connected with the nonperturbed normal velocity $\check{u}_{0,\nu}$. Previously we denoted a similar expression by κ, but in gas dynamics κ is the standard notation for the coefficient of heat conductivity.

Using (2.130) and (2.131), equation (2.115) can be easily transformed into

$$\alpha\check{\rho}_0 v_0 - \check{P}_0 = |\nabla\varphi|\Big(\zeta_0 + \frac{4}{3}\eta_0\Big)\frac{\partial v_0}{\partial\tau}, \tag{2.132}$$

but the equation (2.116) still has a cumbersome form. To simplify the formulas, we introduce the new unknown functions

$$\omega = \frac{\alpha - v_0}{\alpha}, \qquad \psi = 1 + \frac{\check{\theta}_0}{\check{T}_0}. \tag{2.133}$$

It is clear that ω, ψ must possess the property

$$\omega \to 1, \quad \psi \to 1 \qquad \text{as} \quad \tau \to -\infty. \tag{2.134}$$

Now, using (2.112) and (2.131), one can easily establish that

$$\check{P}_0 = \check{\rho}_0\check{T}_0\Big(\frac{\psi}{\omega} - 1\Big), \quad \check{R}_0 = \check{\rho}_0\frac{1-\omega}{\omega}, \tag{2.135}$$

and after some algebra the equations (2.132), (2.116) become

$$\omega - 1 + \frac{1}{\gamma M^2}\Big(\frac{\psi}{\omega} - 1\Big) = \delta\frac{\partial\omega}{\partial\tau}, \tag{2.136}$$

$$\omega^2 - 1 + \frac{2}{(\gamma-1)M^2}(\psi - 1) = 2\delta\omega\frac{\partial\omega}{\partial\tau}. \tag{2.137}$$

Here $\delta = |\nabla\varphi|(\zeta_0 + \frac{4}{3}\eta_0)(\alpha\check{\rho}_0)^{-1}$, and the quantity

$$M = \sqrt{\frac{\alpha^2}{\gamma\check{T}_0}}$$

is the Mach number computed ahead of the wave front.

One can easily deduce from (2.136), (2.137) that the limiting values of ω and ψ as $\tau \to \pm\infty$ are 1, and

$$\omega^+ = \frac{2 + (\gamma-1)M^2}{(\gamma+1)M^2}, \quad \psi^+ = \frac{((\gamma-1)M^2 + 2)(\gamma(2M^2-1)+1)}{(\gamma+1)^2M^2}. \tag{2.138}$$

Next, using (2.137), we can write ψ as a function of ω and ω_τ. Putting the expression thus obtained into (2.136), we obtain the model equation

$$2\delta\omega\frac{\partial\omega}{\partial\tau} + (\gamma+1)(1-\omega)(\omega - \omega^+) = 0. \tag{2.139}$$

Integrating (2.139), one can easily obtain that ω is the solution of the transcendental equation

$$\frac{1-\omega}{(\omega-\omega^+)^{\omega^+}} = e^{\beta(\tau+\varphi_1)}, \tag{2.140}$$

where $\beta = (M^2-1)/M^2\delta$ and $\varphi_1 = \varphi_1(x)$ is the "constant" of integration.

In its turn, for the auxiliary function ψ we obtain the formula

$$\psi = 1 + (\gamma-1)(1-\omega)\Big(\gamma\frac{M^2}{2} + 1 - \omega\Big). \tag{2.141}$$

It is not hard to see that under the condition $M > 1$, i.e., for supersonic front motion, the function ω determined by (2.140) will be infinitely differentiable and stabilizing exponentially fast for $\tau \to \pm\infty$. Moreover, the properties (2.134), which were an assumption so far, are actually fulfilled and the limiting values ω^+, ψ^+ are attained as $\tau \to +\infty$, i.e., in the domain behind the shock wave front.

Finally, by elementary calculations we obtain

$$\begin{gathered} 0 < \omega^+ < 1, \quad \psi^+ > \omega^+, \\ \psi^+ - 1 = 2\frac{(\gamma-1)(\gamma M^2+1)(M^2-1)}{(\gamma+1)^2 M^2} > 0. \end{gathered} \tag{2.142}$$

Combined with (2.133) and (2.135), these imply the inequalities

$$\check{R}_0^+ > 0, \quad \check{P}_0^+ > 0, \quad \check{\theta}_0^+ = \check{T}_0(\psi^+ - 1) > 0,$$

which mean that the solution we have constructed is a compression wave indeed. Thus all our assumptions are fulfilled.

It remains to observe that for $\tau \to \pm\infty$, the function ω tends to ω^+ and 1 as $\exp(\mp\lambda^\pm\tau)$ with $\lambda^+ = \beta/\omega^+$ and $\lambda^- = \beta$, respectively. If the Mach number M is close to one, i.e., $M^2 = 1 + m$ with $m \ll 1$, then we find from (2.140) that up to $O(m^2)$,

$$\omega = 1 - \frac{2m}{\gamma+1}\frac{1}{1+e^{-\beta_0(\tau+\varphi_1)}}, \quad \beta_0 = \frac{m}{\delta}.$$

In this case $\lambda^+ = \lambda^- = \beta_0$ with accuracy $O(m^2)$.

To summarize, we have found the solution to the model equations (2.114)–(2.116) (defined on the surface $\mathcal{T} \times \mathbb{R}^1_\tau$) in the form

$$\begin{aligned} \check{U}_0(\tau,x) &= \check{U}_0^+(x)\chi(\tau,x), \quad & \check{R}_0(\tau,x) &= \check{R}_0^+(x)\chi_1(\tau,x), \\ \check{\theta}_0(\tau,x) &= \check{\theta}_0^+(x)\chi_2(\tau,x), \quad & \check{P}_0(\tau,x) &= \check{P}_0^+(x)\chi_3(\tau,x), \end{aligned} \tag{2.143}$$

where

$$\begin{gathered} \chi = \frac{1-\omega(\tau,x)}{1+\omega^+(x)}, \quad \chi_1 = \frac{\omega^+(x)}{\omega(\tau,x)}\chi, \\ \check{U}_0^+ = 2\check{\alpha}\frac{M^2-1}{(\gamma+1)M^2}\nu, \quad \check{R}_0^+ = 2\check{\rho}_0\frac{M^2-1}{2+(\gamma-1)M^2}, \\ \chi_2 = \frac{\gamma-1}{\psi^+(x)-1}(1-\omega(\tau,x))\Big(\frac{1}{2}\gamma M^2 + 1 - \omega(\tau,x)\Big), \\ \chi_3 = \frac{\gamma+1}{2\gamma(M^2-1)}\Big(\frac{\psi(\tau,x)}{\omega(\tau,x)} - 1\Big), \\ \check{\theta}_0^+(x) = \check{T}_0(\psi^+(x)-1), \quad \check{P}_0^+ = 2\check{p}_0\frac{\gamma(M^2-1)}{\gamma+1}. \end{gathered} \tag{2.144}$$

It can be easily seen that χ, χ_1, χ_2, χ_3 tend to zero as $\tau \to -\infty$ and to 1 as $\tau \to +\infty$, and $\check{U}_0^+$, $\check{R}_0^+$, $\check{\theta}_0^+$, $\check{R}_0^+$ satisfy the Rankine–Hugoniot conditions (2.128).

Define the extension of functions (2.143) outside the surface $\mathcal{T} \times R_\tau^1$ as follows:

$$(2.145)\qquad \begin{aligned} U_0(\tau,x,t) &= U_0^+(x,t)\chi(\tau,x), \quad & R_0(\tau,x,t) &= R_0^+(x,t)\chi_1(\tau,x),\\ \theta_0(\tau,x,t) &= \theta_0^+(x,t)\chi_2(\tau,x), \quad & P_0(\tau,x,t) &= P_0^+(x,t)\chi_3(\tau,x),\end{aligned}$$

where U_0^+, R_0^+, θ_0^+, P_0^+ are smooth functions, so far arbitrary.

Let us turn to construction of the higher order term of the asymptotic expansion, which will enable us to determine U_0^+, R_0^+, θ_0^+, P_0^+. For that, consider the terms of order ε^0 which result from substitution of (2.111) into the first equation (2.107) and from previous constructions. After some simple calculations we arrive at the relation

$$(2.146)\qquad |\nabla\varphi|\frac{\partial}{\partial\tau}\{(V_\nu - u_{0,\nu} - U_{0,\nu})R_1 - (\rho_0 + R_0)U_{1,\nu}\} = \mathcal{F}_\rho(\tau,x,t) + O(\varepsilon),$$

where

$$\begin{aligned}\mathcal{F}_\rho = |\nabla\varphi|\frac{\partial}{\partial\tau}(u_{1,\nu}R_0 + \rho_1 U_{0,\nu}) - \chi_1\frac{\partial R_0^+}{\partial t} - \frac{\partial\rho_0}{\partial t}\\ - \langle\nabla, \rho_0 u_0 + \rho_0 U_0^+\chi + R_0^+(u_0 + U_0^+\chi)\chi_1\rangle - |\nabla\varphi| G(\tau,x).\end{aligned}$$

Here

$$G(\tau,x) = \tau\frac{\partial}{\partial\tau}\left\{\chi_1\frac{\partial}{\partial t}((V_\nu - u_{0,\nu})R_0^+)\bigg|_{t=\varphi} - \chi\frac{\partial}{\partial t}((\rho_0 + R_0^+\chi_1)U_{0,\nu}^+)\bigg|_{t=\varphi}\right\}.$$

Obviously, for (2.146) to hold it is necessary that the right-hand side $\mathcal{F}_\rho$ tends to zero as $\tau \to \pm\infty$. Therefore, taking the limits as $\tau \to \pm\infty$, we obtain the equations of continuity for the background ρ_0, u_0:

$$(2.147)\qquad \frac{\partial\rho_0}{\partial t} + \langle\nabla, \rho_0 u_0\rangle = 0,$$

as well as for the functions R_0^+, U_0^+:

$$(2.148)\qquad \frac{\partial R_0^+}{\partial t} + \langle\nabla, \rho_0 U_0^+ + R_0^+(u_0 + U_0^+)\rangle = 0.$$

We assume here that the functions ρ_0, u_0 satisfy (2.147) for all $x \in \mathbb{R}^3$. However the functions R_0^+, U_0^+, which enter into the right-hand side of (2.146) in the form of coefficients χ and χ_1, satisfy (2.148) only in the domain $D_\varepsilon^- = \{(x,t),\ t > 0,\ \varphi(x) < t - \varepsilon^\delta,\ \delta \in (0,1)\}$, i.e., actually behind the front. Outside this domain R_0^+, U_0^+ can be defined as arbitrary infinitely differentiable functions. We will make similar assumptions regarding the functions p_0, T_0 and P_0^+, θ_0^+.

Now the right-hand side of (2.146) becomes

$$(2.149)\qquad \mathcal{F}_\rho = -\frac{\partial}{\partial\tau}\{\langle U_0, \nabla\varphi_1\rangle(\rho_0 + R_0) + \langle u_0, \nabla\varphi_1\rangle R_0\} + \mathcal{F}_\rho'$$

with

$$\begin{aligned}\mathcal{F}'_\rho = {} & |\nabla\varphi|\frac{\partial}{\partial\tau}(u_{1,\nu}R_0 + \rho_1 U_{0,\nu}) - (\chi - \chi_1)\langle\nabla, \rho_0 U_0^+\rangle \\ & - \chi_1(\chi - 1)\langle\nabla, R_0^+ U_0^+\rangle - \{\rho_0\langle U_0^+, \nabla\chi(\tau', x)\rangle \\ & + R_0^+\langle u_0, \nabla\chi_1(\tau', x)\rangle + R_0^+\langle U_0^+, \nabla(\chi\chi_1)(\tau', x)\rangle\}\Big|_{\tau'=\tau+\varphi_1} \\ & - |\nabla\varphi| G(\tau, x).\end{aligned}$$

It is easily seen that the function $\mathcal{F}_\rho$ as in (2.149) decreases exponentially fast as $\tau \to \pm\infty$. This allows us to expand all functions in (2.146) by the Taylor formula with respect to t at the point $t = \varphi(x)$ and to employ the equality $t - \varphi = \varepsilon\tau$. Then, neglecting the terms of order $O(\varepsilon)$ in (2.146) and integrating with respect to τ, we obtain the equality

$$(\alpha - v_0)\check{R}_1 - (\check{\rho}_0 + \check{R}_0)\check{U}_{1,\nu} = F_\rho(\tau, x), \tag{2.150}$$

where

$$F_\rho = -V_\nu\{\alpha\langle\nu, \nabla\varphi_1\rangle + \langle\check{u}_0, \nabla\varphi_1\rangle\}\check{R}_0 + F'_\rho \tag{2.151}$$

with

$$F'_\rho = V_\nu \int_{-\infty}^{\tau} \mathcal{F}'_\rho(\tau', x, \varphi(x))\, d\tau'.$$

Now we carry out a similar treatment of the terms of order ε^0 resulting from substitution of (2.111) into the second and third equations (2.107). Letting $\tau \to \pm\infty$, we obtain the following equations for the background functions:

$$\begin{aligned}& \frac{\partial}{\partial t}(\rho_0 u_{0i}) + \langle\nabla, \rho_0 u_0 u_{0i}\rangle + \frac{\partial p_0}{\partial x_i} = 0, \quad i = 1, 2, 3, \\ & \frac{\partial}{\partial t}\Big(\frac{1}{2}\rho_0 u_0^2 + \frac{1}{\gamma - 1}p_0\Big) + \Big\langle\nabla, \Big(\frac{1}{2}\rho_0 u_0^2 + \frac{\gamma}{\gamma - 1}p_0\Big)u_0\Big\rangle = 0,\end{aligned} \tag{2.152}$$

as well as equations which complement (2.148) to a complete system:

$$\begin{aligned}& \frac{\partial}{\partial t}(\rho_0 U^+_{0\,i} + R_0^+ u_{0i}) + \Big\langle\nabla, \{(\rho_0 + R_0^+)(u_0 + U_0^+)(u_{0i} + U^+_{0\,i}) \\ & \qquad - \rho_0 u_0 u_{0i}\}\Big\rangle + \frac{\partial P_0^+}{\partial x_i} = 0, \quad i = 1, 2, 3, \\ & \frac{\partial}{\partial t}\Big\{\frac{1}{2}(\rho_0 + R_0^+)(u_0 + U_0^+)^2 - \rho_0 u_0^2 + \frac{1}{\gamma - 1}P_0^+\Big\} \\ & \qquad + \Big\langle\nabla, \Big\{\Big(\frac{1}{2}(\rho_0 + R_0^+)(u_0 + U_0^+)^2 + \frac{\gamma}{\gamma - 1}(p_0 + P_0^+)\Big)(u_0 + U_0^+) \\ & \qquad - \Big(\frac{1}{2}\rho_0 u_0^2 + \frac{\gamma}{\gamma - 1}p_0\Big)u_0\Big\}\Big\rangle = 0.\end{aligned} \tag{2.153}$$

Moreover, if we consider the first correction in the description of the structure of the solution in a neighborhood of the front in each map of the surface Γ_t, we obtain the equations

$$
\begin{aligned}
&\alpha\check{\rho}_0\check{U}_1 - \nu\check{P}_1 = |\nabla\varphi|\frac{\partial}{\partial\tau}\Big\{\eta_0\check{U}_1 + \Big(\zeta_0 + \frac{\eta_0}{3}\Big)\check{U}_{1,\nu}\nu\Big\} + F_u(\tau,x),\\
&\alpha\check{\rho}_0(\alpha - v_0)\{(\check{u}_{0,\nu} + v_0)\check{U}_{1,\nu} + \check{u}_{0,\xi_i}\check{U}_{1,\xi_i}\} - \frac{V_\nu}{\gamma-1}\check{P}_1\\
&\quad + \frac{\gamma}{\gamma-1}(\alpha - v_0)\check{P}_1 = |\nabla\varphi|\frac{\partial}{\partial\tau}\Big\{\Big(\zeta_0 + \frac{4}{3}\eta_0\Big)(\check{u}_{0,\nu} + v_0)\check{U}_{1,\nu}\Big)\\
&\quad + \eta_0\check{u}_{0,\xi_i}\check{U}_{1,\xi_i}\Big\} + F_T(\tau,x),
\end{aligned}
\tag{2.154}
$$

which complement (2.150) to a complete system.

Here ξ_i, $i = 1, 2$, are locally defined unit vectors orthogonal to ν; the vectors u_{0,ξ_i}, U_{1,ξ_i} are projections of u_0, U_1 on ξ_i, and F_u, F_T are smooth functions converging exponentially fast to zero as $\tau \to -\infty$ and to $F_u^+(x)$ and $F_T^+(x)$ respectively as $\tau \to +\infty$.

We do not state here the explicit form of the right-hand sides F_u, F_T, since they are very cumbersome. Let us only point out that F_u^+ and F_T^+ involve terms depending on $\nabla\varphi_1$, similar to those in (2.151).

When deriving (2.154) we used the equality

$$
\check{R}_1 = \frac{\alpha\check{\rho}_0}{(\alpha - v_0)^2}\check{U}_{1,\nu} + (\alpha - v_0)^{-1}F_\rho,
$$

which follows from (2.131) and (2.150).

Thus we have obtained an analog of equations (2.114)–(2.116). It is clear that, letting $\tau \to \pm\infty$ in (2.150), (2.154), we obtain the linearized Rankine–Hugoniot conditions (2.127), (2.128), i.e.,

$$
\begin{aligned}
[\rho_1(V_\nu - u_\nu) - \rho u_{1\nu}] &= f_\rho^+(x),\\
[\rho_1(V_\nu - u_\nu)^2 - 2\rho(V_\nu - u_\nu)u_{1\nu} + p_1] &= f_{u,\nu}^+(x),\\
[(V_\nu - u_\nu)u_{1\nu} - w_1] &= f_T^+(x),\\
[u_{1\xi_i}] &= f_{u,\xi_i}^+(x), \quad i = 1,2.
\end{aligned}
\tag{2.155}
$$

The subscript 1 indicates here the terms of order ε, the functions without subscript form the leading term of the solution, and $f_\rho^+(x), \dots, f_{u,\xi_i}^+(x)$ are functions expressible in terms of $F_\rho^+, \dots, F_{u,\xi_i}^+$ and the leading term of asymptotics. In particular,

$$
f_\rho^+ = -F_\rho^+, \quad f_{u,\xi_i}^+ = -\frac{(\gamma+1)M^2}{\alpha\check{\rho}_0}(2 + (\gamma - 1)M^2)F_{u,\xi_i}^+,
$$

and $F_{u,\xi_i}^+ = \langle\xi_i, F_u^+\rangle$.

The right-hand sides of (2.155) involve the "constant" of integration $\varphi_1(x)$ of equation (2.139), which has not been defined before. Since conditions (2.155) are posed on the surface Γ_t specified in advance, the presence of a free function φ_1 is necessary for them to hold.

Moreover, note that the first correction (of order ε) to velocity has, in general, a nonzero jump of the component tangential to Γ_t This is quite natural since conditions (2.155) are posed on Γ_t, whereas the true position of the front deviates from Γ_t by $O(\varepsilon)$.

Now we return to equations (2.150), (2.154) describing the structure of the first correction to the solution in a neighborhood of the front. Multiplying the first equation (2.154) by the vectors ν and ξ_i, $i = 1, 2$, we obtain the scalar equations

$$\alpha\check{\rho}_0\check{U}_{1,\nu} - \check{P}_1 = |\nabla\varphi|\Big(\zeta_0 + \frac{4}{3}\eta_0\Big)\frac{\partial\check{U}_{1,\nu}}{\partial\tau} + F_{u,\nu}, \tag{2.156}$$

$$\alpha\check{\rho}_0\check{U}_{1,\xi_i} = |\nabla\varphi|\eta_0\frac{\partial\check{U}_{1,\xi_i}}{\partial\tau} + F_{u,\xi_i}, \quad i = 1, 2. \tag{2.157}$$

Recall that the functions F_u, F_ρ, F_T stabilize as $\tau \to \pm\infty$ exponentially fast. This implies that equations (2.157) are solvable in the same class of functions. Thus the tangential components of the first correction are uniquely determined by the right-hand side F_{u,ξ_i}, i.e., actually, by the leading term of the asymptotic expansion.

Now consider (2.156) and the second equation (2.154). Define the auxiliary functions

$$\omega_1 = \frac{\alpha - \check{U}_{1,\nu}}{\alpha}, \quad \psi_1 = 1 + \frac{\check{\theta}_1}{\check{T}_0}. \tag{2.158}$$

Then by transformations similar to those made above these equations can be reduced to the following form:

$$\begin{aligned} \omega_1 - 1 + \frac{1}{\gamma M^2\omega}\Big(\psi_1 - 1 - \frac{\omega_1 - 1}{\omega}\psi\Big) &= \delta\frac{\partial\omega_1}{\partial\tau} + F_\omega(\tau, x), \\ \omega(\omega_1 - 1) + \frac{1}{(\gamma-1)M^2}(\psi_1 - 1) &= \delta\frac{\partial}{\partial\tau}(\omega\omega_1) + F_\psi(\tau, x), \end{aligned} \tag{2.159}$$

where the right-hand sides F_ω, F_ψ are expressible in terms of F_T, F_ρ, and $F_{u,\nu}$, and have the property that $F_\omega, F_\psi \to 0$ as $\tau \to -\infty$.

It is easily seen that (2.159) is a nonhomogeneous linearization of equations (2.136), (2.137). Hence for determination of the function ω_1 we obtain the linearized equation (2.139), i.e.,

$$2\delta\frac{\partial}{\partial\tau}(\omega\omega_1) + (\gamma+1)(1 + \omega^+ - 2\omega)(\omega_1 - 1) = g(\tau, x). \tag{2.160}$$

It is not hard to see that equation (2.160) without any additional conditions is solvable in the class of bounded functions stabilizing exponentially fast as $\tau \to \pm\infty$. It is also clear that (2.160) allows us to find the function ω_1 up to the term $\varphi_2(x)\partial\omega/\partial\tau$, where φ_2 is an arbitrary "constant" of integration, and $\partial\omega/\partial\tau$ is the kernel of the operator in the left-hand side of (2.160).

The subsequent constructions follow the same lines as in the above scheme. We only point out that arbitrariness in the choice of functions φ_2 is necessary for fullfillment of linearized Rankine–Hugoniot conditions which will arise in the course of obtaining the higher order terms of the asymptotics.

Summarizing, we see that the construction of the asymptotic solution of shock wave type for the equations of gas dynamics does not differ methodologically from the corresponding contructions for the Korteweg–de Vries–Burgers equation in the case of negligibly small dispersion (see §1.2). Namely, construction of the leading term of the asymptotics reduces actually to computation of the dynamics of the limiting shock wave (as $\varepsilon \to 0$), i.e., to solving the equations

$$
\begin{aligned}
&\frac{\partial \rho^\pm}{\partial t} + \langle \nabla, \rho^\pm u^\pm \rangle = 0, \\
&\frac{\partial}{\partial t}(\rho^\pm u_i^\pm) + \langle \nabla, \rho^\pm u^\pm u_i^\pm \rangle + \frac{\partial p^\pm}{\partial x_i} = 0, \quad i = 1, 2, 3, \\
&\frac{\partial}{\partial t}\Big(\frac{1}{2}\rho^\pm (u^\pm)^2 + \frac{1}{\gamma - 1}p^\pm\Big) + \Big\langle \nabla, \Big(\frac{1}{2}\rho^\pm (u^\pm)^2 + \frac{\gamma}{\gamma - 1}p^\pm\Big)u^\pm \Big\rangle = 0,
\end{aligned}
\tag{2.161}
$$

in the domains $D^\pm$ ahead of and behind the shock wave front respectively, which are considered jointly with the Rankine–Hugoniot conditions (2.127), (2.128) on the front $\Gamma_t = \{x, \varphi(x) = t\}$. Here $D^+ = \{(x,t),\ t > 0,\ \varphi(x) > t\}$, $D^- = \{(x,t),\ t > 0,\ \varphi(x) < t\}$, and equations (2.161) follow from (2.147), (2.148), (2.152), (2.153) if we denote by $\rho^+ = \rho_0, \dots, p^+ = p_0$ the nonperturbed state of the medium ahead of the front and by $\rho^- = \rho_0 + R_0^+, \dots, p^- = p_0 + P_0^+$ the state of the medium after the passage of the front.

If we know the solution to the system (2.127), (2.128), (2.161), then it is easy to construct the leading term of the asymptotics for the problem with small viscosity. Namely, we have to extend the solutions $\rho^\pm, \dots, p^\pm$ into the domains $D^\mp$ to a distance of order $O(\varepsilon^\delta)$, $\delta \in (0,1)$, which can always be done for a smooth front and sufficiently small ε, and then to match them by means of auxiliary functions $\chi, \dots, \chi_3$. In particular, in this way we will obtain the following representation for the leading term $u_{as}^0 = u_0 + U_0$ of the asymptotic expansion for the velocity (2.111):

$$
u_{\text{as}}^0 = \widetilde{u}^+(x,t) + (\,\widetilde{u}^-(x,t) - \widetilde{u}^+(x,t))\chi(\tau, x), \tag{2.162}
$$

where $\widetilde{u}^\pm$ are the extensions of $u^\pm$ into the domains $D^\mp$, and $\tau = (t - \varphi)/\varepsilon$. Note that by the above construction the component of the vector $\lim_{\varepsilon \to 0} u_{\text{as}}^0$ normal to Γ_t has a jump on the front, while the tangent components, being continuous, have, in general, discontinuous first normal derivatives. Hence formula (2.162) describes a regularization of a function of Heaviside type for the normal component u_{as}^0 and of a function of type τ_+ for tangent components.

Note in conclusion that the Hugoniot type conditions which appear when we describe the dynamics of the front without solving the equations (2.161) are different from the sequence of Rankine–Hugoniot conditions ((2.127), (2.128), (2.155), etc.) which we have obtained. The former conditions follow from the expansion with respect to smoothness of the solution to the limiting (for $\varepsilon \to 0$) problem, and they form an infinite linked chain [**166**], whereas the conditions (2.127), (2.128), (2.155), etc., are obtained from expansion of the solution with respect to a small viscosity parameter, and this sequence of conditions has a triangular structure. However, an infinite chain of Hugoniot type conditions can be restated in the form of Rankine–Hugoniot conditions (2.127), (2.128) if we turn from the local description of the front motion to the global one, which involves solving the equations (2.161) in the domains $D^\pm$.

2.6. The case of finite and small Prandtl numbers

Now we proceed to the physically more meaningful problem on the dynamics of a smoothed shock wave in a medium with finite Prandtl numbers Pr. This case is characteristic for usual gases; it corresponds to equations of gas dynamics with function $\kappa/\Pr$ of the same order as η, ζ. To simplify formulas, we let $\Pr = 1$ and assume, as before, that the equalities (2.110) hold. This seemingly minor change in the problem setting (we assume that $\kappa/\Pr = O(1)$ rather than $\kappa/\Pr = O(\varepsilon)$) can drastically affect the structure of the solution. The thing is that when dealing with the heat conductivity effects in the leading term, the basic model equation describing the profile of the solution in a neighborhood of the wave front is of second order rather than first as in (2.139). Accordingly, when constructing the correction terms of asymptotics we will get a nonhomogeneous linearized model equation $L\omega_1 = g_1$ of second order. In a general situation such an equation can have a bounded solution only for a specific right-hand side which must satisfy the corresponding orthogonality condition. We have repeatedly used this in previous sections, and it was this orthogonality condition from which we derived the parameters of the solution that had not been defined before. However, for the equations of gas dynamics all "free" parameters of the solution, i.e., the amplitudes of the jumps, the geometry and the velocity of the front of the discontinuity of the limiting solution (as $\varepsilon \to 0$), are determined from the Rankine–Hugoniot conditions (2.128). Accordingly, if we assume the natural structure (2.111) of the asymptotic solution, we get into a situation where we have no free parameters to satisfy the orthogonality conditions. Formally this leads to appearance of secular terms, and it actually means that there exists no solution of type (2.111).

Luckily, the above considerations refer to the general case, while for nonlinear equations a lot depends on the specific features of a particular problem. We will show that for the gas dynamics equations with $\Pr \sim 1$ the kernel of the adjoint operator to L is in fact trivial in the class of bounded functions. As a consequence, no additional condition appears, so that our asymptotic procedure remains valid. To establish this fact a detailed study of the model equation is needed, which is the reason for separate treatment of the cases $\Pr \gg 1$ and $\Pr \sim 1$.

For completeness we will also consider the case of small Prandtl numbers, which arises, e.g., in non-magnetized plasma [**251**]. The assumption $\Pr \ll 1$ implies that the viscous terms are negligible in the treatment of the leading term of the asymptotics. For consistency of notation we will write the gas dynamics equations for $\Pr \ll 1$ in the form (2.107) as before, assuming, however, that

$$\eta = \varepsilon\eta', \quad \zeta = \varepsilon\zeta', \tag{2.163}$$

where η' and ζ' are functions of order $O(1)$. Of course, then the parameter ε is inversely proportional to the square root of the Reynolds number Re, so that (2.163) and the norming made in (2.107) mean that the numbers Re and Pr are related by

$$\mathrm{Re} \sim \Pr^{-2}. \tag{2.164}$$

We will consider the case $\Pr \ll 1$ under the additional assumption that the Mach number (ahead of the front) is sufficiently small,

$$1 < M^2 < \frac{1}{\gamma}\frac{3\gamma - 1}{3 - \gamma}. \tag{2.165}$$

The reason for this condition will be discussed later. The general case will be treated in §5.4.

Now let us construct the asymptotic solution of smoothed shock wave type in the case of a finite Prandtl number. Representing the solution in the form (2.111) and making the same calculations as in the previous section, we arrive at model equations which differ from (2.114)–(2.116) only by the additional term responsible for heat conductivity. For the reader's convenience and subsequent references we state these equations again:

$$\check{R}_0 - \langle \nabla\varphi, \check{\rho}_0 \check{U}_0 + \check{R}_0(\check{u}_0 + \check{U}_0)\rangle = 0, \tag{2.166}$$

$$\begin{aligned} &\check{\rho}_0\check{U}_0 + \check{R}_0(\check{u}_0 + \check{U}_0) - \langle \nabla\varphi, (\check{u}_0 + \check{U}_0)\rangle(\check{\rho}_0\check{U}_0 + \check{R}_0(\check{u}_0 + \check{U}_0)) \\ &\quad - \langle \nabla\varphi, \check{U}_0\rangle \check{\rho}_0\check{u}_0 - \nabla\varphi \check{P}_0 \\ &\quad = |\nabla\varphi|^2 \frac{\partial}{\partial\tau}\Big(\eta_0\check{U}_0 + \Big(\zeta_0 + \frac{\eta_0}{3}\Big)\langle\nabla\varphi, \check{U}_0\rangle\nabla\varphi\Big), \end{aligned} \tag{2.167}$$

$$\begin{aligned} &\frac{1}{2}(\check{\rho}_0 + \check{R}_0)(\check{u}_0 + \check{U}_0)^2 + \frac{1}{\gamma-1}\check{P}_0 - \Big\langle \nabla\varphi, \Big(\frac{1}{2}(\check{\rho}_0 + \check{R}_0)(\check{u}_0 + \check{U}_0)^2 \\ &\quad + \frac{\gamma}{\gamma-1}(\check{p}_0 + \check{P}_0)\Big)(\check{u}_0 + \check{U}_0)\Big\rangle \\ &\quad = \frac{1}{2}\frac{\partial}{\partial\tau}\Big(\eta_0|\nabla\varphi|^2(\check{u}_0 + \check{U}_0)^2 + \Big(\zeta_0 + \frac{\eta_0}{3}\Big)\big(\langle\nabla\varphi, \check{u}_0 + \check{U}_0\rangle\big)^2\Big) \\ &\quad + \frac{1}{2}\check{\rho}_0\check{u}_0^2 - \Big\langle \nabla\varphi, \Big(\frac{1}{2}\check{\rho}_0\check{u}_0^2 + \frac{\gamma}{\gamma-1}\check{p}_0\Big)\check{u}_0\Big\rangle + \frac{\partial}{\partial\tau}\Big(\kappa_0|\nabla\varphi|^2\frac{\partial\theta_0}{\partial\tau}\Big). \end{aligned} \tag{2.168}$$

Obviously, by letting $\tau \to \pm\infty$ we arrive at the standard Rankine–Hugoniot conditions (2.127), (2.128) for shock waves.

Now we make the substitutions (2.130) and (2.133)–(2.135). After some transformations we obtain (2.136) and an analog of (2.137):

$$\omega^2 - 1 + \frac{2}{(\gamma-1)M^2}(\psi - 1) = 2\delta\omega\frac{\partial\omega}{\partial\tau} + \delta_1\frac{\partial\psi}{\partial\tau}, \tag{2.169}$$

where

$$\delta_1 = 2\frac{\delta\gamma}{M^2}\frac{\kappa_0}{\zeta_0 + \frac{4}{3}\eta_0}.$$

With the aid of (2.136) the derivative $\partial\omega/\partial\tau$ can be easily excluded from (2.169). Then we obtain the basic model equations

$$\omega - 1 + \mu\Big(\frac{\psi}{\omega} - 1\Big) = \delta\frac{\partial\omega}{\partial\tau}, \tag{2.170}$$

$$\mu_1(\psi - 1) + (\omega - 1)(1 - \omega + 2\mu) = \delta_1\frac{\partial\psi}{\partial\tau}, \tag{2.171}$$

where

$$\mu = \frac{1}{\gamma M^2}, \quad \mu_1 = \frac{2\mu}{\gamma - 1}.$$

Taking the limits as $\tau \to \pm\infty$ in (2.170) and (2.171) and assuming that ω and ψ stabilize for large $|\tau|$, we obtain that the limiting values of ω and ψ are equal to 1 as

$\tau \to -\infty$ and to the quantities ω^+, ψ^+ defined in (2.138) as $\tau \to \infty$. However, now we cannot integrate the model equations in an explicit form, and hence we have to prove the existence and uniqueness of the required solution. This will be done later, and we only remark here that there will arise the necessary and sufficient conditions which are given by the same assumptions

$$M > 1, \quad \gamma > 1, \tag{2.172}$$

as in the case $\mathrm{Pr} \gg 1$. The results of numerical solution of the system (2.170), (2.171) are presented in Figure 2.13.

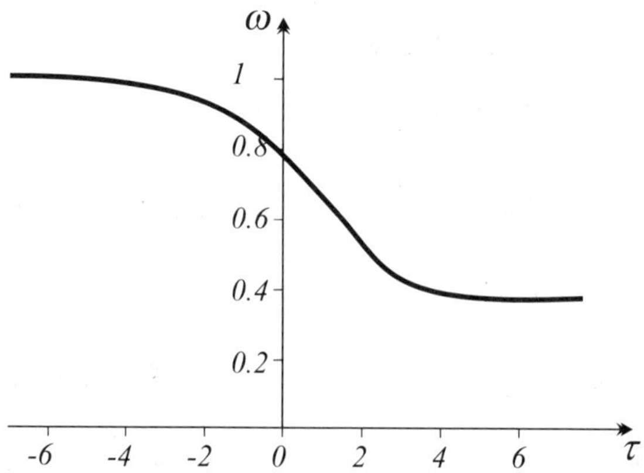

FIGURE 2.13. The shape of $\omega(\tau)$ computed for $\gamma = 1.4$, $M = 2$, and $\delta = \delta_1 = 2$.

If we go on constructing the asymptotics, we arrive at the solution on the surface $\mathcal{T} \times \mathbb{R}^1_\tau$ similar to (2.143):

$$\begin{aligned} \check{U}_0(\tau, x) &= \check{U}_0^+(x)\chi(\tau, x), & \check{R}_0(\tau, x) &= \check{R}_0^+(x)\chi_1(\tau, x), \\ \check{\theta}_0(\tau, x) &= \check{\theta}_0^+(x)\chi_2(\tau, x), & \check{P}_0(\tau, x) &= \check{P}_0^+(x)\chi_3(\tau, x), \end{aligned} \tag{2.173}$$

where

$$\chi_2 = \frac{\psi - 1}{\psi^+ - 1},$$

and all other functions are computed by the same formulas as in (2.143). Recall that $\chi, \chi_1, \chi_2, \chi_3$ tend to zero as $\tau \to -\infty$ and to 1 as $\tau \to \infty$.

Let us write, as before, the extension of the functions (2.143) in the form (2.145) and turn to construction of the higher order term of asymptotics. It is easily seen that, collecting the terms of order ε^0 resulting from substitution of (2.111) into (2.107), we obtain the equations of ideal gas dynamics (2.147), (2.152) for the functions ρ_0, u_0, p_0, T_0 ahead of the front, as well as the equations (2.148), (2.153) for the functions R_0^+, U_0^+, P_0^+, θ_0^+ in the domain behind the front of the shock wave.

Further, to describe the structure of the solution in a neighborhood of the front we again obtain the equations (2.150), (2.154) with the only difference that the right-hand side of the second equation (2.154) now contains the additional terms

$$|\nabla\varphi|\frac{\partial}{\partial\tau}\left\{\kappa_0\frac{\partial\check{\theta}_1}{\partial\tau}+\check{\kappa}_1\frac{\partial\check{\theta}_0}{\partial\tau}-\kappa_0|\nabla\varphi|^{-2}(2\langle\nabla\varphi,\nabla\theta_0\rangle+\Delta\varphi\theta_0)\Big|_{t=\varphi}\right\}$$
$$+|\nabla\varphi|\kappa_0\tau\frac{\partial^2}{\partial\tau^2}\frac{\partial\theta_0}{\partial t}\Big|_{t=\varphi}.$$

Clearly, by letting $\tau \to \pm\infty$ in these equations we arrive at the same linearized Rankine–Hugoniot conditions (2.155) as for $\mathrm{Pr} \gg 1$. Thus, as was to be expected, the auxiliary problem for constructing the leading term is given by the standard equations arising in the computation of the shock wave, while that for the higher order terms is given by the corresponding linearized equations. As a consequence, the functions φ and φ_1 are defined and the leading term of the asymptotics contains no free parameters.

Nevertheless, in order to prove that the asymptotic procedure is valid, we have to establish solvability of the linearized model equations in the class of bounded functions stabilizing for $\tau \to \pm\infty$.

For $\mathrm{Pr} \sim 1$, instead of (2.159) we have the system

$$\begin{aligned}\delta\frac{\partial\omega_1}{\partial\tau}&=\frac{\mu}{\omega}\Big(\psi_1-1-\frac{\psi}{\omega}(\omega_1-1)\Big)+F_\omega(\tau,x),\\ \delta_1\frac{\partial\psi_1}{\partial\tau}&=\mu_1(\psi_1-1)+2(\omega_1-1)(1+\mu-\omega)+F_\psi(\tau,x).\end{aligned}\tag{2.174}$$

Here the auxiliary functions ω_1 and ψ_1 are related to $\check{U}_{1,\nu}$ and $\check{\theta}_1$ by (2.158), and the right-hand sides stabilize at infinity:

$$\begin{aligned}F_\omega, F_\psi &\to 0 &&\text{as}\quad \tau\to-\infty,\\ F_\omega\to F_\omega^+(x),\ F_\psi &\to F_\psi^+(x) &&\text{as}\quad \tau\to+\infty,\end{aligned}\tag{2.175}$$

provided that the solution to (2.170), (2.171) satisfies the boundary conditions

$$\begin{aligned}\omega,\ \psi &\to 1 &&\text{as}\quad \tau\to-\infty,\\ \omega\to\omega^+(x),\ \psi&\to\psi^+(x) &&\text{as}\quad \tau\to+\infty,\end{aligned}\tag{2.176}$$

with ω and ψ stabilizing at an exponential rate.

It is impossible to analyze the equations (2.174) without a detailed study of the behavior of ω and ψ. Hence we return to the treatment of equations (2.170), (2.171) completed with boundary conditions (2.176).

Let us rewrite these equations in the form which is standard for the theory of dynamical systems,

$$\frac{dx}{dt}=P(x,y),\qquad \frac{dy}{dt}=Q(x,y).\tag{2.177}$$

Here $x=\omega$, $y=\psi$, $t=\tau$, and

$$P=\frac{1}{\delta}\Big(x-1+\mu\Big(\frac{y}{x}-1\Big)\Big),\quad Q=\frac{1}{\delta_1}\big(\mu_1(y-1)+(x-1)(1+2\mu-x)\big).\tag{2.178}$$

Obviously, the right-hand sides of (2.178) are infinitely differentiable outside the line $x = 0$. Moreover, one can easily establish that the system (2.177) has two equilibrium points, $(1,1)$ and (x^+, y^+), where in view of (2.138)

$$x^+ = \frac{\gamma-1}{\gamma+1}(1+\gamma\mu_1), \quad y^+ = 1 + 4\frac{(1+\mu)(1-\gamma\mu)}{\mu_1(1+\gamma)^2}. \tag{2.179}$$

Let us investigate the point $(1,1)$. Denoting $x' = 1 - x$ and $y' = y - 1$, we rewrite (2.177) as

$$\begin{aligned} \frac{dx'}{dt} &= ax' + by' + f_1(x', y'), \\ \frac{dy'}{dt} &= cx' + dy' + f_2(x', y'), \end{aligned} \tag{2.180}$$

where

$$a = \frac{1-\mu}{\delta}, \quad b = -\frac{\mu}{\delta}, \quad c = -\frac{2\mu}{\delta_1}, \quad d = \frac{\mu_1}{\delta_1},$$

and the functions f_1, f_2 along with their first derivatives vanish at the point $(0,0)$. One can easily derive that the matrix A related to the linear part of (2.180) is nonsingular, and its determinant equals

$$\Delta_1 = \frac{2\mu(1-\gamma\mu)}{\delta\delta_1(\gamma-1)} > 0,$$

provided that conditions (2.172) are fulfilled. Therefore this equilibrium state is simple. Further, one can easily calculate the eigenvalues of the matrix A:

$$\begin{aligned} \lambda_1^- &= \frac{1}{2\delta_1}\left\{\mu_1 + \frac{\delta_1}{\delta}(1-\mu) - \sqrt{D}\right\} > 0, \\ \lambda_2^- &= \frac{1}{2\delta_1}\left\{\mu_1 + \frac{\delta_1}{\delta}(1-\mu) + \sqrt{D}\right\} > 0, \end{aligned} \tag{2.181}$$

where

$$D = \left(\mu_1 - \frac{\delta_1}{\delta}(1-\mu)\right)^2 + 8\frac{\delta_1}{\delta}\mu^2 > 0.$$

Thus the point $(1,1)$ is a knot to which the trajectories converge as $t \to -\infty$ (i.e., actually, as $\tau \to -\infty$).

For the subsequent analysis we need to know the directions along which the trajectories of the system (2.177) converge to the point $(1,1)$. For their determination we reduce (2.180) to canonical form. Define the new functions $X = x' + py'$, $Y = x' + qy'$ with $p = (\lambda_1^- - a)/c$, $q = (\lambda_2^- - a)/c$. Then we obtain

$$\begin{aligned} \frac{dX}{dt} &= \lambda_1^- X + \widetilde{f}_1(X, Y), \\ \frac{dY}{dt} &= \lambda_2^- Y + \widetilde{f}_2(X, Y), \end{aligned} \tag{2.182}$$

where

$$\widetilde{f}_i = f_i\left(\frac{qX - pY}{k}, \frac{Y - X}{k}\right) \quad \text{and} \quad k = \frac{\lambda_2^- - \lambda_1^-}{c}.$$

Using formulas (2.181), one can easily calculate that

$$(2.183)\qquad \begin{aligned} p &= \frac{1}{4\mu}\Big(-\mu_1 + \frac{\delta_1}{\delta}(1-\mu) + \sqrt{D}\Big) > 0, \\ q &= -\frac{1}{4\mu}\Big(\mu_1 - \frac{\delta_1}{\delta}(1-\mu) + \sqrt{D}\Big) < 0. \end{aligned}$$

It is well known (see, e.g., [**8**]) that as $t \to -\infty$ all trajectories of the system (2.182), except for two of them, converge to the knot $(0,0)$ along the line $Y = 0$. The two remaining trajectories converge to the knot along the line $X = 0$. Therefore, returning to the variables x, y, we obtain that there are only two trajectories of the system (2.177) converging as $t \to -\infty$ to the knot $(1,1)$ along the line

$$(2.184)\qquad \gamma_Y = \Big\{(x,y),\ y = 1 + \frac{x-1}{p}\Big\},$$

whereas the other trajectories converge to the knot along the line

$$(2.185)\qquad \gamma_X = \Big\{(x,y),\ y = 1 - \frac{x-1}{|q|}\Big\}.$$

Now we consider the equilibrium state (x^+, y^+). Substituting $x = x^+ + x'$ and $y = y^+ - y'$, we reduce the system (2.177) to the standard form (2.180), where

$$a = \frac{1}{\delta}\Big(2 - \frac{\mu+1}{x^+}\Big), \quad b = -\frac{\mu}{\delta x^+}, \quad c = -2\frac{1+\mu-x^+}{\delta_1}, \quad d = \frac{\mu_1}{\delta_1}.$$

For the determinant of the matrix A we now obtain the inequality

$$\Delta_2 = -2\frac{\mu(M^2-1)}{\delta\delta_1(\gamma-1)M^2x^+} < 0.$$

Hence A has real different eigenvalues, so that the state we consider is a saddle point. Denote

$$(2.186)\qquad \begin{aligned} \lambda^+ = \frac{1}{2}\Bigg\{ &-\sqrt{4|\Delta_2| + \Big(\frac{\mu_1}{\delta_1} + \frac{1}{\delta}\Big(2 - \frac{\mu+1}{x^+}\Big)\Big)^2} \\ &+ \Big(\frac{\mu_1}{\delta_1} + \frac{1}{\delta}\Big(2 - \frac{\mu+1}{x^+}\Big)\Big)\Bigg\} < 0; \end{aligned}$$

this eigenvalue is related to the separatrix entering the point (x^+, y^+) as $t \to +\infty$.

Now, on the phase plane (x, y) we draw the curves

$$\Gamma_P = \{(x,y),\ y = x(\mu+1-x)/\mu\}$$

and

$$\Gamma_Q = \{(x,y),\ y = 1 + (x-1)(x-1-2\mu)/\mu_1\},$$

which correspond to isoclines $P(x,y) = 0$ and $Q(x,y) = 0$ outside the line $x = 0$. It is easily seen that the curves Γ_P and Γ_Q divide the plane (x, y) into five domains:

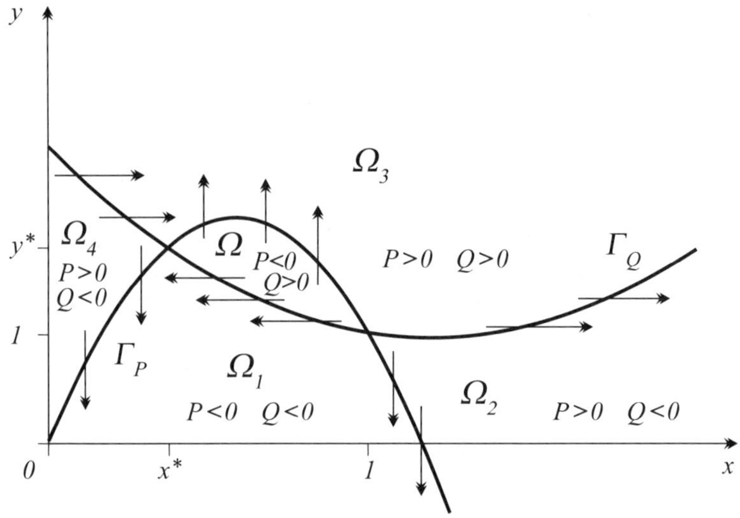

FIGURE 2.14. Isoclines Γ_P and Γ_Q and the domains of constant sign of dy/dx.

Ω, bounded by the intervals Γ_P and Γ_Q, and the external domains $\Omega_1, \ldots, \Omega_4$ (see Figure 2.14).

One can easily establish that the trajectories of the system (2.177) satisfy the inequalities

$$\begin{aligned} &\frac{dy}{dx} < 0 \qquad \text{for } (x,y) \in \Omega \cup \Omega_2 \cup \Omega_4, \\ &\frac{dy}{dx} > 0 \qquad \text{for } (x,y) \in \Omega_1 \cup \Omega_3. \end{aligned} \tag{2.187}$$

The domains of constant sign of the derivative dy/dx so situated and the fact that there are only two simple equilibrium points, the knot $(1,1)$ and the saddle-point (x^+, y^+), imply that the system (2.177) has no closed trajectories. This property and the stability of the knot $(1,1)$ as $t \to -\infty$ entail the existence of a single integral curve joining the saddle point (x^+, y^+) (for $t = +\infty$) and the knot $(1,1)$ (for $t = -\infty$).

An example of phase trajectories of the system (2.177) computed for some particular values of γ, M and δ, δ_1 is shown in Figure 2.15.

Further, denote by $T_P = \{(x,y),\ y = 1-(1-\mu)(x-1)/\mu\}$ and $T_Q = \{(x,y),\ y = 1-2\mu(x-1)/\mu_1\}$ the tangent lines to Γ_P and Γ_Q at the point $(1,1)$. By comparison of T_P and T_Q with the line γ_Y given by (2.184) it is easily seen that γ_Y passes through the point $(1,1)$ lying in the domains Ω_1 and Ω_3. Therefore our integral curve cannot get into the point $(1,1)$ as $t \to -\infty$ along the direction γ_Y, thus entering this point along the direction γ_X (see Figure 2.16).

Now we return to the initial functions ω, ψ and variable τ. The above analysis proves the existence of a unique solution to equations (2.170), (2.171) satisfying the boundary conditions (2.176). We have obtained also the following behavior of

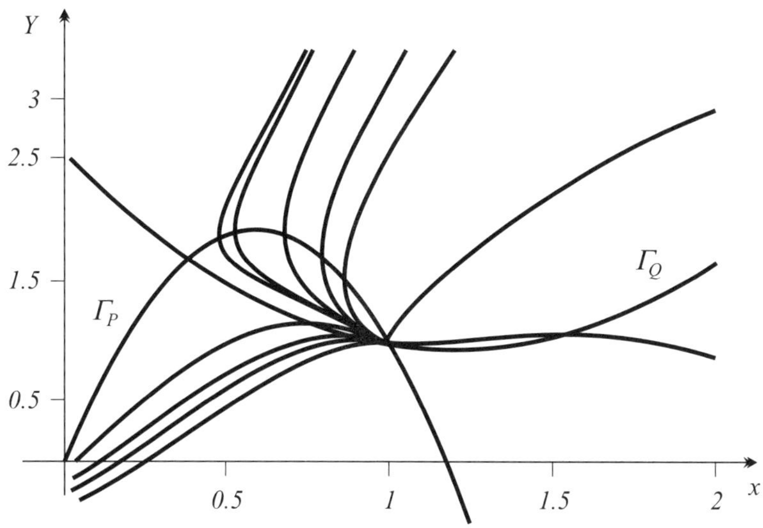

FIGURE 2.15. Phase trajectories for $\gamma = 1.4$, $M = 2$, and $\delta = \delta_1 = 2$.

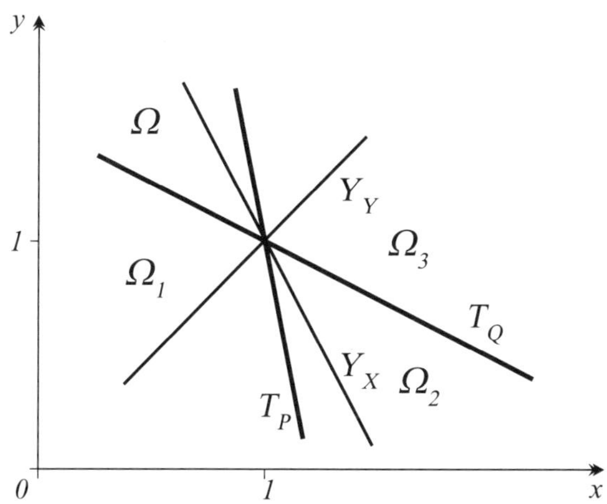

FIGURE 2.16. Directions of convergence to the knot $(1, 1)$ as $t \to \infty$.

ω and ψ at infinity:

$$
\begin{aligned}
&1-\omega,\ \psi-1 \sim e^{\lambda_1^- \tau} &&\text{as}\quad \tau \to -\infty,\\
&\omega-\omega^+,\ \psi^+-\psi \sim e^{-|\lambda^+|\tau} &&\text{as}\quad \tau \to +\infty,
\end{aligned}
\tag{2.188}
$$

where the exponents λ_1^- and λ^+ are defined by (2.181) and (2.186).

Now we consider the equations (2.174). It will be expedient to rewrite this system as a single equation:

$$
\delta_2 \frac{\partial^2}{\partial \tau^2}(\omega\omega_1) + \frac{\partial}{\partial \tau}(\zeta\omega_1) + \frac{\gamma+1}{\gamma-1}(2\omega - 1 - \omega^+)(\omega_1 - 1) = g,
\tag{2.189}
$$

where

$$\delta_2 = \frac{\delta\delta_1}{\mu}, \qquad \zeta = \delta_1\frac{1+\mu}{\mu} - 2\left(\frac{\delta_1}{\mu} + \frac{\delta}{\gamma-1}\right)\omega,$$

and the right-hand side g has properties similar to (2.175). One can readily establish that (2.189) obtains as the linearization of the equation

$$\begin{aligned}\delta_2\frac{\partial}{\partial\tau}\left(\omega\frac{\partial\omega}{\partial\tau}\right) + \frac{\partial}{\partial\tau}\left\{\delta_1\frac{1+\mu}{\mu}\omega - \left(\frac{\delta_1}{\mu} + \frac{\delta}{\gamma-1}\right)\omega^2\right\}\\ + \frac{\gamma+1}{\gamma-1}(\omega-1)(\omega-\omega^+) = 0,\end{aligned} \tag{2.190}$$

which is an immediate consequence of (2.170) and (2.171). Hence the homogeneous equation corresponding to (2.189) has the solution $y_1 = \partial\omega/\partial\tau$. Next, rewrite (2.189) in the form

$$L\omega_1 \overset{\text{def}}{=} \delta_2\frac{\partial^2}{\partial\tau^2}(\omega\omega_1) + \frac{\partial}{\partial\tau}(\zeta\omega_1) + \xi\omega_1 = g_1, \tag{2.191}$$

where

$$\xi = \frac{\gamma+1}{\gamma-1}(2\omega - 1 - \omega^+), \qquad g_1 = g + \xi.$$

Obviously, y_1 satisfies the equation $Ly = 0$. In order to find the second solution, observe that the substitution

$$\omega_1 = ue^z,$$

with $z = z(\tau, x)$ satisfying the equation

$$\frac{\partial z}{\partial\tau} = -\frac{1}{2\delta_2\omega}\left(\zeta + \delta_2\frac{\partial\omega}{\partial\tau}\right), \tag{2.192}$$

reduces the operator L to a symmetric form:

$$L_1 u \overset{\text{def}}{=} \delta_2\frac{\partial}{\partial\tau}\left(\omega\frac{\partial u}{\partial\tau}\right) + q(\tau, x)u.$$

Setting $u_1 = y_1\exp(-z)$, we obtain that $L_1 u_1 = 0$. Now the second solution u_2 to this equation can be easily found by standard formulas (see, e.g., [**81**]). Letting $y_2 = u_2\exp(z)$, it is readily seen that y_2 increases exponentially as $\tau \to -\infty$.

It is not hard to establish now (see also §1.6) that the $v_i = y_i\exp(-2z)$, $i = 1, 2$, with z satisfying (2.192) solve the equation $L^+ v = 0$, where

$$L^+ = \delta_2\omega\frac{\partial^2}{\partial\tau^2} - \zeta\frac{\partial}{\partial\tau} + \xi.$$

It remains to invoke the relations (2.188) to obtain that v_1 increases as an exponential function with exponent $-\lambda_2^-\tau$ as $\tau \to -\infty$. Since y_2 is an increasing function, this implies that v_2 also increases as $\tau \to -\infty$. Thus the operator L^+ has no kernel in the class of bounded functions, so that the equations (2.174) prove to be solvable without any additional conditions. By establishing this property we complete the treatment of the case $\Pr \sim 1$, since the higher order terms of the asymptotics are constructed similarly to the first correction.

Now we consider the problem on the motion of a smoothed shock wave for small Prandtl numbers (2.164). Clearly, the above constructions remain valid with the only difference that the viscous terms are carried over into higher order terms of the asymptotic expansion. As a consequence, we obtain the limiting problem (as $\varepsilon \to 0$), which is now given by the equations of ideal gas dynamics and the Rankine–Hugoniot conditions (2.155), while the wave profile in a neighborhood of the front is described by, instead of (2.170), (2.171), the equations

$$\begin{gathered} \omega - 1 + \mu\Big(\frac{\psi}{\omega} - 1\Big) = 0, \\ \mu_1(\psi - 1) + (\omega - 1)(1 - \omega + 2\mu) = \delta_1 \frac{\partial \psi}{\partial \tau}, \end{gathered} \tag{2.193}$$

with boundary conditions (2.176).

On excluding ψ the system (2.193) reduces to the equation

$$\frac{\partial \omega}{\partial \tau} = \frac{\mu(\gamma+1)}{\delta_1(\gamma-1)} \frac{(1-\omega)(\omega-\omega^+)}{1+\mu-2\omega}. \tag{2.194}$$

It is easily seen that, unlike the equation (2.139) whose singular point $\omega^* = 0$ lies outside the interval of interest $[\omega^+, 1]$, the singular point $\omega^* = (1+\mu)/2$ of (2.194) can belong to this interval. Hence we impose the additional condition $\omega^* < \omega^+$. Using the fact that $\gamma < 3$ in actual media, it is not hard to calculate that this condition implies the inequality (2.165). Then (2.194) can be easily integrated, and we arrive at the implicit formula

$$\frac{1-\omega}{(\omega-\omega^+)^a} = e^{\beta_1(\tau+\varphi_1(x))}, \tag{2.195}$$

where

$$\begin{gathered} a = \frac{2\omega^+ - 1 - \mu}{1-\mu} = \gamma \frac{(3-\gamma)(M_T^2 - M^2)}{(\gamma+1)(\gamma M^2 - 1)}, \quad M_T^2 = \frac{1}{\gamma}\frac{3\gamma-1}{3-\gamma}, \\ \beta_1 = \frac{\gamma+1}{\gamma-1}\frac{\mu(1-\omega^+)}{\delta_1(1-\mu)} = \frac{\alpha\check{\rho}_0(M^2-1)}{\gamma\kappa_0|\nabla\varphi|(\gamma-1)(\gamma M^2-1)}. \end{gathered} \tag{2.196}$$

It is obvious from (2.196) that whenever (2.165) holds the exponent a will be positive, and hence (2.195) describes the required solution. In this case all subsequent construcions differ from those carried out above only in minor details.

However, when the strict inequality (2.165) fails, formula (2.195) describes an exponentially increasing function ω. Obviously, such a solution has no physical meaning. Proceeding formally, one could try to remedy the situation by keeping the viscous term in the model equations—in other words, by treating the equations (2.170), (2.171) with small viscosity $\delta \sim O(\varepsilon)$ instead of (2.194). But the above analysis of the system (2.170), (2.171), and especially formula (2.188), shows that in this case the solution will involve the dependence on τ/ε. At the same time all the above constructions, in particular the derivation of (2.170), (2.171), rely on the assumption that the singular dependence on the small parameter is determined only by the fast variable $\tau = (t-\varphi)/\varepsilon$. Thus we again arrive at a contradiction.

This means that in the case where (2.164) holds and (2.165) fails the asymptotic solution cannot be represented as a one-phase solution (2.111). In fact, in the problem under consideration there are three scales, $l_1 \sim \varepsilon^2$, $l_2 \sim \varepsilon$, and $l_3 \sim 1$. Accordingly, a solution can be constructed only in the form of a multiphase asymptotic expansion. The reason for the contradictions which appear when neglecting small-scale effects may be explained as follows. Taking into consideration the intermediate scale l_2 leads to a partial balance between the nonlinearity and diffusion effects. However the width of the transition zone remains large, and the nonlinearity effect in the equation of motion may lead to twisting the wave profile inside the transition zone. This cannot happen only if the wave moves sufficiently slowly, i.e., (2.165) is fulfilled. Taking small-scale diffusion effects into consideration leads to a balance, and hence to the existence of the required solution. On the contrary, when neglecting the diffusion terms we arrive at an imbalance which destroys the solution.

A detailed discussion of the physical aspects of the problems discussed here can be found in [**251**]. Construction of a three-scale solution will be carried out in §5.4.

2.7. Temperature boundary layer for domains with smooth boundary

So far we have dealt with asymptotic solutions whose fast variation was localized about some surfaces which themselves were to be determined. However, there is a class of meaningful problems where the position of these surfaces is known in advance. The best-known example of such a situation is the problem on the boundary layer where one seeks a solution rapidly varying near the outer boundary (the so-called boundary layer) and slowly varying inside the domain. Discussion of various aspects of the boundary layer problem can be found, e.g., in [**42, 250**] (one-dimensional case) and in [**27, 28, 42, 65, 84, 192–194, 231, 249, 252, 253**], (multidimensional case). These papers contain also an extensive bibliography on this subject.

From the viewpoint of our preceding constructions, the boundary layer asymptotics is given by a function of the form $u(S/\varepsilon, x, t, \varepsilon)$, where, as before, u is a function of fast variation only in a small neighborhood of the set $\Gamma = \{(x,t),\ S = 0\}$, but now $u(\tau, x, t, \varepsilon)$ is considered only on the semiaxis $\tau \geq 0$ and the set Γ coincides with the outer boundary. It is clear that such solutions can be constructed in the framework of the methodology developed above, but some details of the calculations have to be altered.

Consider the equations of gas dynamics, taking heat effects into consideration, which are written in the non-divergence form

$$(2.197)\qquad \begin{aligned} &\rho\Big(\frac{\partial u_{(i)}}{\partial t} + \langle u, \nabla\rangle u_{(i)}\Big) + \frac{\partial p}{\partial x_i} = \varepsilon^2 \frac{\partial}{\partial x_j}\sigma_{ij}, \quad i = 1,2,3,\\ &\frac{\partial \rho}{\partial t} + \operatorname{div} \rho u = 0,\\ &\rho c_p\Big(\frac{\partial T}{\partial t} + \langle u, \nabla T\rangle\Big) = \varepsilon^2 D + E\Big(\frac{\partial p}{\partial t} + \langle u, \nabla p\rangle\Big). \end{aligned}$$

Here, as before, $x = (x_1, x_2, x_3)$, $u = (u_{(1)}, u_{(2)}, u_{(3)})$ is the velocity, p is the pressure, ρ is the density, and T is the temperature, all these functions being reduced to a dimensionless form; ε is related to the Reynolds number Re as $\varepsilon = 1/\sqrt{\mathrm{Re}}$, σ_{ij}

is the stress tensor,

$$\sigma_{ij} = \Big(\frac{\partial u_{(i)}}{\partial x_j} + \frac{\partial u_{(j)}}{\partial x_i}\Big) + \Big(\zeta - \frac{2}{3}\eta\Big)\delta_{ij}\frac{\partial u_{(k)}}{\partial x_k},$$

where δ_{ij} is Kronecker's delta, repeated indices indicate the summation, and the normalized coefficients of viscosity $\eta > 0$, $\zeta > 0$ are smooth functions of density and temperature; further, c_p is the coefficient of specific heat,

$$D = \operatorname{div}(\kappa\nabla T) + \frac{\eta}{2}\Big(\frac{\partial u_{(i)}}{\partial x_j} + \frac{\partial u_{(j)}}{\partial x_i} - \frac{2}{3}\delta_{ij}\frac{\partial u_{(k)}}{\partial x_k}\Big)^2 + \zeta\Big(\frac{\partial u_{(k)}}{\partial x_k}\Big)^2,$$

and E is the Eckert number. In (2.197) we assume that the Prandtl number is proportional to one and include the coefficient of proportionality into the coefficient of heat conductivity $\kappa = \kappa(\rho, T) > 0$. Moreover, we assume that the gas is ideal, i.e., that

$$P = \rho T. \tag{2.198}$$

We will consider equations (2.197) in a domain with smooth boundary $\partial\Omega$ and show that construction of the boundary layer type solution reduces in the leading term to the Prandtl equations. These equations are widely known [**84, 231**]; however, their derivation (as given, e.g., in the book [**231**]) relies, on a deep insight into the nature of processes in fluids. We will derive these equations from a formal mathematical standpoint and will indicate the method of deriving model equations for higher order corrections.

It should be pointed out that the Prandtl equations play a fundamental role in the theory of hydro- and gas dynamics. Apart from the well-known boundary layer situation, similar equations arise in the study of coherent structures (including magnetohydrodynamics) [**146, 147, 154, 155, 159, 163**] as well as in the problem of constructing localized solutions of stretched vortex type for the Navier–Stokes equations [**224–228**]. The model equations in these problems contain nontrivial additional terms, but they have basically the same structure as the Prandtl equations.

Let us proceed to construct the asymptotic solution of boundary layer type for equations (2.197).

We set the following initial and boundary conditions:

$$u\big|_{t=0} = u^0(x,\varepsilon), \quad \rho\big|_{t=0} = \rho^0(x,\varepsilon), \quad T\big|_{t=0} = T^0(x,\varepsilon), \tag{2.199}$$

$$u\big|_{\partial\Omega} = 0, \qquad T\big|_{\partial\Omega} = T_b(x',t), \tag{2.200}$$

where $u^0, \dots, T_b$ are given functions whose properties will be specified later, and $x' \in \partial\Omega$.

We will seek the asymptotic solution to the problem (2.197)–(2.200) in the form of an expansion similar to (2.111):

$$\begin{aligned}
u &= u_0(x,t) + U_0\Big(\frac{S}{\varepsilon}, x, t\Big) + \varepsilon\Big(u_1(x,t) + U_1\Big(\frac{S}{\varepsilon}, x, t\Big)\Big) + \dots,\\
\rho &= \rho_0(x,t) + R_0\Big(\frac{S}{\varepsilon}, x, t\Big) + \varepsilon\Big(\rho_1(x,t) + R_1\Big(\frac{S}{\varepsilon}, x, t\Big)\Big) + \dots,\\
T &= T_0(x,t) + \Theta_0\Big(\frac{S}{\varepsilon}, x, t\Big) + \varepsilon\Big(T_1(x,t) + \Theta_1\Big(\frac{S}{\varepsilon}, x, t\Big)\Big) + \dots,\\
p &= p_0(x,t) + P_0\Big(\frac{S}{\varepsilon}, x, t\Big) + \varepsilon\Big(p_1(x,t) + P_1\Big(\frac{S}{\varepsilon}, x, t\Big)\Big) + \dots,
\end{aligned} \tag{2.201}$$

where the u_i, ρ_i, T_i, p_i, $U_i(\tau, x, t)$, $R_i(\tau, x, t)$, $\Theta_i(\tau, x, t)$, $P_i(\tau, x, t)$ are smooth functions such that

$$U_i, \dots, P_i \to 0 \quad \text{as} \quad \tau \to \infty \tag{2.202}$$

together with all their derivatives, and $S = S(x)$ is a smooth function satisfying the condition

$$S\big|_{\partial\Omega} = 0, \quad |\nabla S|\big|_{\partial\Omega} \neq 0. \tag{2.203}$$

We assume for definiteness that $S > 0$ for $x \in \Omega$.

Substitute (2.201) into (2.197) and (2.198), and pass to the limit as $\tau \to \infty$. Equating to zero the coefficients of corresponding powers of ε, we obtain the equations of ideal gas dynamics for determination of the leading term of the "background" $u_0, \dots, T_0$:

$$\begin{aligned}
&\rho_0\Big(\frac{\partial u_0}{\partial t} + \langle u_0, \nabla\rangle u_0\Big) + \nabla p_0 = 0,\\
&\frac{\partial \rho_0}{\partial t} + \operatorname{div}(\rho_0 u_0) = 0,\\
&\rho_0 c_p\Big(\frac{\partial T_0}{\partial t} + \langle u_0, \nabla T_0\rangle\Big) - E\Big(\frac{\partial p_0}{\partial t} + \langle u_0, \nabla p_0\rangle\Big) = 0,\\
&p_0 = \rho_0 T_0,
\end{aligned} \tag{2.204}$$

while for the higher order terms $u_i, \dots, T_i$, $i \geq 1$, we obtain nonhomogeneous linearized equations. It is well known that the system (2.204) allows us to impose the boundary condition of impenetrability

$$u_0^{\perp}\big|_{\partial\Omega} = 0, \tag{2.205}$$

where $u_0^{\perp}$ denotes the component of the vector u_0 normal to $\partial\Omega$. We complete (2.204), (2.205) with the initial conditions

$$u_0\big|_{t=0} = \overline{u^0}(x), \quad \rho_0\big|_{t=0} = \overline{\rho^0}(x), \quad T_0\big|_{t=0} = \overline{T^0}(x), \tag{2.206}$$

where $\overline{u^0}$, $\overline{\rho^0}$, $\overline{T^0}$ denote the weak limits as $\varepsilon \to 0$ of the initial functions u^0, ρ^0, T^0 in (2.199). Then we obtain the well-posed problem (2.204)–(2.206) for determining $u_0, \dots, T_0$, and by setting similar initial and boundary conditions for linearized equations we can determine recursively all regular terms $u_i, \dots, T_i$ of (2.201).

We will assume that the solution to (2.204)–(2.206) remains smooth within the time interval $[0, t_0]$. Clearly, this implies the existence of smooth functions $u_i, \dots, T_i$, $i \geq 1$, for $t \in [0, t_0]$.

In the literature on fluid mechanics the background terms of the expansion (2.201) are usually regarded as "outer flow". Our assumption on the existence of a smooth solution to (2.204)–(2.206) is nothing but the assumption that the outer flow is inviscid.

Now we turn to the more significant step, namely, to determination of boundary layer components of the solution. Substitute (2.201) into the equation of continuity, collect the terms according to the order in ε, and equate to zero the coefficient of ε^{-1}. Since S does not depend on time, we obtain the relation

$$\frac{\partial}{\partial\tau}\{(\rho_0+R_0)\langle\nabla S, u_0+U_0\rangle\}=0 \pmod{O(\varepsilon)}. \tag{2.207}$$

By (2.203) the vector $\nabla S|_{\partial\Omega}$ is directed along the normal to the boundary, and by (2.205) $\langle\nabla S, u_0\rangle|_{\partial\Omega}=0$. Therefore, taking (2.202) into account, we have

$$\frac{\partial}{\partial\tau}(R_0\langle\nabla S, u_0\rangle)\big|_{\tau=S/\varepsilon}=0 \pmod{O(\varepsilon)}, \tag{2.208}$$

uniformly in $x\in\Omega$. Thus, integrating (2.207) and taking the terms of order $O(\varepsilon)$ into the higher order corrections, we obtain the equality

$$(\rho_0+R_0)\langle\nabla S, U_0\rangle=0.$$

Since the density does not vanish anywhere, this implies the geometric condition

$$\langle\nabla S, U_0\rangle=0. \tag{2.209}$$

Consider now the terms of order ε^{-1} arising when (2.201) is substituted into the equation of motion

$$(\rho_0+R_0)\frac{\partial}{\partial\tau}\{\langle u_0+U_0,\nabla S\rangle(u_0+U_0)\}+\nabla S\frac{\partial P_0}{\partial\tau}=0 \pmod{O(\varepsilon)}.$$

Taking (2.208) and (2.209) into account, we immediately obtain

$$\frac{\partial P_0}{\partial\tau}=0, \tag{2.210}$$

i.e., the pressure in its leading term is a slowly varying function.

A similar treatment of the equation of energy shows that by (2.205), (2.209) it is fulfilled within $O(1)$. Therefore the only condition on the boundary layer part of the temperature which appears at the first step of the asymptotic procedure is related to the state equation. Namely, (2.198) and (2.210) immediately imply the relation

$$\Theta_0=-\frac{T_0R_0}{\rho_0+R_0}. \tag{2.211}$$

Now we consider the terms of order ε^0 which result when substituting the expansion (2.201) into (2.197). Since $\langle\nabla S, u_1\rangle|_{\partial\Omega}=0$, after some routine calculations we obtain the relations

$$
\begin{aligned}
&(\rho_0 + R_0)\Big\{\frac{\partial}{\partial t}(u_0 + U_0) + \langle u_0 + U_0, \nabla\rangle(u_0 + U_0) \\
&\qquad + \Big(\frac{1}{\varepsilon}\langle u_0, \nabla S\rangle + \langle U_1, \nabla S\rangle\Big)\frac{\partial U_0}{\partial \tau}\Big\} \\
&\qquad + \nabla p_0 + \nabla S\frac{\partial P_1}{\partial \tau} - |\nabla S|^2\frac{\partial}{\partial \tau}\Big(\eta_0\frac{\partial U_0}{\partial \tau}\Big) = 0 \ (\operatorname{mod} O(\varepsilon)), \\
&\frac{\partial}{\partial t}(\rho_0 + R_0) + \langle \nabla, (\rho_0 + R_0)(u_0 + U_0)\rangle \\
&\qquad + \frac{\partial}{\partial \tau}\Big\{\Big(\frac{1}{\varepsilon}\langle \nabla S, u_0\rangle + \langle \nabla S, U_1\rangle\Big)(\rho_0 + R_0)\Big\} = 0 \ (\operatorname{mod} O(\varepsilon)), \\
&c_p(\rho_0 + R_0)\Big\{\frac{\partial}{\partial t}(T_0 + \Theta_0) + \langle u_0 + U_0, \nabla(T_0 + \Theta_0)\rangle \\
&\qquad + \Big(\frac{1}{\varepsilon}\langle \nabla S, u_0\rangle + \langle U_1, \nabla S\rangle\Big)\frac{\partial \Theta_0}{\partial \tau}\Big\} \\
&\qquad - |\nabla S|^2\frac{\partial}{\partial \tau}\Big(\kappa_0\frac{\partial \Theta_0}{\partial \tau}\Big) - |\nabla S|^2\eta_0\Big|\frac{\partial U_0}{\partial \tau}\Big|^2 \\
&\qquad - E\Big(\frac{\partial p_0}{\partial t} + \langle u_0 + U_0, \nabla p_0\rangle\Big) = 0 \ (\operatorname{mod} O(\varepsilon)).
\end{aligned} \tag{2.212}
$$

Here $\kappa_0 = \kappa(\rho_0 + R_0, T_0 + \Theta_0)$, $\eta_0 = \eta(\rho_0 + R_0, T_0 + \Theta_0)$, and we write $0 \ (\operatorname{mod} O(\varepsilon))$ in the right-hand side to emphasize that the terms of order $O(\varepsilon)$ in the left-hand side may be neglected. One of the terms of this kind is, e.g., the expression

$$
(\rho_0 + R_0)u_0^{\perp}\frac{\partial U_0}{\partial \nu}
$$

(where ν is the inner normal to $\partial\Omega$), which enters into the term $(\rho_0 + R_0)\langle u_0, \nabla\rangle U_0$ in the first equation (2.212). We leave such terms in the left-hand sides of (2.212) in order to simplify the formulas. For the same reason we take no consideration of (2.204) in (2.212). Recall, however, that $\varepsilon^{-1}\langle u_0, \nabla S\rangle \partial U_0/\partial\tau|_{\tau=S/\varepsilon}$ has magnitude $O(1)$ uniformly in $x \in \overline{\Omega}$.

It is easily seen that the system (2.212) contains seven unknown scalar functions, viz., three components of the vector U_0, $\langle \nabla S, U_1\rangle$, R_0, P_1, and Θ_0. If we complete (2.212) with equalities (2.209) and (2.211), we obtain seven scalar equations.

In fact, the unknown function P_1 can be easily excluded from this system. Namely, take the scalar product of the first equation (2.212) by the vector ∇S. Using the relations

$$
\begin{aligned}
R_0\Big\langle \nabla S, \frac{\partial u_0}{\partial t}\Big\rangle &= O(\varepsilon), \\
\langle U_0, \nabla\rangle\langle \nabla S, u_0\rangle &= O(\varepsilon), \\
R_0\langle u_0, \nabla\rangle\langle \nabla S, u_0\rangle &= O(\varepsilon),
\end{aligned}
$$

and neglecting the terms of order $O(\varepsilon)$, we obtain the equality

$$
|\nabla S|^2\frac{\partial P_1}{\partial \tau} = (\rho_0 + R_0)\langle u_0 + U_0, S_{xx}(u_0 + U_0)\rangle - \rho_0\langle u_0, S_{xx}u_0\rangle, \tag{2.213}
$$

where S_{xx} denotes the Hessian of the function S, i.e., the matrix with entries $\partial^2 S/\partial x_i \partial x_j$. The boundary condition $P_1 \to 0$ as $\tau \to \infty$ allows us to determine P_1 uniquely as a function of R_0 and U_0.

Let us set the initial and boundary conditions

$$\begin{gathered} U_0\big|_{t=0} = U_0^0(\tau, x), \quad R_0\big|_{t=0} = R_0^0(\tau, x), \quad \Theta_0\big|_{t=0} = \Theta_0^0(\tau, x), \\ U_0\big|_{\tau=0,\, x\in\partial\Omega} = -u_0\big|_{x\in\partial\Omega}, \quad \Theta_0\big|_{\tau=0,\, x\in\partial\Omega} = T_b(x,t) - T_0\big|_{x\in\partial\Omega}, \\ U_0 \to 0, \quad \Theta_0 \to 0 \qquad \text{as} \quad \tau \to \infty, \end{gathered} \tag{2.214}$$

where U_0^0, R_0^0, Θ_0^0 are functions such that

$$\begin{gathered} u^0(x,\varepsilon) - \overline{u^0}(x) - U_0^0\Big(\frac{S}{\varepsilon}, x\Big) = O(\varepsilon), \\ \rho^0(x,\varepsilon) - \overline{\rho^0}(x) - R_0^0\Big(\frac{S}{\varepsilon}, x\Big) = O(\varepsilon), \\ T^0(x,\varepsilon) - \overline{T^0}(x) - \Theta_0^0\Big(\frac{S}{\varepsilon}, x\Big) = O(\varepsilon), \end{gathered} \tag{2.215}$$

and u^0, ρ^0, T^0 are the initial conditions from (2.199). The relations (2.215) show that the initial values u^0, ρ^0, T^0 may involve a singular dependence on ε only in the form $\tau = S/\varepsilon$, and they must stabilize as $\tau \to \infty$. It is also clear that the initial data must satisfy the equalities (2.209) and (2.211) within $O(\varepsilon)$. Moreover, we will assume that the initial and boundary conditions are consistent with each other. Then the problem (2.212), (2.214) and (2.209), (2.211) is well posed and, together with (2.204)–(2.206), constitutes a closed system for determination of the leading term of the asymptotics. Furthermore, we will show later on that equations (2.212) represent one possible form of Prandtl equations for the temperature boundary layer.

Construction of the higher order terms of the asymptotics is carried out along the same lines. We only mention the following point. As usual, to determine the boundary layer corrections $U_i - \nabla S\langle \nabla S, U_i\rangle/|\nabla S|^2$, $\langle \nabla S, U_{i+1}\rangle$, R_i, P_{i+1}, Θ_i we obtain nonhomogeneous linearized equations. However, unlike the previous problems, they are considered only on the semiaxis $\tau > 0$. Hence no additional conditions for solvability of these equations appear. This entails that the only conditions on S are (2.203), so that some arbitrariness remains in the choice of the phase S. This observation agrees with the choice of S as the distance to $\partial\Omega$ along the inner normal, which is standard for the boundary layer theory.

We will show that the model equations (2.212) coincide with the Prandtl equations $\operatorname{mod} O(\varepsilon)$. Prior to the treatment of the general case we consider a simple example.

Let Ω be the half-space $x_2 > 0$ and let the solution be independent of the variable x_3. Denote $x = x_1$, $y = x_2$ and assume that the outer flow is laminar, i.e., stationary, and parallel to the boundary $y = 0$, so that $u_0 = (W(x), 0)^T$, $\rho_0 = \rho_0(x)$, $T_0 = T_0(x)$, $p_0 = p_0(x)$, where $(\cdot,\cdot)^T$ denotes a column-vector. Set $S = y$ and observe that in this case $U_0 = (U_0^1, 0)^T$, $\langle \nabla S, U_1\rangle = U_1^2$, where U_0^1 is the longitudinal (along the x-axis) and U_1^2 the normal (with respect to the x-axis) component of the corresponding vectors. We will use the notation $\mathcal{U} = W + U_0^1$ and $\mathcal{V} = U_1^2$ for the longitudinal and normal components of velocity; let, moreover,

$\mathcal{R} = \rho_0 + R_0$, $\mathcal{T} = T_0 + \Theta_0$. Obviously, in our example

$$\langle u_0 + U_0, \nabla\rangle(u_0 + U_0) = \mathcal{U}\frac{\partial \mathcal{U}}{\partial x}, \quad \langle \nabla, (\rho_0 + R_0)(u_0 + U_0)\rangle = \frac{\partial}{\partial x}(\mathcal{R}\mathcal{U}),$$
$$\langle u_0 + U_0, \nabla(T_0 + \Theta_0)\rangle = \mathcal{U}\frac{\partial \mathcal{T}}{\partial x}, \quad \langle u_0 + U_0, \nabla p_0\rangle = \mathcal{U}\frac{\partial p_0}{\partial x}.$$

Moreover, $\langle u_0, \nabla S\rangle = 0$ and (2.213) implies that $P_1 = 0$. Hence the equations (2.212) in this particular case become

$$\begin{gathered}
\mathcal{R}\Big(\frac{\partial \mathcal{U}}{\partial t} + \mathcal{U}\frac{\partial \mathcal{U}}{\partial x} + \mathcal{V}\frac{\partial \mathcal{U}}{\partial \tau}\Big) + \frac{\partial p_0}{\partial x} = \frac{\partial}{\partial \tau}\Big(\eta_0 \frac{\partial \mathcal{U}}{\partial \tau}\Big),\\
\frac{\partial \mathcal{R}}{\partial t} + \frac{\partial}{\partial x}(\mathcal{R}\mathcal{U}) + \frac{\partial}{\partial \tau}(\mathcal{R}\mathcal{V}) = 0,\\
c_p\mathcal{R}\Big(\frac{\partial \mathcal{T}}{\partial t} + \mathcal{U}\frac{\partial \mathcal{T}}{\partial x} + \mathcal{V}\frac{\partial \mathcal{T}}{\partial \tau}\Big) = \frac{\partial}{\partial \tau}\Big(\kappa_0 \frac{\partial \mathcal{T}}{\partial \tau}\Big) + \eta_0\Big(\frac{\partial \mathcal{U}}{\partial \tau}\Big)^2 + E\mathcal{U}\frac{\partial p_0}{\partial x},\\
p_0 = \mathcal{R}\mathcal{T}.
\end{gathered} \tag{2.216}$$

Thus we have obtained standard Prandtl equations for the temperature boundary layer in the case of a planar boundary (see, e.g., [**231**]), where $\tau = y/\varepsilon$ is the stretched normal variable and the pressure p_0 is a known function.

In the general case we have to pass to a specific coordinate system related to the geometry of the surface $\partial\Omega$. Namely, let the functions

$$x^1 = x^1(x'), \quad x^2 = x^2(x') \tag{2.217}$$

locally determine the curvilinear coordinates on $\partial\Omega$. Denote

$$x^3 = S(x) \tag{2.218}$$

and pass from the Cartesian coordinates x_1, x_2, x_3 to the curvilinear coordinates x^1, x^2, x^3. We will use the standard notation

$$e_j = \frac{\partial x_1}{\partial x^j}\mathbf{i} + \frac{\partial x_2}{\partial x^j}\mathbf{j} + \frac{\partial x_3}{\partial x^j}\mathbf{k},$$
$$e^j = \frac{\partial x^j}{\partial x_1}\mathbf{i} + \frac{\partial x^j}{\partial x_2}\mathbf{j} + \frac{\partial x^j}{\partial x_3}\mathbf{k}\Big|_{x_i = x_i(x^1,x^2,x^3)}$$

for contravariant and covariant basis vectors (with $\mathbf{i}, \mathbf{j}, \mathbf{k}$ being the basis of the Cartesian coordinate system), and let

$$g_{ij} = \langle e_i, e_j\rangle, \quad g^{ij} = \langle e^i, e^j\rangle$$

be the corresponding components of the metric tensors. Recall that $\langle e_i, e^j\rangle = \delta_i^j$, i.e., $\langle e_i, e^j\rangle = 1$ for $i = j$ and 0 for $i \neq j$, and that the vectors e_j and e^j (the so-called reciprocal bases) are related by the equations

$$e_j = g_{ji}e^i, \quad e^j = g^{ji}e_i.$$

We will assume for definiteness that x^1, x^2, x^3 is a right-hand coordinate system, i.e., the Jacobian of the change of coordinates is positive:

$$\frac{\partial(x_1, x_2, x_3)}{\partial(x^1, x^2, x^3)} = \sqrt{\det[g_{ij}]} = \sqrt{g} > 0.$$

The coordinate system x^1, x^2, x^3 is called monocline if $g^{i3} = g_{i3} = 0$ for $i \neq 3$.

For an arbitrary vector F, we will use the decomposition

$$F = e_i F^i,$$

where the contravariant coordinates F^i are given by

$$F^i = \langle e^i, F \rangle.$$

Further, we will need the Christoffel symbols of the second kind

$$\Gamma^k_{ij} = \frac{1}{2} g^{kl} \Big(\frac{\partial g_{il}}{\partial x^j} + \frac{\partial g_{jl}}{\partial x^i} - \frac{\partial g_{ij}}{\partial x^l} \Big). \tag{2.219}$$

Recall that the Christoffel symbols are symmetric with respect to the subscripts, i.e., $\Gamma^k_{ij} = \Gamma^k_{ji}$, and that along with (2.219) we can also use the formulas

$$\Gamma^k_{ij} = \Big\langle e^k, \frac{\partial e_i}{\partial x^j} \Big\rangle = -\Big\langle e_i, \frac{\partial e^k}{\partial x^j} \Big\rangle.$$

The Christoffel symbols arise in a natural way when differentiating vectors. Namely, since

$$\frac{\partial}{\partial x_i} = \frac{\partial x^k}{\partial x_i} \frac{\partial}{\partial x^k},$$

we have $\nabla = e^k \partial/\partial x^k$. Therefore

$$\langle f, \nabla \rangle F = f^k \frac{\partial}{\partial x^k}(e_i F^i) = f^k \Big(e_i \frac{\partial F^i}{\partial x^k} + F^i \frac{\partial e_i}{\partial x^k} \Big) = e_i f^k \Big(\frac{\partial F^i}{\partial x^k} + \Gamma^i_{jk} F^j \Big). \tag{2.220}$$

The expression in parentheses in the right-hand side of (2.220) is referred to as the covariant derivative and is denoted by $F^i_{,k}$:

$$F^i_{,k} = \frac{\partial F^i}{\partial x^k} + \Gamma^i_{jk} F^j.$$

In the above notation the expression for the contravariant vector $\langle f, \nabla \rangle F$ takes a simple form:

$$\langle f, \nabla \rangle F = e_i f^k F^i_{,k}.$$

It remains to recall the formula

$$\frac{\partial}{\partial x^i} \log \sqrt{g} = \Gamma^k_{ik},$$

which enables us to calculate that

$$\operatorname{div} F = \frac{1}{\sqrt{g}} \frac{\partial}{\partial x^i} (\sqrt{g} F^i).$$

Let us rewrite the equations (2.212) in the new coordinates (2.217), (2.218). First of all, since $x^3 = S$, the equality (2.209) means that $U_0^3 = 0$. Denote

$$\begin{aligned} \mathcal{U}^\alpha &= u_0^\alpha + U_0^\alpha, \quad \alpha = 1,2, \qquad \mathcal{U}^3 = u_0^3 + \varepsilon U_1^3, \\ \mathcal{R} &= \rho_0 + R_0, \qquad \mathcal{T} = T_0 + \Theta_0. \end{aligned} \tag{2.221}$$

Here we use the fact that $\langle \nabla S, U_1 \rangle = \langle e^3, U_1 \rangle = U_1^3$; hence (2.221) is a natural analog of the notation in the previous example. However, here we do not normalize the normal component $\mathcal{U}^3$ of the velocity.

We will demonstrate in detail the change from Cartesian to curvilinear coordinates for the first equation (2.212). Using the auxiliary formulas stated above, we obtain the relations

$$\begin{aligned} & e_i \mathcal{R} \Big\{ \frac{\partial}{\partial t}(u_0^i + U_0^i) + (u_0^k + U_0^k)(u_{0,k}^i + U_{0,k}^i) + \frac{1}{\varepsilon}\mathcal{U}^3 \frac{\partial U_0^i}{\partial \tau} \Big\} \\ & \quad + e_i g^{ik} \frac{\partial p_0}{\partial x^k} + e_i g^{i3} \frac{\partial P_1}{\partial \tau} \\ & \quad - e_i g^{33} \frac{\partial}{\partial \tau}\Big(\eta_0 \frac{\partial U_0^i}{\partial \tau} \Big) = 0 \ (\operatorname{mod} O(\varepsilon)), \quad i = 1,2,3. \end{aligned} \tag{2.222}$$

The specific choice of our coordinates (2.217), (2.218) implies that, uniformly in $x^3 \geq 0$,

$$g^{\alpha 3} \frac{\partial P_1}{\partial \tau} \Big|_{\tau = x^3/\varepsilon} = O(\varepsilon), \quad \alpha = 1,2.$$

Neglecting these terms, we see that the first two equations (2.222) do not depend on the pressure correction P_1, while the third equation, which does not contain second derivatives of U_0 (since $U_0^3 = 0$), can be used for determination of P_1 and leads to a formula similar to (2.213) (see also below). Therefore we can use the first two equations to obtain the leading term. Since $\partial U_0^\alpha / \partial \tau = \partial \mathcal{U}^\alpha / \partial \tau$ for $\alpha = 1,2$, we obtain

$$\begin{aligned} & \mathcal{R} \Big\{ \frac{\partial \mathcal{U}^\alpha}{\partial t} + (u_0^k + U_0^k)(u_{0,k}^\alpha + U_{0,k}^\alpha) + \frac{1}{\varepsilon}\mathcal{U}^3 \frac{\partial \mathcal{U}^\alpha}{\partial \tau} \Big\} \\ & \qquad + g^{\alpha k} \frac{\partial p_0}{\partial x^k} - g^{33} \frac{\partial}{\partial \tau}\Big(\eta_0 \frac{\partial \mathcal{U}^\alpha}{\partial \tau} \Big) = 0 \ (\operatorname{mod} O(\varepsilon)). \end{aligned} \tag{2.223}$$

Consider now the expression

$$I^\alpha = (u_0^k + U_0^k)(u_{0,k}^\alpha + U_{0,k}^\alpha).$$

Since $U_0^3 = 0$, we easily obtain

$$I^\alpha = (u_0^\beta + U_0^\beta)(u_{0,\beta}^\alpha + U_{0,\beta}^\alpha) + u_0^3 u_{0,3}^\alpha + u_0^3 U_{0,3}^\alpha,$$

where the last term in the right-hand side is of order $O(\varepsilon)$, so that it can be neglected. Here and subsequently we assume that the indices denoted by greek letters take only the values 1 and 2. Denote

$$F^\alpha_{,\beta'} = \frac{\partial F^\alpha}{\partial x^\beta} + \Gamma^\alpha_{\beta\gamma} F^\gamma, \tag{2.224}$$

so that $F^\alpha_{,\beta'}$ differs from the covariant derivative $F^\alpha_{,\beta}$ by absence of the term $\Gamma^\alpha_{\beta 3}F^3$. Obviously, $U^\alpha_{0,\beta} = U^\alpha_{0,\beta'}$ and $U^\alpha_{0,3} = U^\alpha_{0,3'}$. Next,

$$U^\beta_0 u^\alpha_{0,\beta} = U^\beta_0 u^\alpha_{0,\beta'} + O(\varepsilon),$$

since $\Gamma^\alpha_{\beta 3}U^\beta_0 u^3_0 = O(\varepsilon)$ uniformly in $x^3 \geq 0$. Hence, up to $O(\varepsilon)$, I^α can be rewritten as

$$I^\alpha = \mathcal{U}^\beta \mathcal{U}^\alpha_{,\beta'} + u^3_0(u^\alpha_{0,3} + \Gamma^\alpha_{3\beta}u^\beta_0).$$

Finally, we point out the uniform estimate

$$R_0 u^3_0(u^\alpha_{0,3} + \Gamma^\alpha_{3\beta}u^\beta_0) = O(\varepsilon).$$

Therefore, neglecting the terms of order $O(\varepsilon)$ (i.e., combining them with the higher order terms), we can rewrite the relations (2.223) in the form of the equations

$$\begin{aligned}\mathcal{R}\left\{\frac{\partial \mathcal{U}^\alpha}{\partial t} + \mathcal{U}^\beta \mathcal{U}^\alpha_{,\beta'} + \frac{1}{\varepsilon}\mathcal{U}^3\frac{\partial \mathcal{U}^\alpha}{\partial \tau}\right\} + \rho_0 u^3_0(u^\alpha_{0,3} + \Gamma^\alpha_{3\beta}u^\beta_0) + g^{\alpha k}\frac{\partial p_0}{\partial x^k} \\ = g^{33}\frac{\partial}{\partial \tau}\left(\eta_0\frac{\partial \mathcal{U}^\alpha}{\partial \tau}\right), \quad \alpha = 1, 2.\end{aligned} \tag{2.225}$$

A similar treatment of the remaining relations (2.212) leads to the equations

$$\frac{\partial \mathcal{R}}{\partial t} + \frac{1}{\sqrt{g}}\frac{\partial}{\partial x^\alpha}(\sqrt{g}\mathcal{R}\mathcal{U}^\alpha) + \frac{1}{\varepsilon}\frac{\partial}{\partial \tau}(\mathcal{R}\mathcal{U}^3) + + \frac{u^3_0}{\sqrt{g}}\frac{\partial}{\partial x^3}(\sqrt{g}\rho_0) + \mathcal{R}\frac{\partial u^3_0}{\partial x^3} = 0, \tag{2.226}$$

$$\begin{aligned}& c_p\mathcal{R}\left\{\frac{\partial \mathcal{T}}{\partial t} + \mathcal{U}^\alpha\frac{\partial \mathcal{T}}{\partial x^\alpha} + \frac{1}{\varepsilon}\mathcal{U}^3\frac{\partial \mathcal{T}}{\partial \tau}\right\} \\ &= g^{33}\frac{\partial}{\partial \tau}\left(\kappa_0\frac{\partial \mathcal{T}}{\partial \tau}\right) + g^{33}g_{\alpha\beta}\eta_0\frac{\partial \mathcal{U}^\alpha}{\partial \tau}\frac{\partial \mathcal{U}^\beta}{\partial \tau} \\ &\quad + E\left(\frac{\partial p_0}{\partial t} + \mathcal{U}^\alpha\frac{\partial p_0}{\partial x^\alpha} + u^3_0\frac{\partial p_0}{\partial x^3}\right) - c_p\rho_0 u^3_0\frac{\partial T_0}{\partial x^3}.\end{aligned} \tag{2.227}$$

The system (2.225)–(2.227) is valid for all $x^3 \geq 0$ for which the coordinates x^1, x^2, x^3 exist (up to intersection of inner normals to $\partial\Omega$). One can easily establish that as $\tau \to \infty$, which corresponds to $x^3 \geq c\varepsilon^{1-\delta}$ for $\varepsilon \to 0$ and $\delta > 0$, the equations (2.225)–(2.227) turn into

$$\begin{aligned}&\rho_0\left\{\frac{\partial u^\alpha_0}{\partial t} + u^k_0 u^\alpha_{0,k}\right\} + g^{\alpha k}\frac{\partial p_0}{\partial x^k} = 0, \quad \alpha = 1, 2, \\ &\frac{\partial \rho_0}{\partial t} + \frac{1}{\sqrt{g}}\frac{\partial}{\partial x^k}(\sqrt{g}\rho_0 u^k_0) = 0, \\ &c_p\rho_0\left\{\frac{\partial T_0}{\partial t} + u^k_0\frac{\partial T_0}{\partial x^k}\right\} = E\left\{\frac{\partial p_0}{\partial t} + u^k_0\frac{\partial p_0}{\partial x^k}\right\}.\end{aligned} \tag{2.228}$$

This system is nothing but the equations of ideal gas dynamics (2.204) for u^1_0, u^2_0, ρ_0, p_0, T_0 rewritten in the curvilinear coordinates (the complete system is obtained from (2.228) by letting $\alpha = 1, 2, 3$).

There are particular cases where equations (2.225)–(2.227) take a simpler form. For instance, let the normal component of the outer velocity be identically equal to zero, $u_0^3 \equiv 0$, and let the coordinate system x^1, x^2, x^3 be monocline uniformly in x^3, i.e., $g^{\alpha 3} = 0$ for $\alpha = 1, 2$. Let us normalize the variable x^3 by defining S as a solution to the eikonal equation

$$|\nabla S| = 1, \quad S\big|_{\partial\Omega} = 0. \tag{2.229}$$

Then $g^{33} = 1$, and equations (2.225)–(2.227) become

$$\begin{aligned}
&\mathcal{R}\Big\{\frac{\partial \mathcal{U}^\alpha}{\partial t} + \mathcal{U}^\beta \mathcal{U}^\alpha_{,\beta'} + \frac{1}{\varepsilon}\mathcal{U}^3 \frac{\mathcal{U}^\alpha}{\partial \tau}\Big\} + g^{\alpha\beta}\frac{\partial p_0}{\partial x^\beta} = \frac{\partial}{\partial \tau}\Big(\eta_0 \frac{\partial \mathcal{U}^\alpha}{\partial \tau}\Big),\\
&\frac{\partial \mathcal{R}}{\partial t} + \frac{1}{\sqrt{g}}\frac{\partial}{\partial x^\alpha}(\sqrt{g}\mathcal{R}\mathcal{U}^\alpha) + \frac{1}{\varepsilon}\frac{\partial}{\partial \tau}(\mathcal{R}\mathcal{U}^3) = 0,\\
&c_p\mathcal{R}\Big\{\frac{\partial \mathcal{T}}{\partial t} + \mathcal{U}^\alpha \frac{\partial \mathcal{T}}{\partial x^\alpha} + \frac{1}{\varepsilon}\mathcal{U}^3 \frac{\partial \mathcal{T}}{\partial \tau}\Big\}\\
&\quad = \frac{\partial}{\partial \tau}\Big(\kappa_0 \frac{\partial \mathcal{T}}{\partial \tau}\Big) + \eta_0 g_{\alpha\beta}\frac{\partial \mathcal{U}^\alpha}{\partial \tau}\frac{\partial \mathcal{U}^\beta}{\partial \tau} + E\Big(\frac{\partial p_0}{\partial t} + \mathcal{U}^\alpha \frac{\partial p_0}{\partial x^\alpha}\Big).
\end{aligned} \tag{2.230}$$

This is the form of the Prandtl equations as they are presented in the book [**84**].

A form of (2.225)–(2.227) similar to (2.230) will be obtained if the Prandtl equations are considered only in the domain $0 < x^3 \le \mu$ such that

$$|u_0^3| \ll 1. \tag{2.231}$$

Obviously, for $\mu \ll 1$ our coordinate system is close to a monocline one, i.e., $|g^{\alpha 3}| \ll 1$. Letting $\tau = x^3/\varepsilon$, we return from the extended set of variables τ, x^1, x^2, x^3, t to the variables x^1, x^2, x^3, t. Then, neglecting the terms of order $O(|u_0^3| + |g^{\alpha 3}| + \varepsilon)$ in the first equation, the terms of order $O(\varepsilon)$ in the second, and the terms of order $O(|u_0^3| + \varepsilon)$ in the third, we arrive at the equations

$$\begin{aligned}
&\mathcal{R}\Big\{\frac{\partial \mathcal{U}^\alpha}{\partial t} + \mathcal{U}^\beta \mathcal{U}^\alpha_{,\beta'} + \mathcal{U}^3 \frac{\partial \mathcal{U}^\alpha}{\partial x^3}\Big\} + g^{\alpha\beta}\frac{\partial p_0}{\partial x^\beta} = \varepsilon^2 \frac{\partial}{\partial x^3}\Big(\eta_0 \frac{\partial \mathcal{U}^\alpha}{\partial x^3}\Big),\\
&\frac{\partial \mathcal{R}}{\partial t} + \frac{1}{\sqrt{g}}\frac{\partial}{\partial x^k}(\sqrt{g}\mathcal{R}\mathcal{U}^k) = 0,\\
&c_p\mathcal{R}\Big\{\frac{\partial \mathcal{T}}{\partial t} + \mathcal{U}^k \frac{\partial \mathcal{T}}{\partial x^k}\Big\} = \varepsilon^2 \frac{\partial}{\partial x^3}\Big(\kappa_0 \frac{\partial \mathcal{T}}{\partial x^3}\Big)\\
&\quad + \varepsilon^2 \eta_0 g_{\alpha\beta}\frac{\partial \mathcal{U}^\alpha}{\partial x^3}\frac{\partial \mathcal{U}^\beta}{\partial x^3} + E\Big(\frac{\partial p_0}{\partial t} + \mathcal{U}^\alpha \frac{\partial p_0}{\partial x^\alpha}\Big),
\end{aligned} \tag{2.232}$$

which are the equations (2.230) written in a different form.

All functions here depend on the variables $x^3/\varepsilon, x^1, x^2, x^3, t$, and we assume that (2.229) holds.

Equations (2.232) are more suitable for numerical computations than (2.230), since they contain fewer independent variables. It should be remembered, however, that for $x^3 \ge c\varepsilon^{1-\delta}$ (which corresponds to passage to the limit as $\tau \to \infty$) the equations (2.232) differ from the gas dynamics equations (2.228). Namely, the first equation (2.232) does not contain the terms

$$\rho_0 u_0^3 (2u_0^\beta \Gamma_{3\beta}^\alpha + u_0^3 \Gamma_{33}^\alpha) + g^{\alpha 3} \frac{\partial p_0}{\partial x^3},$$

and the third equation does not contain the term $Eu_0^3 \partial p_0 / \partial x^3$. In other words, outside a small neighborhood of the boundary the equations (2.232) are valid only approximately, up to $O(|u_0^3| + |g^{\alpha 3}|)$.

It remains to consider the relation (2.222) for $i = 3$. Making the same estimates as in deriving (2.225), one can easily see that (2.222) for $i = 3$ reduces, up to $O(\varepsilon)$, to the equality

$$g^{33} \frac{\partial P_1}{\partial \tau} = \Gamma_{\alpha\beta}^3 \{\rho_0 u_0^\alpha u_0^\beta - \mathcal{R}\mathcal{U}^\alpha \mathcal{U}^\beta\}, \tag{2.233}$$

which is just a different form of (2.213). Thus the basic model equations (2.212), which arise when constructing the asymptotics of boundary layer type, turn out to be the Prandtl equations. These equations have been thoroughly investigated, and in the literature one can find numerous examples of their application to specific problems [**84, 192, 193, 231**], which allows us to avoid a detailed discussion of the system (2.212).

As was pointed out above, there is a certain arbitrariness in the choice of the phase S. We will use it to consider one possible alternative form of the Prandtl equations. Namely, define the phase $S = S(x,t)$ as the solution to the problem

$$\frac{\partial S}{\partial t} + \langle u_0, \nabla S \rangle = 0, \quad S\big|_{t=0} = S^0(x), \tag{2.234}$$

where S^0 is some smooth function satisfying the conditions (2.203), and u^0 is the solution to the problem (2.204)–(2.206). This choice of S is admissible because for all $x \in \overline{\Omega}$ and $t > 0$ for which the velocity remains smooth, the solution to (2.234) satisfies conditions (2.203).

It is easily shown that whenever the component of velocity u_0 normal to $\partial\Omega$ either vanishes or remains negligibly small at all points (x,t) of interest and $S^0(x)$ is the distance to $\partial\Omega$ along the inner normal, the phase S remains equal to S^0. Thus in this case we are still in the same setting as above.

On the other hand, if the normal component of u_0 is not negligibly small, then the phase S will be moved along the flow u_0. Consider, for example, the two-dimensional case and let Ω be the half-plane $\{(x,y),\ y > 0,\ x \in \mathbb{R}^1\}$. Let the longitudinal component of u_0 be equal to some function $W(x,y,t)$, and let the vertical component be equal to $y^m t$ with $m > 1$. Set $S^0 = y$. Then the solution to (2.234) is given by the function

$$S = \frac{y}{(1 + \lambda t^2 y^{m-1})^{1/(m-1)}} \approx \frac{y}{1 + \lambda t^2 y^{m-1}/(m-1)}, \tag{2.235}$$

where $\lambda = 1/2(m-1)$. For finite t the difference between S and y is, of course, negligible. However, for large t the denominator of (2.235) means that the boundary layer is extended because of vertical displacement.

Thus by defining the phase as the solution to (2.234) we explain (partially) the change of the width of the boundary layer domain owing to the shift along u_0. This is also true for statinary problems (see, e.g., [**231**]), where the role of time is played by the distance z measured from the point $z = 0$. In this case the initial condition is set for $z = 0$, and the variation in the width of the boundary layer is determined (only partially) by the profile of the outer flow u_0.

The asymptotic solution for such a choice of S has the same form (2.201) as before and is constructed along the same lines. The only difference is that now we have to take into consideration the dependence of S on time. Omitting the intermediate calculations, we present the basic model equations which are obtained now instead of (2.212):

$$
\begin{gathered}
(\rho_0 + R_0)\Big\{\frac{\partial}{\partial t}(u_0 + U_0) + \langle u_0 + U_0, \nabla\rangle(u_0 + U_0) + \langle U_1, \nabla S\rangle\frac{\partial U_0}{\partial \tau}\Big\} \\
+ \nabla p_0 + \nabla S\frac{\partial P_1}{\partial \tau} = |\nabla S|^2\frac{\partial}{\partial \tau}\Big(\eta_0\frac{\partial U_0}{\partial \tau}\Big), \\
\frac{\partial}{\partial t}(\rho_0 + R_0) + \langle \nabla, (\rho_0 + R_0)(u_0 + U_0)\rangle + \frac{\partial}{\partial \tau}((\rho_0 + R_0)\langle \nabla S, U_1\rangle) = 0, \\
c_p(\rho_0 + R_0)\Big\{\frac{\partial}{\partial t}(T_0 + \Theta_0) + \langle u_0 + U_0, \nabla\rangle(T_0 + \Theta_0) + \langle U_1, \nabla S\rangle\frac{\partial \Theta_0}{\partial \tau}\Big\} \\
= |\nabla S|^2\frac{\partial}{\partial \tau}\Big(\kappa_0\frac{\partial \Theta_0}{\partial \tau}\Big) + |\nabla S|^2\eta_0\Big|\frac{\partial U_0}{\partial \tau}\Big|^2 + E\Big(\frac{\partial p_0}{\partial t} + \langle u_0 + U_0, \nabla p_0\rangle\Big).
\end{gathered}
\tag{2.236}
$$

Here u_0, ρ_0, p_0, and T_0 constitute the solution to the problem (2.204)–(2.206), the boundary layer velocity U_0 satisfies the geometric condition (2.209), and the pressure in the leading term is a slowly varying function, i.e., $P_0 = 0$.

As is easily seen, equations (2.236) differ from (2.212) only by absence of the terms $\varepsilon^{-1}\langle \nabla S, u_0\rangle$. This is what was to be expected, because the effect of these terms is accounted for by the choice of the phase S as the solution to equation (2.234). Thus we again obtain the Prandtl equations as the model equations for the leading term of the asymptotics.

In order to take the geometry of the boundary into account, rewrite equations (2.236) in curvilinear coordinates. Since now S depends on time, it is natural that the tangent coordinates x^1, x^2 should also depend on t. Namely, let the functions $x_0^1 = x_0^1(x')$ and $x_0^2 = x_0^2(x')$ determine the local coordinates on the surface $\partial\Omega$. Set $x_0^3 = S^0(x)$ and let x_0^1, x_0^2, x_0^3 determine a right-hand system of local coordinates. Define the functions $x^j = x^j(x, t)$ as the solution to the problem

$$
\frac{\partial x^j}{\partial t} + \langle u_0, \nabla x^j\rangle = 0, \quad x^j\big|_{t=0} = x_0^j(x), \qquad j = 1, 2, 3.
\tag{2.237}
$$

It will be shown below that if the function u_0 remains smooth for $t \le t_0$, then for such t the triple x^1, x^2, x^3 locally determines a right-hand coordinate system. The boundary $\partial\Omega$ then turns into the plane $x^3 = 0$, but the orthogonality between x^i and x^j (if it holds for $t = 0$) is not preserved for $t > 0$.

The passage to the coordinates x^1, x^2, x^3 in equation (2.236) is carried out in the same way as before. The main difference is that now an additional term of the form $(u_0^i + U_0^i)\partial e_i/\partial t$ appears. To calculate it we will use the formula

$$\begin{aligned}(2.238)\quad &\Big(\frac{\partial}{\partial t} + \langle u_0, \nabla\rangle\Big)F(x^1(x,t), x^2(x,t), x^3(x,t), t) \\ &= \Big\{e_i\Big(\frac{\partial F^i(x^1,x^2,x^3,t)}{\partial t} + F^j(x^1,x^2,x^3,t)u^i_{0,j}(x^1,x^2,x^3,t)\Big)\Big\}\Big|_{x^k=x^k(x,t)},\end{aligned}$$

which is valid for any vector $F = e_iF^i$. We will demonstrate the derivation of this formula later on.

We will also use the formula for differentiation of a scalar function f:

$$(2.239)\quad \begin{aligned}&\Big(\frac{\partial}{\partial t} + \langle u_0, \nabla\rangle\Big)f(x^1(x,t),x^2(x,t), x^3(x,t), t) \\ &\qquad = \frac{\partial f(x^1,x^2,x^3,t)}{\partial t}\Big|_{x^k=x^k(x,t)},\end{aligned}$$

which is a consequence of the fact that $x^k(x,t)$ is constant along the vector field u_0.

Omitting technical details, we state the analog of equations (2.225)–(2.227) for functions $\mathcal{U}^\alpha = u_0^\alpha + U_0^\alpha$, $\alpha = 1,2$, $U^3 = \langle\nabla S, U_1\rangle$, and $\mathcal{R} = \rho_0 + R_0$, $\mathcal{T} = T_0 + \Theta_0$, which depend on the variables τ, x^1, x^2, x^3, t:

$$(2.240)\quad\begin{aligned}&\mathcal{R}\Big\{\frac{\partial\mathcal{U}^\alpha}{\partial t} + \mathcal{U}^\beta\mathcal{U}^\alpha_{,\beta'} + \mathcal{U}^3\frac{\partial\mathcal{U}^\alpha}{\partial\tau} - L_{u_0}\mathcal{U}^\alpha\Big\} \\ &\quad + g^{\alpha k}\frac{\partial p_0}{\partial x^k} + g^{\alpha 3}\frac{\partial P_1}{\partial\tau} + \rho_0u_0^3(u^\alpha_{0,3} + \Gamma^\alpha_{3\beta}u_0^\beta) \\ &= g^{33}\frac{\partial}{\partial\tau}\Big(\eta_0\frac{\partial\mathcal{U}^\alpha}{\partial\tau}\Big),\quad \alpha = 1,2, \\ &\frac{\partial\mathcal{R}}{\partial t} + \frac{1}{\sqrt{g}}\frac{\partial}{\partial x^\alpha}(\sqrt{g}\mathcal{R}\mathcal{U}^\alpha) + \frac{\partial}{\partial\tau}(\mathcal{R}\mathcal{U}^3) = 0, \\ &c_p\mathcal{R}\Big\{\frac{\partial\mathcal{T}}{\partial t} + \mathcal{U}^\alpha\frac{\partial\mathcal{T}}{\partial x^\alpha} + \mathcal{U}^3\frac{\partial\mathcal{T}}{\partial\tau}\Big\} \\ &= g^{33}\frac{\partial}{\partial\tau}\Big(\kappa_0\frac{\partial\mathcal{T}}{\partial\tau}\Big) + g^{33}g_{\alpha\beta}\eta_0\frac{\partial\mathcal{U}^\alpha}{\partial\tau}\frac{\partial\mathcal{U}^\beta}{\partial\tau} + E\Big(\frac{\partial p_0}{\partial t} + \mathcal{U}^\alpha\frac{\partial p_0}{\partial x^\alpha}\Big).\end{aligned}$$

Here $\partial P_1/\partial\tau$ is a function of $\mathcal{U}$ and $\mathcal{R}$ which is obtained from (2.233), and $L_{u_0}\mathcal{U}^\alpha$ means the Lie derivative along the vector field u_0, i.e.,

$$(2.241)\quad L_{u_0}\mathcal{U}^\alpha = u_0^\beta\frac{\partial\mathcal{U}^\alpha}{\partial x^\beta} - \mathcal{U}^\beta\frac{\partial u_0^\alpha}{\partial x^\beta},$$

with summation over the index β taking only values 1 and 2.

It is easily seen that equations (2.240) differ from (2.225)–(2.227) by terms $\mathcal{R}L_{u_0}\mathcal{U}^\alpha$ and $g^{\alpha 3}\partial P_1/\partial\tau$ which appear in the equation of motion and by absence of drift terms

$$\frac{u_0^3}{\sqrt{g}}\frac{\partial}{\partial x^3}(\sqrt{g}\rho_0) + \mathcal{R}\frac{\partial u_0^3}{\partial x^3}$$

in the equation of continuity and

$$u_0^3\left(E\frac{\partial p_0}{\partial x^3} - c_p\rho_0\frac{\partial T_0}{\partial x^3}\right)$$

in the equation of energy. Moreover, in (2.240) the normal component of velocity $\mathcal{U}^3$, unlike (2.225)–(2.227), is of purely boundary layer nature, since it does not involve u_0^3. Note that (2.233) and (2.241) do not depend on the normal component u_0^3 of the external flow and that the following equality holds:

$$g^{\alpha k}\frac{\partial p_0}{\partial x^k} + \rho_0 u_0^3(u_{0,3}^\alpha + \Gamma_{3\beta}^\alpha u_0^\beta) = -\rho_0\left(\frac{\partial u_0^\alpha}{\partial t} + u_0^\beta u_{0,\beta'}^\alpha\right).$$

Hence equations (2.240) involve neither the normal component of u_0 nor differentiation with respect to the slow normal variable x^3.

Thus defining the curvilinear coordinates as the solution to (2.237) allows us to take into account the drift along the normal component of the flow. However in (2.240) there remain terms responsible for the drift along the tangent components. Note that we do not discuss here the diffusion effects, which, as is well known, substantially affect the width of the boundary layer as well. Moreover, it is to be stressed that equations (2.236) and (2.240) are again Prandtl equations, but written in a specific coordinate system.

It remains to justify the use of the solution to (2.237) for new coordinates, and to derive formula (2.238).

For sufficiently small t the first claim immediately follows from our assumption about smoothness of the function u_0 for $t \in [0, t_0]$. Indeed, if $u_0 \in C^\infty$ then the solutions of the linear equations (2.237) exist for $t \leq t_0$ and are smooth functions.

For $t = 0$ the Jacobian $\sqrt{g}|_{t=0}$ does not vanish by the choice of the functions $x_0^j(x)$, $j = 1, 2, 3$. Hence the inequality $\sqrt{g} > 0$ will continue to hold for $t \leq t_1$ with $t_1 \leq t_0$ small enough.

Let us derive formula (2.238) for $t \leq t_1$.

To this end we first calculate the derivatives of the basis vectors. Applying the operator ∇ to (2.237), we obtain the equalities

$$\frac{d}{dt}e^j = -e^i u_{0,i}^j, \quad j = 1, 2, 3.$$

Further, since in the three-dimensionl space $e_1 = \sqrt{g}e^2 \times e^3$, we have

$$\frac{de_1}{dt} = \frac{e_1}{\sqrt{g}}\frac{\partial}{\partial t}\sqrt{g} - \sqrt{g}\{e^i \times e^3 u_{0,i}^2 + e^2 \times e^i u_{0,i}^3\}.$$

After some simple transformations and calculation of derivatives of e_2 and e_3 we obtain

$$\frac{de_j}{dt} = e_j\left(\frac{1}{\sqrt{g}}\frac{\partial}{\partial t}\sqrt{g} - \operatorname{div} u_0\right) + e_i u_{0,j}^i, \quad j = 1, 2, 3. \tag{2.242}$$

The next step consists in proving the equality

$$\frac{\partial}{\partial t}\sqrt{g} = \sqrt{g}\operatorname{div} u_0, \tag{2.243}$$

which is an analog of the Liouville theorem as applied to equations (2.237). For this we calculate the derivative of a coefficient of the metric tensor g_{ik}. Using (2.242), we have

$$\begin{aligned}\frac{d}{dt}g_{ik}(x^1(x,t),&x^2(x,t),x^3(x,t),t)\\ &= \left\{2\mu g_{ik} + g_{kl}u^l_{0,i} + g_{il}u^l_{0,k}\right\}\Big|_{x^k=x^k(x,t)},\end{aligned} \tag{2.244}$$

where

$$\mu = \frac{1}{\sqrt{g}}\frac{\partial}{\partial t}\sqrt{g} - \operatorname{div} u_0,$$

and differentiation in the right-hand side of (2.244) is performed with respect to independent variables x^1, x^2, x^3, and t.

Now (2.244) and (2.239) imply the equality

$$\begin{aligned}\frac{1}{\sqrt{g}}\frac{\partial}{\partial t}\sqrt{g} = 3\mu + \frac{1}{g}\{u^l_{0,n_1}&[g_{n_1l}g_{n_2n_3}g_{n_3n_3} + g_{n_2l}g_{n_1n_3}g_{n_2n_3}\\ + g_{n_3l}g_{n_1n_2}g_{n_2n_2} - g_{n_3l}g_{n_2n_2}g_{n_1n_3} &- g_{n_1l}g^2_{n_3n_2} - g_{n_2l}g_{n_1n_2}g_{n_3n_3}]\},\end{aligned} \tag{2.245}$$

where the summation extends over indices l, n_k taking values $1, 2, 3$, with tuples (n_1, n_2, n_3) obtainable by cyclic permutations of the tuple $(1, 2, 3)$. With some calculations, one can show that the expression in braces in the right-hand side of (2.245) equals $gu^l_{0,l}$. Then (2.245) combined with this fact implies (2.243). In its turn, (2.243) enables us to rewrite (2.242) in the final form

$$\frac{de_j}{dt} = e_i u^i_{0,j}, \quad j = 1, 2, 3. \tag{2.246}$$

Now, substituting the vector F written as the sum $e_i F^i$ and using (2.239) and (2.246), we easily obtain formula (2.238).

Finally, observe that by (2.243) the Jacobian $\sqrt{g}$ does not vanish for all $t_1 \leq t_0$. This means that $t_1 = t_0$, i.e., the functions $x^j = x^j(x,t)$ determine a curvilinear coordinate system such that the formula (2.238) remains valid on the entire time interval $[0, t_0]$ on which the "background" u_0 remains smooth.

CHAPTER 3

Asymptotic Solutions of Some Pseudodifferential Equations and Dynamical Systems with Small Dispersion

In this chapter we continue the treatment of problems which contain new mathematical aspects as compared to those considered before. Namely, we will show that the method developed above can also be used to construct asymptotic soliton-type solutions of pseudodifferential equations and dynamical systems. This will be exemplified by the Whitham equation and by the chains of the Toda and Langmuir equations. Equations of the latter type will also be reduced to pseudodifferential equations.

3.1. Propagation of interior soliton-type waves in a fluid

To describe the of characteristic effects on the water, such as cusps and overturning of waves, Whitham [**262**] proposed the integrodifferential equation

$$\frac{\partial u}{\partial t} + (\rho_1 + \rho_2 u)\frac{\partial u}{\partial x} - \frac{\partial}{\partial x}\widehat{\mathcal{A}}u = 0. \tag{3.1}$$

Here $u = u(x,t)$, $x \in \mathbb{R}^1$, $t > 0$, is the unknown function describing the amplitude of the wave, $\rho_1, \rho_2 \neq 0$ are constants, and the operator $\widehat{\mathcal{A}}$ acts by the formula

$$\widehat{\mathcal{A}}u = \int_{-\infty}^{\infty} A(x-s)u(s,t)\,ds,$$

where the kernel $A(x)$ may be a fairly general function. By selecting various dispersion relationships, i.e., various Fourier images of the kernel $A(x)$, equation (3.1) yields the Burgers, Korteweg–de Vries, Benjamin–Ono and other equations. As a particular case of the Whitham equation (3.1), one obtains the so-called "intermediate", or Joseph equation [**93**]:

$$\begin{aligned} &\frac{\partial u}{\partial t} + (\rho_1 + \rho_2 u)\frac{\partial u}{\partial x} \\ &\quad + \rho_3 V.P.\int_{-\infty}^{\infty}\left\{\coth\frac{\pi(x-y)}{\varepsilon H} - \operatorname{sign}(x-y)\right\}\frac{\partial^2 u(y,t)}{\partial y^2}\,dy = 0, \end{aligned} \tag{3.2}$$

which describes inner waves in a stratified fluid. Here the symbol $V.P.$ indicates, as usual, that the integral is meant in the sense of principal value, $\rho_3 \neq 0$ is a constant, $H > 0$ is a constant characterizing the depth of nonperturbed fluid, and

$\varepsilon > 0$ is the parameter characterizing the magnitude of dispersion. We will treat (3.1) for small dispersion, i.e., assuming that ε is a small parameter. Moreover, we will assume for definiteness that $\rho_2 > 0$, $\rho_3 > 0$ (which can always be achieved by a suitable change of variables).

To simplify the subsequent calculations, we will employ another representation of the integral term in the left-hand side of (3.2). Namely, it is not hard to check that (3.2) can be rewritten in the form (3.1) with a pseudodifferential operator $\widehat{\mathcal{A}} = \widehat{\mathcal{A}}(-i\varepsilon\partial/\partial x)$ which acts by the formula

$$\widehat{\mathcal{A}}\Big(-i\varepsilon\frac{\partial}{\partial x}\Big)u(x,t) = \frac{1}{2\pi}\int_{-\infty}^{\infty}\int_{-\infty}^{\infty} e^{i\xi(x'-x)}a(\varepsilon\xi)u(x',t)\,dx'\,d\xi, \tag{3.3}$$

where $i = \sqrt{-1}$ and the symbol $a(\xi)$ of the operator $\widehat{\mathcal{A}}(-i\partial/\partial x)$ has the form

$$a(\xi) = \rho_3\xi H\{\coth(\xi H) - (\xi H)^{-1}\}.$$

Of course, (3.3) is nothing but the standard representation of the pseudodifferential operator

$$\widehat{\mathcal{A}}\Big(-i\frac{\partial}{\partial x}\Big)u = \mathcal{F}^{-1}_{\xi\to x}\big[a(\xi)\mathcal{F}_{x\to\xi}[u(x)]\big],$$

where $\mathcal{F}_{x\to\xi}$ stands for the Fourier transform and $\mathcal{F}^{-1}_{\xi\to x}$ for the inverse Fourier transform (see, e.g., [**241, 246**]).

Solvability of the class (3.1) of integrodifferential equations, which comprises (3.2) for $\varepsilon = \text{const} > 0$, was established in [**180, 181**]. Moreover, the exact solution of equation (3.2), which describes the motion of a solitary wave in the case of a constant background, is known [**168**]. Our aim is to construct an asymptotic solution describing the dynamics of a distorted solitary wave in the case of a smooth variable background.

First we state the formula for the leading term of the asymptotics:

$$\begin{aligned} u = u_0(x,t) + A(t)\{&\cosh^2(\beta(t)\tau) \\ &+ \sigma^2(t)\sinh^2(\beta(t)\tau)\}^{-1}\big|_{\tau=(x-\varphi(t)/\varepsilon)+\varphi_1(t)} + O(\varepsilon) \end{aligned} \tag{3.4}$$

(see [**161, 162**]), and discuss the result thus obtained.

In (3.4) u_0 is the background perturbation of the fluid, which is defined as the solution of the degenerate equation (3.2) (as $\varepsilon \to 0$),

$$\frac{\partial u_0}{\partial t} + (\rho_1 + \rho_2 u_0)\frac{\partial u_0}{\partial x} = 0. \tag{3.5}$$

We assume that the solution to the Hopf equation (3.5) remains smooth for $t \in [0,T]$ and that $u_0(x,0)$ together with all its derivatives with respect to x tends to zero as $|x| \to \infty$ faster than any positive power of $1/|x|$. Clearly, the solution $u_0(x,t)$ possesses the same property for all $t \in [0,T]$.

Further, we find the function $\varphi(t)$, which determines the motion of the front $\Gamma_t = \{x,\ x = \varphi(t)\}$ of the soliton, as well as the auxiliary function $\delta = \delta(t)$, as the solution to the equations

$$\frac{\partial\varphi}{\partial t} = \rho_1 + \rho_2 u_0(\varphi,t) + \rho_3(1 - \delta\cot\delta), \tag{3.6}$$

$$\log\Big(\frac{\sin\delta}{\delta}\Big) + \delta\cot\delta - \frac{\rho_2}{\rho_3}u_0(\varphi,t) = E, \tag{3.7}$$

where the constant $E > 0$ is determined by the initial conditions for (3.2). It is assumed that $0 < \delta(t) < \pi$ for all $t \in [0, T]$.

Finally, the functions $A(t)$, $\sigma(t)$, and $\beta(t)$ involved in (3.4) can be expressed in terms of the solution of (3.6), (3.7) as

$$\sigma = \tan\left(\frac{\delta}{2}\right), \quad A = 2\frac{\rho_3}{\rho_2}\delta\sigma, \quad \beta = \frac{\delta}{2H}, \tag{3.8}$$

and the phase shift $\varphi_1(t)$ is a smooth function to be determined later.

As is easily seen, formula (3.4) describes a distorted solitary wave with amplitude $A(t)$ and velocity $d\varphi/dt$ moving over the smooth background $u_0(x,t)$. Clearly, if $u_0 = \text{const}$, then (3.7) implies that δ is also a constant. In this case φ is a linear function of t, and the first two terms of (3.4) form an exact solution of (3.2).

Further, differentiating (3.6), (3.7) and using (3.5), we can obtain the formula for the acceleration $d^2\varphi/dt^2$ of the wave front:

$$\frac{d^2\varphi}{dt^2} = -\rho_2\rho_3 \frac{(1-\delta\cot\delta)^2}{\delta^2 + (1-\delta\cot\delta)^2} \frac{\partial u_0}{\partial x}\bigg|_{x=\varphi(t)}.$$

We see from this formula that the wave slows down when moving "upwards" (where the background u_0 increases) and accelerates when moving "downwards".

Finally, for small depth $H \ll 1$, expanding the functions involved in (3.4) and (3.6)–(3.8) in H, it is not hard to see that equations (3.6), (3.7) become, up to $O(H^2)$,

$$\begin{aligned}
&\frac{d\varphi}{dt} = \rho_1 + \rho_2 u_0(\varphi, t) + \frac{\rho_2}{3} A_1(t),\\
&A_1(t) + 2u_0(\varphi, t) = \text{const},
\end{aligned}$$

and the soliton-type component of the solution (3.4) within the same accuracy is given by the formula

$$A_1(t)\cosh^{-2}\left(\frac{1}{2H}\sqrt{A_1\frac{\rho_2}{\rho_3}}\,\tau\right).$$

Thus for small depth the soliton-type solution of (3.2) coincides up to $O(H^2)$ with the similar solution to the Korteweg–de Vries equation with constant coefficients (see §1.1).

Now we construct the asymptotic solution to (3.2). According to our method we write the solution in the form of the standard expansion

$$u = u_0(x,t) + V_0(\tau, t) + \varepsilon\big(u_1(x,t) + V_1(\tau, x, t)\big) + \dots, \tag{3.9}$$

where, as usual, $\tau = (x - \varphi(t))/\varepsilon$, and the $u_i(x,t)$, $V_i(\tau,x,t)$ are smooth functions with $V_0 \in S$, i.e., $V_0 \to 0$ as $\tau \to \pm\infty$, while $V_j \in \mathcal{H}$, i.e., $V_j(\tau,x,t) \to 0$ as $\tau \to \infty$ and $V_j(\tau,x,t) \to V_j^-(x,t)$ as $\tau \to -\infty$, $j = 1, 2, \dots, N$. Moreover, we will assume that the u_i belong to the Schwartz space uniformly in t.

We will restrict ourselves to construction of the leading term of asymptotics. However, as in previous problems, in order to derive the Hugoniot type conditions (3.6), (3.7) and to determine the phase shift φ_1 we will have to analyze additional conditions which are inferred from construction of higher order terms. Hence we will consider only three terms in the expansion (3.9), i.e., we take $N = 2$.

First we determine the background perturbations $u_i(x,t)$. For that, consider the Taylor expansion of the symbol $a(\varepsilon\xi)$ about the point $\varepsilon = 0$:

$$a(\varepsilon\xi) = (\varepsilon\xi)^2 a_2 + (\varepsilon\xi)^3 \mathcal{A}_3(\varepsilon\xi),$$

where $a_2 = \frac{1}{3}\rho_3 H^2$ and $\mathcal{A}_3$ is the remainder term.

Denote

$$U_2 = \sum_{k=0}^{2} \varepsilon^k u_k(x,t). \tag{3.11}$$

Putting U_2 into (3.2) and using the expansion (3.10), we obtain the equation

$$\frac{\partial U_2}{\partial t} + (\rho_1 + \rho_2 U_2)\frac{\partial U_2}{\partial x} = -\varepsilon^2 a_2 \frac{\partial^3 U_2}{\partial x^3} + i\varepsilon^3\Big(-i\frac{\partial}{\partial x}\Big)^4 \widehat{\mathcal{A}}_3\Big(-i\varepsilon\frac{\partial}{\partial x}\Big)U_2, \tag{3.12}$$

where the operator $\widehat{\mathcal{A}}_3$ with symbol $\mathcal{A}_3$ acts by the formula similar to (3.3).

Now, substituting (3.11) into (3.12) and equating to zero the coefficients of ε^j, $j = 0,1,2$, we obtain equation (3.5) for the determination of u_0, and the nonhomogeneous linearized Hopf equations

$$\frac{\partial u_k}{\partial t} + \frac{\partial}{\partial x}\{(\rho_1 + \rho_2 u_0)u_k\} = f_k(x,t), \quad k = 1,2, \tag{3.13}$$

for the determination of u_k, $k = 1,2$. Here $f_1 = 0$, and f_2 is a polynomial in u_0, u_1 and their derivatives.

We will assume that all functions $u_k(x,0)$ are infinitely differentiable and tend to zero exponentially as $|x| \to \infty$. Then, in view of our assumption on u_0, the functions $u_k(x,t)$ for $t \in [0,T]$ remain smooth and exponentially decreasing at infinity.

Next, it is clear from the examples considered before that the process of obtaining the higher order rapidly varying terms of the expansion (3.9) gives rise to functions of the form

$$V_k(\tau,x,t) = V_k^-(x,t)\eta(\tau,t) + \omega_k(\tau,t), \quad k = 1,2, \tag{3.14}$$

where $\omega_k \to 0$ as $\tau \to \pm\infty$, while $\eta \to 1$ as $\tau \to -\infty$ and $\eta \to 0$ as $\tau \to \infty$.

In turn, the tail perturbations $V_k^-(x,t)$ are a sort of background functions, which, along with u_k, can be determined by the regular theory of perturbations. Clearly, by the same arguments we obtain equations for V_k^- similar to (3.13):

$$\frac{\partial V_k^-}{\partial t} + \frac{\partial}{\partial x}\{(\rho_1 + \rho_2 u_0)V_k^-\} = f_k^-(x,t), \quad k = 1,2, \tag{3.15}$$

which are considered, however, for $x < \varphi(t)$. We will assume that the $V_k^-(x,0) \in C^\infty$ and tend to zero exponentially fast as $x \to -\infty$.

Now we construct the soliton components of the solution. Substituting (3.9) into (3.2), taking the Taylor expansion in x about the point $x = \varphi(t)$, and setting the coefficient of ε^{-1} equal to zero, we obtain the following equation for the determination of $V_0(\tau, t)$:

$$-\varphi_t \frac{\partial V_0}{\partial \tau} + \big(\rho_1 + \rho_2(u_0(\varphi, t) + V_0)\big)\frac{\partial V_0}{\partial \tau} - \widehat{\mathcal{A}}_\tau \frac{\partial}{\partial \tau} V_0 = 0, \tag{3.16}$$

where the operator $\widehat{\mathcal{A}}_\tau$ acts by the formula

$$\widehat{\mathcal{A}}_\tau \frac{\partial}{\partial \tau} V(\tau, t) = \frac{1}{2\pi}\int_{-\infty}^{\infty}\int_{-\infty}^{\infty} e^{i\xi(\tau'-\tau)} a(\xi) \frac{\partial}{\partial \tau'} V(\tau', t)\, d\tau'\, d\xi.$$

Since (3.16) can be regarded as a stationary Whitham equation (3.2) (with coefficient $\rho_1 + \rho_2 u_0(\varphi, t) - \varphi_t$ in place of ρ_1) and $V_0 \to 0$ as $\tau \to \pm\infty$, using the results of [**168**] we can easily find the solution of (3.16) in an explicit form. Namely, we obtain the formula

$$V_0 = A\big\{\cosh^2 \beta(\tau + \varphi_1) + \sigma^2 \sinh^2 \beta(\tau + \varphi_1)\big\}^{-1}, \tag{3.17}$$

as well as the first Hugoniot type condition (3.6). As usual, φ_1 denotes here the "constant" of integration and A, β, and σ are related by (3.8). Obviously, the formulas given above are well defined only for $\delta \in (0, \pi)$, and for these δ the amplitude A will be positive.

Next, we use the fact that the functions $V_k^-(x, t)$ satisfy the equations (3.15). Then, on carrying out the standard procedure, we obtain the following equations for the determination of $\check{V}_k = V_k(\tau, x, t)|_{x=\varphi(t)}$:

$$\frac{\partial}{\partial \tau}\{(\rho_1 - \varphi_t + \rho_2(u_0(\varphi, t) + V_0))\check{V}_k\} - \widehat{\mathcal{A}}_\tau \frac{\partial}{\partial \tau}\check{V}_k = \mathcal{F}_k(\tau, t), \quad k = 1, 2, \tag{3.18}$$

where the $\mathcal{F}_k \in \mathcal{S}$ are polynomials in $\tau, u_0, V_0, \dots, V_{k-1}, u_k$, and their derivatives for $t = \varphi$.

Using (3.14), we can rewrite (3.18) in the form

$$\frac{\partial}{\partial \tau}\{(\rho_1 - \varphi_t + \rho_2(\check{u}_0 + V_0))\omega_k - \widehat{\mathcal{A}}_\tau \omega_k\} = \mathcal{F}_k^1(\tau, t), \tag{3.19}$$

where

$$\mathcal{F}_k^1 = \mathcal{F}_k - \check{V}_k^- \frac{\partial}{\partial \tau}\{(\rho_1 - \varphi_t + \rho_2(\check{u}_0 + V_0))\eta\} + \check{V}_k^- \widehat{\mathcal{A}}_\tau \frac{\partial \eta}{\partial \tau}. \tag{3.20}$$

Obviously, $\mathcal{F}_k^1 \in \mathcal{S}$ if $\mathcal{F}_k \in \mathcal{S}$.

On integrating (3.19) with respect to τ and choosing the "constant" of integration to fulfill the condition $\omega_k \to 0$ as $\tau \to \infty$, we obtain

$$\big(-\rho_3(1 - \delta \cot \delta) + \rho_2 V_0\big)\omega_k - \widehat{\mathcal{A}}_\tau \omega_k = -\int_\tau^\infty \mathcal{F}_k^1(\tau', t)\, d\tau'. \tag{3.21}$$

Observe that the operator in the left-hand side of (3.21) is symmetric on functions from $\mathcal{S}$ and $\partial V_0/\partial\tau$ belongs to its kernel. Hence we obtain a necessary condition for solvability of (3.21):

$$\int_{-\infty}^{\infty} \mathcal{F}_k^1(\tau,t)V_0(\tau,t)\,d\tau = 0, \quad k=1,2. \tag{3.22}$$

Next we note that the operator in the left-hand side of (3.21) maps any function in $\mathcal{S}$ into $\mathcal{S}$ (see, e.g., [**241, 246**]). This immediately implies a second necessary condition for solvability of (3.21) in $\mathcal{S}$:

$$\int_{-\infty}^{\infty} \mathcal{F}_k^1(\tau,t)\,d\tau = 0, \tag{3.23}$$

since for any function $\mathcal{F}_k^1 \in \mathcal{S}$ (3.23) implies that the right-hand side of (3.21) belongs to $\mathcal{S}$.

Let us transform conditions (3.22) and (3.23). First of all, using (3.20) and integrating by parts, we can easily obtain the formula

$$\begin{aligned}\int_{-\infty}^{\infty} V_0\mathcal{F}_k^1\,d\tau &= \int_{-\infty}^{\infty} V_0\mathcal{F}_k\,d\tau \\ &\quad + \check{V}_k^- \int_{-\infty}^{\infty} \eta\Big\{(\rho_1 - \varphi_t + \rho_2(\check{u}_0 + V_0))\frac{\partial V_0}{\partial\tau} - \mathcal{A}_\tau\frac{\partial V_0}{\partial\tau}\Big\}\,d\tau.\end{aligned} \tag{3.24}$$

Since V_0 satisfies (3.16), the second integral in the right-hand side of (3.24) vanishes, and we see that the conditions (3.22) and

$$\int_{-\infty}^{\infty} V_0\mathcal{F}_k\,d\tau = 0 \tag{3.25}$$

are equivalent.

Consider condition (3.23). In view of the arbitrariness in determination of the function $\eta(\tau,t)$ we can choose it so that

$$\frac{\partial\eta}{\partial\tau} = cV_0, \quad \eta\to 0 \quad \text{as} \quad \tau\to\infty,$$

where the constant $c = -\sigma\beta/\delta A$ is chosen from the condition $\eta\to 1$ as $\tau\to-\infty$. Then (3.23) and (3.20) imply the equality

$$\int_{-\infty}^{\infty} \mathcal{F}_k\,d\tau + \check{V}_k^-\Big\{\rho_1 - \varphi_t + \rho_2\check{u}_0 + c\int_{-\infty}^{\infty} \mathcal{A}_\tau V_0\,d\tau\Big\} = 0. \tag{3.26}$$

To evaluate the last integral in (3.26) we can use the formula

$$\int_{-\infty}^{\infty} \mathcal{A}_\tau V_0\,d\tau = (\rho_1 - \varphi_t + \rho_2\check{u}_0)\int_{-\infty}^{\infty} V_0\,d\tau + \frac{\rho_2}{2}\int_{-\infty}^{\infty} V_0^2\,d\tau, \tag{3.27}$$

which follows from (3.16) on integrating twice. One can easily calculate that

$$\int_{-\infty}^{\infty} V_0\,d\tau = 4H\frac{\rho_3}{\rho_2}\delta, \quad \int_{-\infty}^{\infty} V_0^2\,d\tau = 8H\Big(\frac{\rho_3}{\rho_2}\Big)^2(1-\delta\cot\delta)\delta.$$

Now we obtain from this formula and (3.6) that the right-hand side of (3.27) vanishes. Hence (3.26) and (3.6) entail the equality

$$\check{V}_k^- = \frac{1}{\rho_3(1-\delta\cot\delta)}\int_{-\infty}^{\infty}\mathcal{F}_k(\tau,t)\,d\tau. \tag{3.28}$$

Thus we have obtained two necessary conditions for solvability of equation (3.21) in $\mathcal{S}$ and hence for solvability of equation (3.18) in $\mathcal{H}$. We reduced the first of them, namely (3.22), to the form (3.25), which precisely coincides with the corresponding orthogonality condition (1.101) for the Korteweg–de Vries equation. It is obvious that we can satisfy this condition, because (3.25) for $k=1$ involves the phase $\varphi(t)$, which is unspecified so far, and (3.25) for $k=2$ contains the "constant" of integration $\varphi_1(t)$. Let us also mention that the solution of (3.18) surely contains an arbitrary "constant" of integration $\varphi_{k+1}(t)$, since the function $\varphi_{k+1}V_{0\tau}$ satisfies the homogeneous equation corresponding to (3.18). The second necessary condition, namely (3.23), was reduced to (3.28). We can satisfy this condition using the arbitrariness in specifying the boundary value of the function $V_k^-(x,t)$ on the curve $x=\varphi(t)$. Formula (1.103), which is similar to (3.28), also appeared when we constructed the soliton-type asymptotics of the Korteweg–de Vries equation.

However, in contrast with the problems treated before, we cannot guarantee that conditions (3.23) and (3.25) will be sufficient.

Hence we adopt the assumption that under conditions (3.22) and (3.23) the equation (3.21) with $\mathcal{F}_k$ decreasing exponentially fast as $\tau\to\pm\infty$ has a bounded solution decreasing to zero as $t\to\infty$. In this respect the technique which enables us to find an equation for the phase shift φ_1 without determination of V_1 (see below and §1.1) becomes extremely important. Note also that when finding φ_1 we will obtain a partial validation of our hypothesis about sufficiency of (3.23) and (3.25) for solvability of (3.21) in the required class of functions. Namely, we will establish convergence of the integrals

$$\int_{-\infty}^{\infty} V_0V_1\,d\tau \quad\text{and}\quad \int_{-\infty}^{\infty} V_0V_1V_{1\tau}\,d\tau.$$

Now we are going to derive an equation complementing (3.6) to a complete system. To this end, consider the necessary condition (3.25) for $k=1$. The function $\mathcal{F}_1$ is given by the formula

$$\mathcal{F}_1 = -\left\{\frac{\partial V_0}{\partial t} + \rho_2\Big(V_0+\tau\frac{\partial V_0}{\partial\tau}\Big)\frac{\partial u_0}{\partial x} + \rho_2 u_1\frac{\partial V_0}{\partial \tau}\right\}\Bigg|_{t=\varphi}. \tag{3.29}$$

Substituting this expression into (3.25) and evaluating the integral, we obtain the equation

$$\frac{d}{dt}(\beta\alpha) + \frac{\rho_2}{\rho_3}\beta\frac{du_0(\varphi,t)}{dt} = 0, \tag{3.30}$$

where $\alpha = 1-\delta\cot\delta$. Since $\beta=\delta/2H$, this equation can be easily integrated and we obtain the second Hugoniot condition (3.7).

Next, one can easily calculate, using (3.28) with $k=1$ and (3.29), that

$$\check{V}_1^-(t) = -4\frac{H}{\alpha\rho_2}\frac{d\delta}{dt}.$$

Now we can complete the construction of the tail perturbations $V_k^-(x,t)$. Namely, we define these functions as the solution to equations (3.15) in the domain $D^- = \{(x,t),\ t>0,\ x<\varphi(t)\}$ completed with the boundary condition

$$V_k^-\big|_{x=\varphi(t)} = \check{V}_k^-(t) \tag{3.31}$$

and the natural initial condition for $t=0$, $x<\varphi(0)$. It is easily seen that for $\delta\in(0,\pi)$ the expression $1-\delta\cot\delta$ is strictly positive. This fact combined with equation (3.6) means that we again obtained a "supersonic" velocity of the soliton motion; hence the problem (3.15), (3.29) is well-posed. Finally, we define an extension of V_k^- as the solution of (3.15) for $\varphi(t)<x\le\varphi(t)+\sqrt{\varepsilon}$ and as an arbitrary smooth function tending to zero exponentially fast as $x\to\infty$ for $x\ge\varphi(t)+\sqrt{\varepsilon}$. Then we obtain the representation (3.14) for $V_k(\tau,x,t)$.

It remains to find the phase shift $\varphi_1(t)$. For that, consider the necessary condition (3.25) for $k=2$. After some algebra we obtain the formula

$$\begin{aligned}\mathcal{F}_2 = -\Bigg\{&\frac{\partial V_1}{\partial t}+\rho_2V_1\frac{\partial u_0}{\partial x}-\left(\frac{\partial V_1^-}{\partial t}+\rho_2V_1^-\frac{\partial u_0}{\partial x}\right)\eta+\rho_2(u_1+V_1)\frac{\partial V_1}{\partial\tau}\\ &+\rho_2V_0\left(\frac{\partial u_1}{\partial x}+\frac{\partial V_1}{\partial x}\right)-\tau\rho_3(1-\delta\cot\delta)\frac{\partial^2V_1}{\partial\tau\partial x}\\ &+\tau\rho_2\left(\frac{\partial u_0}{\partial x}\frac{\partial V_1}{\partial\tau}+\frac{\partial}{\partial\tau}\left(V_0\frac{\partial V_1}{\partial x}\right)+\frac{\partial u_1}{\partial x}\frac{\partial V_0}{\partial\tau}+\frac{\partial^2u_0}{\partial x^2}V_0\right)\\ &-\widehat{\mathcal{A}}_{1\tau}\frac{\partial^2V_1}{\partial\tau\partial x}+\frac{1}{2}\tau^2\rho_2\frac{\partial^2u_0}{\partial x^2}\frac{\partial V_0}{\partial\tau}-\widehat{\mathcal{A}}_\tau\left(\tau\frac{\partial^2V_1}{\partial x\partial\tau}\right)\Bigg\}\Bigg|_{x=\varphi(t)},\end{aligned}\tag{3.32}$$

where

$$\eta = \frac{1}{2}-\frac{1}{\delta}\arctan\left\{\tan\left(\frac{\delta}{2}\right)\tanh(\beta(\tau+\varphi_1))\right\},$$

and $\widehat{\mathcal{A}}_{1\tau}$ is the operator with symbol $i\rho_3H\big(\coth(\xi H)-(\xi H)^{-1}\big)$.

Now we evaluate the integral involved in condition (3.25) for $k=2$ without finding the function V_1 explicitly. For that we will need the equalities

$$\begin{aligned}\rho_2\int_{-\infty}^{\infty}V_0\check{V}_1\frac{\partial\check{V}_1}{\partial\tau} = &\left\{\int_{-\infty}^{\infty}\frac{\partial V_0}{\partial t}V_1\,d\tau-\rho_2\int_{-\infty}^{\infty}\left(u_1V_0+\tau V_0\frac{\partial u_0}{\partial x}\right)\frac{\partial V_1}{\partial\tau}\,d\tau\right\}\Bigg|_{x=\varphi(t)}\\ &+\frac{1}{2}(\check{V}_1^-)^2\rho_3(1-\delta\cot\delta),\end{aligned}\tag{3.33}$$

and

$$\begin{aligned}&\int_{-\infty}^{\infty}V_0\widehat{\mathcal{A}}_\tau\left(\tau\frac{\partial^2V_1}{\partial\tau\partial x}\right)d\tau\Bigg|_{x=\varphi(t)}\\ &=-\left\{\rho_3(1-\delta\cot\delta)\int_{-\infty}^{\infty}\tau V_0\frac{\partial^2V_1}{\partial\tau\partial x}\,d\tau-\frac{\rho_2}{2}\int_{-\infty}^{\infty}\tau V_0^2\frac{\partial^2V_1}{\partial\tau\partial x}\,d\tau\right\}\Bigg|_{x=\varphi(t)},\end{aligned}\tag{3.34}$$

which are obtained if we multiply equation (3.18) for $k = 1$ by $\check{V}_1$ and equation (3.16) integrated once by $\tau \partial^2 V_1/\partial x \partial \tau|_{x=\varphi(t)}$ respectively, and then integrate them with respect to τ.

Now we substitute the function (3.32) into (3.25) for $k = 2$ and employ (3.33) and (3.34). After some transformations we arrive at the relation

$$\begin{aligned} \delta\alpha \frac{d}{dt}\Big\{ \frac{1}{\delta\alpha} \int_{-\infty}^{\infty} V_0 \check{V}_1 \, d\tau \Big\} \\ + \frac{1}{2}\rho_2 \Big\{ \varphi_1 \frac{\partial^2 u_0}{\partial x^2} + \frac{\partial u_1}{\partial x} + \frac{\partial V_1^-}{\partial x} \Big\}\Big|_{x=\varphi(t)} \int_{-\infty}^{\infty} V_0^2 \, d\tau \\ + 32\rho_3 \Big(\frac{H}{\rho_2}\Big)^2 \alpha \delta^{3/2} \frac{d}{dt}\Big(\frac{1}{\alpha^2} \frac{d}{dt} \sqrt{\delta} \Big) = 0. \end{aligned} \tag{3.35}$$

Let us calculate the first integral in the left-hand side of (3.35). For that, consider the equations with parameter q

$$q g_0 + \frac{1}{2}\rho_2 g_0^2 - \widehat{\mathcal{A}}_\tau g_0 = 0, \tag{3.36}$$

$$\widehat{L} g_1 = \Phi_1, \tag{3.37}$$

where

$$\widehat{L} = q + \rho_2 g_0 - \widehat{\mathcal{A}}_\tau, \qquad \Phi_1 = - \int_{\tau}^{\infty} \mathcal{F}_1^1 \, d\tau',$$

with the function V_0 in $\mathcal{F}_1^1$ replaced by g_0.

It is obvious that the solutions, $g_0(\tau, q)$ and $g_1(\tau, q, t)$, of these equations coincide with $V_0(\tau, t)$ and $\omega_1(\tau, t)$ for $q = q^0 \stackrel{\text{def}}{=} \rho_2 u_0(\varphi, t) + \rho_1 - \varphi_t$, and that

$$\frac{\partial g_0}{\partial q}\Big|_{q=q_0} = \frac{1}{\delta}\Big\{ \alpha V_0 + \frac{1}{2}\frac{\rho_2}{\rho_3} V_0^2 + \tau \frac{\partial V_0}{\partial \tau} \Big\} \frac{\partial \delta}{\partial q}\Big|_{q=q_0}.$$

In turn, we find from (3.7) that

$$\frac{\partial \delta}{\partial q}\Big|_{q=q^0} = 2 \sin^2 \delta \{\rho_3 (\sin 2\delta - 2\delta)\}^{-1}.$$

Further, on differentiating (3.36) with respect to q we obtain

$$g_0 = -\widehat{L} \frac{\partial g_0}{\partial q}.$$

Hence, taking (3.37) into account, we have

$$\int_{-\infty}^{\infty} V_0 \omega_1 \, d\tau = - \int_{-\infty}^{\infty} \frac{\partial g_0}{\partial q} \widehat{L} g_1 \, d\tau \Big|_{q=q^0} = - \int_{-\infty}^{\infty} \Phi_1 \frac{\partial g_0}{\partial q} \, d\tau \Big|_{q=q^0}. \tag{3.38}$$

The integral in the right-hand side of (3.38) can be calculated explicitly. It remains to note that by (3.14)

$$\int_{-\infty}^{\infty} V_0 \check{V}_1 \, d\tau = \frac{1}{2} \check{V}_1^- \int_{-\infty}^{\infty} V_0 \, d\tau + \int_{-\infty}^{\infty} V_0 \omega_1 \, d\tau.$$

Thus equation (3.35), after some tedious calculations, can be transformed to the form

$$\frac{d}{dt}\Big\{\mu_1\frac{d}{dt}(\alpha\delta\varphi_1)\Big\}+\mu_2\varphi_1=G(t), \tag{3.39}$$

where $G(t)$ does not depend on φ_1 and vanishes for constant u_0 and u_1,

$$\mu_1=\frac{\alpha_1}{\alpha^2\delta^2},\qquad \mu_2=\frac{1}{2}\rho_2\rho_3\frac{\partial^2 u_0}{\partial x^2}\bigg|_{x=\varphi},$$

$$\alpha=1-\delta\cot\delta,\qquad \alpha_1=\frac{\sin^2\delta-\delta^2}{\sin 2\delta-2\delta}.$$

So, we have obtained the required equation for the correction φ_1 to the phase of the solution to (3.4). Note that for $u_0=u_1=0$ we obtain from (3.39) that $\varphi_1=c_0+c_1t$. In this case the constants c_0 and c_1 are calculated from the initial data for equation (3.1) in the same manner as for the Korteweg–de Vries equation (see §1.1).

3.2. Solution of smoothed shock wave type in the Toda model of the continuum

Consider a discrete system of equations

$$m_k\frac{d^2q_k}{dt^2}=a_{k+1}\big(e^{r_{k+1}(q_{k+1}-q_k)}-1\big)-a_k\big(e^{r_k(q_k-q_{k-1})}-1\big),\qquad k=0,\pm1,\pm2,\dots, \tag{3.40}$$

which is the infinite Toda chain with variable coefficients. The Toda model of the continuum has been explored by many mathematicians and physicists. We will only refer here to the papers [**16, 242–244**] and point out that the equations (3.40) with constant coefficients can be integrated by the method of the inverse scattering problem [**66, 67, 136**].

Let us discuss the physical meaning of equations (3.40). The models of type (3.40) arise in the microscopic description of a continuum as a system of balls of masses m_k which interact nonlinearly. Then the functions $q_k=q_k(t)$ describe the deviations of the balls from the equilibrium positions, while a_k and r_k characterize the elasticity of the medium.

It is easily seen that by taking formal expansions of the exponential functions involved in (3.40) into Taylor series and neglecting the quadratic terms we can rewrite (3.40) as

$$m_k\frac{d^2q_k}{dt^2}=a_{k+1}r_{k+1}(q_{k+1}-q_k)-a_kr_k(q_k-q_{k-1}). \tag{3.41}$$

Assuming that $m_k\ll 1$, which is quite natural, and that $r_k\gg 1$, which means that the coefficient of elasticity is large, we set $m_k=\varepsilon m'(x_k)$, $r_k=r'(x_k)/\varepsilon$, $a_k=\alpha(x_k)$, where ε is a small parameter characterizing the distance between the balls at rest positions, $x_k=k\varepsilon$, and m',r',α are smooth functions. Then the equations (3.41) can be rewritten as a differential-difference scheme

$$\begin{aligned}m'(x_k)\frac{d^2q_k}{dt^2}=\frac{1}{\varepsilon}\Big\{&\alpha(x_{k+1})r'(x_{k+1})\frac{q_{k+1}-q_k}{\varepsilon}\\&-\alpha(x_k)r'(x_k)\frac{q_k-q_{k-1}}{\varepsilon}\Big\},\qquad k=0,\pm1,\pm2,\dots.\end{aligned} \tag{3.42}$$

Making the normalization $u_k = q_k/\varepsilon$ and passing to the limit as $\varepsilon \to 0$, we obtain from (3.42) the usual wave equation

$$m'(x)\frac{\partial^2 u}{\partial t^2} = \frac{\partial}{\partial x}\left(\alpha(x)r'(x)\frac{\partial u}{\partial x}\right), \quad x \in \mathbb{R}^1. \tag{3.43}$$

Thus the Toda chain (3.40) is a kind of microscopic analog of the wave equation.

Of course, the above heuristic arguments can be put into a rigorous form only if the deviations $\Delta_k = q_{k+1} - q_k$ are small and, moreover, the Δ_k vary slowly with k. In this case the wave equation (3.43) provides a correct macroscopic description of the processes, since the behavior of each particular ball is, of course, immaterial.

On the contrary, when the deviations Δ_k are not small, the above scheme does not go through, which brings up the problem of averaging the equations (3.40), i.e., derivation of a macroscopic model.

This is the problem we deal with in this section. In order to formulate the main result, viz., a macroscopic analog of (3.40) which can produce a smoothed shock wave, we formulate our assumptions and introduce auxiliary functions.

Regarding the coefficients m_k, a_k, r_k in equations (3.40), we will assume that for all k we have $r_k a_k \geq \text{const} > 0$ and that there exist smooth functions $\alpha = \alpha(x)$, $\rho = \rho(x)$, $\mu = \mu(x)$, uniformly bounded together with their derivatives, such that for all k

$$m_k = \varepsilon\mu(x_k), \quad r_k = \frac{1}{\varepsilon}\rho\Big(x_k + \frac{\varepsilon}{2}\Big), \quad a_k = \alpha\Big(x_k + \frac{\varepsilon}{2}\Big), \tag{3.44}$$

where $x_k = k\varepsilon$, $\varepsilon \ll 1$ being a small parameter, and for all x

$$\rho(x)\alpha(x) \geq \text{const} > 0. \tag{3.45}$$

These assumptions mean that in the medium the density and the coefficients of elasticity vary slowly.

Define the auxiliary functions $\varphi = \varphi(t)$ and $\beta = \beta(t)$ as the solution to the equations

$$\left(\frac{d\varphi}{dt}\right)^2 = \lambda(\varphi)\frac{\sinh^2\beta}{\beta^2}, \tag{3.46}$$

$$\frac{\alpha(\varphi)}{\rho(\varphi)}(\sinh 2\beta - 2\beta) = E, \tag{3.47}$$

where $E > 0$ is a constant and

$$\lambda(\varphi) = \frac{\alpha\rho}{\mu}\bigg|_{x=\varphi}.$$

Denote by $\mathcal{T}$ the one-dimensional curve in $\mathbb{R}^2_{tx}$ specified by the equation $x = \varphi(t)$, and let ν denote the unit vector in the space of (t, x),

$$\nu = \frac{1}{\sqrt{1+\dot\varphi^2}}\begin{pmatrix}\dot\varphi \\ -1\end{pmatrix}, \tag{3.49}$$

where the dot means differentiation with respect to t. Obviously, ν is a normal to $\mathcal{T}$.

Furthermore, we introduce the auxiliary functions

$$
\begin{aligned}
&l_1 = \frac{\dot\varphi^2 - \lambda(\varphi)}{\sqrt{1+\dot\varphi^2}}, && l_2 = \dot\varphi\frac{\dot\varphi^2 + \lambda(\varphi) + 2}{\sqrt{1+\dot\varphi^2}},\\
&l_3 = 2\frac{\beta}{\mu(\varphi)\rho(\varphi)}\frac{d}{dt}(\dot\varphi\mu(\varphi)), && \delta = \frac{1}{l_1}\Bigl(l_3 + 2l_2\frac{d}{dt}\Bigl(\frac{\beta}{\rho(\varphi)}\Bigr)\Bigr).
\end{aligned}
\tag{3.50}
$$

Finally, denote by $u = u(x,t,\varepsilon)$ the function related to q_k by

$$
u\big|_{x=k\varepsilon} = \frac{1}{\varepsilon}q_k,
\tag{3.51}
$$

and let $v(x,t)$ be the limit of u as $\varepsilon \to 0$.

Our main result consists in deriving the limiting equation describing the motion of the shock wave. Namely, the dynamics of the shock wave front is described by equation $x = \varphi(t)$, and, moreover,

$$
v = v^+(x,t) \quad \text{for } x > \varphi(t), \qquad v = v^-(x,t) \quad \text{for } x < \varphi(t),
\tag{3.52}
$$

where $v^\pm$ solve the wave equation

$$
\mu\frac{\partial^2 v^\pm}{\partial t^2} = \frac{\partial}{\partial x}\Bigl(\alpha\rho\frac{\partial v^\pm}{\partial x}\Bigr)
\tag{3.53}
$$

in the domains $D^+ = \{(x,t),\ t>0,\ x>\varphi(t)\}$, $D^- = \{(x,t),\ t>0,\ x<\varphi(t)\}$, respectively, subject to the following conditions on the surface of discontinuity $\mathcal{T}$:

$$
[v] = 2\frac{\beta}{\rho(\varphi)}, \qquad \Bigl[\frac{\partial v}{\partial \nu}\Bigr] = -\delta.
\tag{3.54}
$$

Here, as usual, the square brackets denote the magnitude of the jump,

$$
[u] = u^+\big|_{\mathcal{T}+0} - u^-\big|_{\mathcal{T}-0},
$$

and $\partial/\partial\nu = \langle \nu, \nabla\rangle$ is the derivative along the normal to $\mathcal{T}$.

In the course of deriving the limiting problem (3.53), (3.54) we will also obtain an intermediate result concerning the asymptotic solution to the chain (3.40). Namely, we will establish that the leading term of the asymptotic solution of shock wave type for the system (3.40) has the form

$$
\begin{aligned}
&q_k = \varepsilon Y(x,t,\varepsilon)\big|_{x=k\varepsilon},\\
&Y(x,t,\varepsilon) = u_0(x,t) + u_0^-(x,t)\chi(\tau,t) + \omega_0(\tau,t)\big|_{\tau=(x-\varphi(t))/\varepsilon+\varphi_1(t)},
\end{aligned}
\tag{3.55}
$$

where

$$
\begin{aligned}
&\chi(\tau,t) = \frac{1}{2}(1-\tanh\beta\tau),\\
&\omega_0(\tau,t) = \frac{1}{\rho(\varphi)}\Bigl\{\log\frac{\cosh\beta(\tau+\frac12)}{\cosh\beta(\tau-\frac12)} - \beta\tanh\beta\tau\Bigr\},
\end{aligned}
\tag{3.56}
$$

the background u_0 is a smooth solution to the wave equation

$$\mu\frac{\partial^2 u_0}{\partial t^2} = \frac{\partial}{\partial x}\Big(\alpha\rho\frac{\partial u_0}{\partial x}\Big), \tag{3.57}$$

and $u_0^- = (v^- - u_0)_\varepsilon$. Here the function v^- introduced above denotes the perturbation to the left of the front, i.e., in D^-, and the subscript ε means the extension of the function $v^- - u_0$ into D^+ at a distance of order $\sqrt{\varepsilon}$.

Obviously, the above formulas imply that in the case where $u_0 = 0$ and the coefficients m_k, r_k, a_k do not depend on k we have $\beta = \text{const}$ and φ is a linear function of t. Then $\delta = 0$, and the limiting problem admits a piecewise constant solution. In this case the function determined by (3.55) equals the exact solution of (3.40) found by the method of the inverse scattering problem [**67, 136**].

Now we proceed to the derivation of the limiting problem (3.53), (3.54). For that, following [**162**], we construct first an asymptotic solution for the chain (3.40). For subsequent constructions it will be convenient to transform the equations (3.40). First of all, using the notation (3.44), (3.51), we pass from the discrete equations (3.40) to the differential-difference equation

$$\begin{aligned}
&\varepsilon^2\mu(x)\frac{\partial^2 u(x,t,\varepsilon)}{\partial t^2}\\
&\quad= \alpha\Big(x+\frac{\varepsilon}{2}\Big)\Big(\exp\Big\{\rho\Big(x+\frac{\varepsilon}{2}\Big)\big(u(x+\varepsilon,t,\varepsilon)-u(x,t,\varepsilon)\big)\Big\}-1\Big)\\
&\quad\quad -\alpha\Big(x-\frac{\varepsilon}{2}\Big)\Big(\exp\Big\{\rho\Big(x-\frac{\varepsilon}{2}\Big)\big(u(x,t,\varepsilon)-u(x-\varepsilon,t,\varepsilon)\big)\Big\}-1\Big)
\end{aligned} \tag{3.58}$$

for function u defined for all $x \in \mathbb{R}^1$. Next, we define the translation operator by the formula

$$e^{\delta\partial/\partial x} f(x,t,\varepsilon) = f(x+\delta,t,\varepsilon)$$

and introduce the operator $\sinh(\delta\partial/\partial x)$,

$$\sinh(\delta\partial/\partial x) = \frac{1}{2}\big(e^{\delta\partial/\partial x} - e^{-\delta\partial/\partial x}\big).$$

Then equation (3.58) can be restated as a pseudodifferential equation:

$$\begin{aligned}
&\varepsilon^2\mu(x)\frac{\partial^2 u(x,t,\varepsilon)}{\partial t^2}\\
&\quad= 2\sinh\Big(\frac{\varepsilon}{2}\frac{\partial}{\partial x}\Big)\Big\{\alpha(x)\Big(\exp\Big\{2\rho(x)\sinh\Big(\frac{\varepsilon}{2}\frac{\partial}{\partial x}\Big)u\Big\}-1\Big)\Big\}.
\end{aligned} \tag{3.59}$$

We will seek an asymptotic solution of shock wave type with accuracy $O(\varepsilon^3)$ in the form of the standard expansion

$$\begin{aligned}
u &= \mathcal{Y}_2\Big(\frac{x-\varphi(t)}{\varepsilon}, x, t, \varepsilon\Big) + O(\varepsilon^3),\\
\mathcal{Y}_2(\tau,x,t,\varepsilon) &= \sum_{j=0}^{2}\varepsilon^j\big(u_j(x,t)+V_j(\tau,x,t)\big),
\end{aligned} \tag{3.60}$$

where, as usual, the φ, u_j, V_j are smooth bounded functions with $V_j \to 0$ as $\tau \to \infty$ and $V_j \to V_j^-(x,t)$ as $\tau \to -\infty$.

It is clear that the function $\mathcal{Y}_2((x-\varphi(t))/\varepsilon, x, t, \varepsilon)$ will be an asymptotic solution $\operatorname{mod} O(\varepsilon^3)$ of (3.59) if the function $\mathcal{Y}_2(\tau, x, t, \varepsilon)$ satisfies the relation

$$\begin{aligned}\dot{\varphi}^2\frac{\partial^2\mathcal{Y}_2}{\partial\tau^2} + 2\varepsilon\widehat{\pi}\frac{\partial\mathcal{Y}_2}{\partial\tau} + \varepsilon^2\frac{\partial^2\mathcal{Y}_2}{\partial t^2} - \frac{2}{\mu}\sinh\Big(\widehat{D} + \frac{\varepsilon}{2}\frac{\partial}{\partial x}\Big)\\ \times\Big\{\alpha(x)\Big(\exp\Big\{2\rho(x)\sinh\Big(\widehat{D}+\frac{\varepsilon}{2}\frac{\partial}{\partial x}\Big)\mathcal{Y}_2\Big\} - 1\Big)\Big\} = O(\varepsilon^3)\end{aligned} \tag{3.61}$$

for all $\tau \in R^1$. Here and subsequently we use the notation

$$\widehat{D} = \frac{1}{2}\frac{\partial}{\partial\tau}, \qquad \widehat{\pi} = -\dot{\varphi}\frac{\partial}{\partial t} - \frac{1}{2}\ddot{\varphi}.$$

To determine the background part of the solution (3.60), i.e., the functions u_j, $j = 0, 1, 2$, we observe that smooth functions fulfill the following relation:

$$\begin{aligned}\sinh\Big(\widehat{D} + \frac{\varepsilon}{2}\frac{\partial}{\partial x}\Big)f(x) &= \sinh\Big(\frac{\varepsilon}{2}\frac{\partial}{\partial x}\Big)f(x)\\ &= \sum_{k=1}^{M} c_k\Big(\frac{\varepsilon}{2}\Big)^k\frac{\partial^k}{\partial x^k}f(x) + O(\varepsilon^{M+1}),\end{aligned} \tag{3.62}$$

where $c_k = 1/k!$ for odd k and $c_k = 0$ for even k. Therefore letting $\tau \to \infty$ in (3.61) and equating to zero the coefficients of like powers of ε, we obtain for determination of u_j the recursive chain of wave equations

$$\mu\frac{\partial^2 u_j}{\partial t^2} = \frac{\partial}{\partial x}\Big(\alpha\rho\frac{\partial u_j}{\partial x}\Big) + f_j(x,t), \tag{3.63}$$

where $f_0 = 0$ and the f_j are polynomials in $u_0, \dots, u_{j-1}$ and their derivatives for $j = 1, 2$.

Next, we represent the fast-varying components of the solution (3.60) in the standard form

$$V_j(\tau, x, t) = V_j^-(x,t)\chi(\tau,t) + \omega_j(\tau,t), \tag{3.64}$$

where $\omega_j \to 0$ as $\tau \to \pm\infty$, and χ is a function such that $\chi \to 0$ as $\tau \to \infty$ and $\chi \to 1$ as $\tau \to -\infty$.

Now we let $\tau \to -\infty$ in (3.61) and repeat the above arguments to obtain the equations for the tail functions V_j^- similar to (3.63):

$$\mu\frac{\partial^2 V_j^-}{\partial t^2} = \frac{\partial}{\partial x}\Big(\alpha\rho\frac{\partial V_j^-}{\partial x}\Big) + f_j^-(x,t), \tag{3.65}$$

where, obviously, $f_0^- = 0$.

Let us proceed to determiantion of functions χ and ω_j, $j = 0, 1, 2$. Clearly, for such functions we cannot use expansions like (3.62) any more. Hence we consider the terms of order ε^0 in (3.61) to obtain the pseudodifferential equation

$$\dot{\varphi}^2\frac{\partial^2\check{V}_0}{\partial\tau^2} - 2\frac{\alpha(\varphi)}{\mu(\varphi)}\sinh\widehat{D}\Big\{e^{2\rho(\varphi)\sinh\hat{D}\check{V}_0} - 1\Big\} = 0. \tag{3.66}$$

As before, we use here the notation $\check{f}$ for the function f on the surface $\mathcal{T}$, i.e., $\check{f} = f(\tau, x, t)|_{x=\varphi(t)}$. Moreover, we observe that the function

$$2\sinh \widehat{D} V(\tau, x) = V(\tau + \tfrac{1}{2}, x) - V(\tau - \tfrac{1}{2}, x)$$

tends to zero exponentially fast as $\tau \to \pm\infty$ if $V(\tau, x)$ stabilizes at an exponential rate as $\tau \to \pm\infty$. This fact enables us to expand the coefficients α, ρ, μ by the Taylor formula around the point $x = \varphi(t)$.

Using the exact solution to the Toda chain with constant coefficients [**67, 136**], we can easily solve (3.66) in an explicit form:

$$\check{V}_0 = \frac{1}{\rho(\varphi)} \left\{ \log \frac{\cosh \beta(\tau + \varphi_1 + \frac{1}{2})}{\cosh \beta(\tau + \varphi_1 - \frac{1}{2})} - \beta \right\}, \tag{3.67}$$

where $\varphi_1 = \varphi_1(t)$ is the "constant" of integration, and the functions $\beta = \beta(t)$ and $\dot{\varphi}$ are related by (3.46).

We split the function $\check{V}_0$ into a soliton component and a component stabilizing as $\tau \to -\infty$. Analyzing the formula (3.67), we obtain the representation

$$\check{V}_0 = \check{V}_0^-(t)\chi(\tau, t) + \omega_0(\tau, t), \tag{3.68}$$

similar to (3.64). Here $\check{V}_0^- = -2\beta/\rho(\varphi)$, and the functions η and ω_0 are defined by (3.56).

Now we define the extension $\check{V}_0$ in the form (3.64) with $j = 0$, which for V_0^- entails the boundary condition

$$V_0^-|_{x=\varphi(t)} = -2\frac{\beta}{\rho(\varphi)}. \tag{3.69}$$

In order to derive the equation (3.47) and to determine the phase shift φ_1 we have to consider higher order terms of asymptotics. Putting the expansion (3.60) into (3.61), taking (3.63) and (3.65) into account, and writing down the Taylor expansions, we obtain the following equations for $\check{V}_j$:

$$\dot{\varphi}^2 \frac{\partial^2 \check{V}_j}{\partial \tau^2} - 4\lambda(\varphi) \sinh \widehat{D} \Big\{ e^{2\rho(\varphi)(\sinh \hat{D}\check{V}_0)} \sinh \widehat{D} \check{V}_j \Big\} = F_j, \tag{3.70}$$

where the F_j are functions which depend on $\tau, t, u_0, V_0, \dots, V_{j-1}, u_j$ and their derivatives at $x = \varphi(t)$. In particular,

$$\begin{aligned} F_1 = -\Bigg\{ & 2\widehat{\pi} \frac{\partial V_0}{\partial \tau} - 2\lambda \sinh \widehat{D} \left(e^{2\rho \sinh \hat{D} V_0} \cosh \widehat{D} \Big(\frac{\partial V_0}{\partial x} + \frac{\partial u_0}{\partial x} \Big) \right) \\ & - \frac{1}{\mu} \cosh \widehat{D} \frac{\partial}{\partial x} \left(\alpha \Big(e^{2\rho \sinh \hat{D} V_0} - 1 \Big) \right) \\ & + \tau \left(\dot{\varphi}^2 \frac{\partial^3 V_0}{\partial \tau^2 \partial x} - \dot{\varphi}^2 \frac{\mu}{\alpha} \frac{\partial^2 V_0}{\partial \tau^2} \frac{\partial}{\partial x} \Big(\frac{\alpha}{\mu} \Big) \right. \\ & \left. - 2\frac{\alpha}{\mu} \sinh \widehat{D} \frac{\partial}{\partial x} e^{2\rho \sinh \hat{D} V_0} \right) \Bigg\} \Bigg|_{x=\varphi(t)}. \end{aligned} \tag{3.71}$$

The differential expression in the left-hand side of (3.70) is symmetric, and it is not hard to see that the functions 1 and $\partial \check{V}_0/\partial\tau$ satisfy the corresponding homogeneous equation. This leads us to necessary conditions for solvability of (3.70):

$$\int_{-\infty}^{\infty} F_j(\tau,t)\,d\tau = 0, \tag{3.72}$$

$$\int_{-\infty}^{\infty} F_j(\tau,t)\frac{\partial \check{V}_0}{\partial\tau}\,d\tau = 0. \tag{3.73}$$

As in §3.1, we will restrict ourselves to the assumption that conditions (3.72), (3.73) are sufficient.

Consider these conditions for $k=1$. Using (3.68) and the explicit form (3.71) of F_1, we can easily integrate in the left-hand side of (3.72) to obtain that (3.72) for $j=1$ means the condition

$$2\dot\varphi\frac{\partial V_0^-}{\partial t} + (\dot\varphi^2+\lambda)\frac{\partial V_0^-}{\partial x}\bigg|_{x=\varphi(t)} = -\left(\ddot\varphi + \frac{\dot\varphi^2}{\mu}\frac{\partial\mu}{\partial x}\right)V_0^-\bigg|_{x=\varphi}. \tag{3.74}$$

Note that the expression in the left-hand side of (3.74) is the derivative of V_0^- along the vector

$$l = \begin{pmatrix} 2\dot\varphi \\ \dot\varphi^2+\lambda\end{pmatrix} \in \mathbb{R}^2_{tx}.$$

Let ν be the unit normal to $\mathcal{T}$ given by (3.49), and let ζ be a unit vector tangent to $\mathcal{T}$. For definiteness, we set

$$\zeta = \frac{1}{\sqrt{1+\dot\varphi^2}}\begin{pmatrix}1\\ \dot\varphi\end{pmatrix}.$$

Then $l = l_1\nu + l_2\zeta$, where the coefficients l_1, l_2 are defined by (3.50). Obviously, by condition (3.69) we have

$$\langle\zeta,\nabla_{t,x}\rangle V_0^-(x,t)\big|_{x=\varphi(t)} = \frac{d}{dt}\big(V_0^-(\varphi(t),t)\big) = -2\frac{d}{dt}\Big(\frac{\beta}{\rho(\varphi)}\Big).$$

Using this equality, we can easily rewrite condition (3.74) as

$$\frac{\partial V_0^-}{\partial\nu}\Big|_{\mathcal{T}} = \delta, \tag{3.75}$$

with $\delta=\delta(t)$ as in (3.50).

Consider now condition (3.73) for $j=1$. We introduce the auxiliary functions

$$r_0 = 2\rho(\varphi)\sinh\widehat{D}\check{V}_0, \tag{3.76}$$

$$\mathcal{F}_1 = -\bigg\{\widehat{\pi}\frac{\partial r_0}{\partial\tau} - \frac{\tau}{\lambda}\frac{\partial^2 r_0}{\partial\tau^2}\frac{\partial\lambda}{\partial x} - \frac{1}{2}\frac{\rho}{\alpha}\frac{\partial}{\partial x}\Big(\frac{\alpha^2}{\mu}\Big)\sinh(2\widehat{D})e^{r_0} - \frac{1}{2}\dot\varphi^2\rho\frac{\partial^2 u_0}{\partial x^2}\frac{\partial^2 r_0}{\partial\tau^2}\bigg\}\bigg|_{x=\varphi}. \tag{3.77}$$

We will also use the following equality, which can be verified directly:

$$\frac{\partial \check{V}_0}{\partial \tau} = 2\frac{\beta^2}{\rho \sinh^2 \beta} \sinh \widehat{D} w,$$

where

$$w = \int_{-\infty}^{\tau} (e^{r_0} - 1)\, d\tau' = \frac{\sinh^2 \beta}{\beta}(\tanh \beta\tau + 1). \tag{3.78}$$

Then condition (3.73) for $j = 1$ can be transformed into

$$\int_{-\infty}^{\infty} w\mathcal{F}_1 \, d\tau = 0, \tag{3.79}$$

since

$$\begin{aligned}\int_{-\infty}^{\infty} F_1 \frac{\partial \check{V}_0}{\partial \tau}\, d\tau &= -2\frac{\beta^2}{\rho \sinh^2 \beta} \int_{-\infty}^{\infty} w \sinh \widehat{D} F_1 \, d\tau \\ &= -2\frac{\beta^2}{\rho \sinh^2 \beta} \int_{-\infty}^{\infty} w\mathcal{F}_1 \, d\tau.\end{aligned}$$

In its turn, after some transformations condition (3.79) becomes

$$\begin{aligned}&\Big\{2\dot{\varphi} \int_{-\infty}^{\infty} w \frac{\partial^2 r_0}{\partial \tau \partial t}\, d\tau + \Big(\ddot{\varphi} - \frac{1}{\lambda}\frac{\partial \lambda}{\partial x}\Big) \int_{-\infty}^{\infty} w \frac{\partial r_0}{\partial \tau}\, d\tau \\ &- \frac{1}{\lambda}\frac{\partial \lambda}{\partial x} \int_{-\infty}^{\infty} \tau \frac{\partial r_0}{\partial \tau}\frac{\partial w}{\partial \tau}\, d\tau + \frac{\rho}{\alpha}\frac{\partial}{\partial x}\Big(\frac{\alpha^2}{\mu}\Big) \int_{-\infty}^{\infty} w \sinh(2\widehat{D}) e^{r_0}\, d\tau \Big\}\Big|_{x=\varphi} = 0.\end{aligned} \tag{3.80}$$

On integrating in (3.80) we reduce this equality to the form

$$\frac{d}{dt}\Big\{\frac{\alpha(\varphi)}{\rho(\varphi)}(\sinh 2\beta - 2\beta)\Big\} = 0.$$

Obviously, this implies (3.47).

Now, in order to derive the limiting problem (3.53), (3.54), it remains to note that $V_0^-(x,t)$ is the tail perturbation arising after the passage of the front of the smoothed shock wave. Then, letting ε tend to zero, we obtain that $v = \lim_{\varepsilon \to 0} u$ fulfills the equalities $v = u_0(x,t) + V_0^-(x,t)$ for $x < \varphi(t)$, $t > 0$, and $v = u_0(x,t)$ for $x > \varphi(t)$, $t > 0$.

Using the notation (3.52), we obtain the problem (3.53), (3.54) from (3.63), (3.65) for $j = 0$ and conditions (3.69), (3.75).

Conversely, setting u_0 to be the solution of (3.63) with $j = 0$ for all $x \in \mathbb{R}^1$, with the aid of solutions of (3.53), (3.54) we define the function $V_0^-(x,t) = v^-(x,t) - u_0(x,t)$ in D^-. It remains to construct the extension of V_0^- as the solution to the wave equation in a small neighborhood of $\mathcal{T}$ and to take an arbitrary smooth extension for $x \geq \varphi(t) + \sqrt{\varepsilon}$.

To complete the construction of the leading term of the asymptotics, it remains to write down the equation for the phase shift φ_1. For this we will evaluate the integral in (3.73) for $j = 2$ without finding the explicit form of $\check{V}_1$.

First of all, we observe that equation (3.70) for $j = 1$ can be transformed into

$$\dot{\varphi}^2 \frac{\partial^2 r_1}{\partial \tau^2} - 4\lambda \sinh^2 \widehat{D}(r_1 e^{r_0}) = \mathcal{F}_1, \tag{3.81}$$

where $r_1 = \sinh \widehat{D}\check{V}_1$ and the functions r_0 and $\mathcal{F}_1$ are defined by (3.76) and (3.77), while condition (3.73) for $j = 2$ can be rewritten as

$$\int_{-\infty}^{\infty} \mathcal{F}_2 w \, d\tau = 0. \tag{3.82}$$

Here w is the function defined by (3.78), and

$$\begin{aligned} \mathcal{F}_2 = \Big\{ &- 2\widehat{\pi} \frac{\partial r_1}{\partial \tau} + 4\lambda\rho \frac{\partial u_0}{\partial x} \sinh^2 \widehat{D}(r_1 e^{r_0}) + 2\lambda \sinh^2 \widehat{D}(r_1^2 e^{r_0}) \\ &+ \frac{\rho}{\alpha}\frac{\partial}{\partial x}\Big(\frac{\alpha^2}{\mu}\Big) \sinh 2\widehat{D}(r_1 e^{r_0}) + 4\tau \frac{\partial \lambda}{\partial x} \sinh^2 \widehat{D}(r_1 e^{r_0}) \\ &+ 2\dot{\varphi}\Big(\frac{\partial \eta}{\partial t} - \rho \frac{\partial^2 V_0^-}{\partial x \partial t}\Big)\frac{\partial \chi}{\partial \tau} + \tau \frac{\partial}{\partial x}\Big(\frac{\rho}{\alpha}\frac{\partial}{\partial x}\Big(\frac{\alpha^2}{\mu}\Big)\Big) \sinh 2\widehat{D} e^{r_0} \\ &+ \tau \frac{\dot{\varphi}^2}{\lambda}\frac{\partial^2 r_0}{\partial \tau^2}\Big(\frac{\partial(\lambda\rho u_{0x})}{\partial x} + \frac{1}{2}\tau \frac{\partial^2 \lambda}{\partial x^2}\Big) + G_1 \Big\}\Big|_{x=\varphi}, \end{aligned} \tag{3.83}$$

where

$$\begin{aligned} G_1 = {} & \frac{\partial(\rho V_0^-)}{\partial x}\Big\{2\lambda(\sinh 2\widehat{D} + 2\tau \sinh^2 \widehat{D})(\chi e^{r_0}) - \tau\dot{\varphi}^2 \frac{\partial^2 \chi}{\partial \tau^2}\Big\} - \frac{\partial^2 r_0}{\partial t^2} \\ &+ \frac{1}{\alpha u_{0x}}\frac{\partial}{\partial x}\Big(\frac{\alpha^2\rho^2 u_{0x}^2}{\mu}\Big) \sinh 2\widehat{D} e^{r_0} + \rho \frac{\partial}{\partial x}\Big(\frac{\alpha_x}{\mu}\Big) \cosh 2\widehat{D}(e^{r_0} - 1), \end{aligned}$$

and the subscript x means differentiation with respect to x.

In obtaining the formula for $\mathcal{F}_2$ we have used the fact that w satisfies the equation

$$\dot{\varphi}^2 \frac{\partial^2 w}{\partial \tau^2} - 4\lambda e^{r_0} \sinh^2 \widehat{D} w = 0. \tag{3.84}$$

Moreover, in (3.83) we denote by η the limiting value of r_0 as $\tau \to -\infty$,

$$\eta = \sqrt{\lambda} \frac{\beta^2 \sinh \beta}{\beta^2 - \sinh^2 \beta} \frac{d}{dt} \log \Big\{ \beta^3 \sinh \beta + \Big(\frac{\alpha\mu}{\rho^3}\Big)\Big|_{x=\varphi} \Big\}.$$

To write down the integral (3.82) we will need the auxiliary equalities

$$4\lambda \int_{-\infty}^{\infty} w \sinh^2 \widehat{D}(r_1^2 e^{r_0}) \, d\tau = -2\dot{\varphi}^2 \int_{-\infty}^{\infty} r_1 \frac{\partial r_1}{\partial \tau} e^{r_0} \, d\tau - \dot{\varphi}^2 \eta^2, \tag{3.85}$$

$$\dot{\varphi}^2 \int_{-\infty}^{\infty} r_1 \frac{\partial r_1}{\partial \tau} e^{r_0} \, d\tau = \int_{-\infty}^{\infty} r_1 e^{r_0} \int_{-\infty}^{\tau} \mathcal{F}_1(\tau', t) \, d\tau' \, d\tau - \frac{1}{2}\lambda \eta^2. \tag{3.86}$$

To derive (3.85), one has to multiply (3.84) by r_1^2, integrate with respect to τ, and make some tedious transformations.

Derivation of (3.86) uses the identity

$$4\int_{-\infty}^{\infty}\frac{\partial\psi}{\partial\tau}\sinh^2\widehat{D}\psi\,d\tau=-\frac{1}{2}\lim_{\tau\to-\infty}\Big(\frac{\partial\psi}{\partial\tau}\Big)^2, \tag{3.87}$$

valid for smooth functions ψ such that $\psi_\tau\to 0$ as $\tau\to\infty$ and $\psi_\tau\to$ const as $\tau\to-\infty$, as well as the equality

$$\dot\varphi^2\frac{\partial r_1}{\partial\tau}=4\lambda\sinh^2\widehat{D}\int_{-\infty}^{\tau}r_1e^{r_0}\,d\tau'+\int_{-\infty}^{\tau}\mathcal{F}_1(\tau',t)\,d\tau', \tag{3.88}$$

which results by integrating (3.81) and using the obvious equality

$$\sinh^2\widehat{D}_\tau\int_{-\infty}^{\tau}f(\tau')\,d\tau'=\int_{-\infty}^{\tau}\sinh^2\widehat{D}_{\tau'}f(\tau')\,d\tau'.$$

It is clear that, on multiplying (3.88) by $r_1\exp(r_0)$, integrating, and applying (3.87), we obtain (3.86).

Using (3.85), (3.86) and the explicit form of F_1,F_2, we can rewrite (3.82) as

$$\begin{aligned}\Big\{2\dot\varphi\frac{d}{dt}\int_{-\infty}^{\infty}w\frac{\partial r_1}{\partial\tau}\,d\tau&+\Big(\ddot\varphi-2\frac{\dot\varphi^2}{\lambda}\frac{\partial\lambda}{\partial x}\Big)\int_{-\infty}^{\infty}w\frac{\partial r_1}{\partial\tau}\,d\tau\\&-\Big(\ddot\varphi-2\frac{\dot\varphi^2}{\lambda}\frac{\partial\lambda}{\partial x}\Big)\int_{-\infty}^{\infty}r_0r_1e^{r_0}\,d\tau+G\Big\}\Big|_{x=\varphi}=0.\end{aligned} \tag{3.89}$$

Here

$$\begin{aligned}G=\Big\{\int_{-\infty}^{\infty}wG_1\,d\tau-\frac{\lambda_x}{\lambda}\int_{-\infty}^{\infty}\tau w\mathcal{F}_1\,d\tau+\frac{\lambda_x\dot\varphi^2}{\lambda\beta}\eta\sinh^2\beta\Big\}\Big|_{x=\varphi}\\-\frac{1}{2}(\dot\varphi^2-1)\eta^2-2\dot\varphi\eta\frac{d}{dt}\Big(\frac{\sinh^2\beta}{\beta}\Big)\end{aligned}$$

denotes the terms that do not depend on r_1.

Denote $\lambda_1=\dot\varphi^2/\lambda\big|_{x=\varphi}$ and observe that we have the equality

$$\lambda_1r_0=4\sinh^2\widehat{D}\int_{-\infty}^{\tau}w_1(\tau',\lambda_1)\,d\tau', \tag{3.90}$$

where $w_1(\tau',\lambda_1)\big|_{\lambda_1=\dot\varphi^2/\lambda}=w(\tau,t)$.

Now we differentiate (3.90) with respect to the parameter λ_1 and employ (3.88). Then we can write down the third integral in (3.89) without finding the explicit form of r_1 as

$$\begin{aligned}\int_{-\infty}^{\infty}r_1r_0e^{r_0}\,d\tau&=\int_{-\infty}^{\infty}r_1e^{r_0}\Big\{-\lambda_1\frac{\partial r_0}{\partial\lambda_1}+4\sinh^2\widehat{D}\int_{-\infty}^{\tau}\frac{\partial w_1}{\partial\lambda_1}\,d\tau'\Big\}\,d\tau\\&=\Big\{\frac{1}{\lambda}\int_{-\infty}^{\infty}\frac{\partial w_1}{\partial\lambda_1}\int_{-\infty}^{\tau}\mathcal{F}_1\,d\tau'\,d\tau+\lambda_1\eta\lim_{\tau\to-\infty}\frac{\partial w_1}{\partial\lambda_1}\Big\}\Big|_{\lambda_1=\dot\varphi^2/\lambda}.\end{aligned} \tag{3.91}$$

Next, multiply (3.88) by w and integrate with respect to τ. After some algebra we obtain the equality

$$\int_{-\infty}^{\infty} w\frac{\partial r_1}{\partial \tau}\,d\tau = -\int_{-\infty}^{\infty} r_1 r_0 e^{r_0}\,d\tau + \dot{\varphi}^{-2}\int_{-\infty}^{\infty} w\int_{-\infty}^{\tau}\mathcal{F}_1(\tau',t)\,d\tau'\,d\tau, \tag{3.92}$$

which, combined with (3.91), enables us to evaluate the remaining integral in the left-hand side of (3.89).

Now, using (3.77), (3.78), (3.91), and (3.92), we obtain after very tedious calculations that (3.89) transforms into the required equation for φ_1:

$$\gamma_1\frac{d^2\varphi_1}{dt^2} - \gamma_2\frac{d\varphi_1}{dt} + \gamma_3\varphi_1 + g = 0. \tag{3.93}$$

Here

$$\gamma_1 = 4\frac{\beta}{\kappa}\sinh^2\beta,$$

$$\gamma_2 = 4\dot{\kappa}_1 + 8\frac{\dot{\lambda}\beta}{\lambda\kappa}\sinh^2\beta + E\frac{\mu^2}{\alpha^4}\frac{d}{dt}\left(\frac{\alpha^3\rho}{\mu^2}\Big|_{x=\varphi}\right),$$

$$\begin{aligned}
\gamma_3 = {}& 2\lambda^{3/2}\frac{d}{dt}\left(\frac{\kappa_1\dot{\lambda}}{\lambda^{5/2}}\right) - 2\lambda\kappa\kappa_1\frac{d^2}{dt^2}(\lambda^{-1}) + \frac{1}{2}E\dot{\lambda}\frac{\mu\rho\kappa^2}{\alpha^3\beta^2}\frac{d}{dt}\left(\frac{\alpha}{\rho}\Big|_{x=\varphi}\right)\\
&+ E\frac{\rho}{\alpha}\left\{\frac{d}{dt}\left(\frac{\mu}{\alpha}\Big|_{x=\varphi}\frac{d}{dt}\left(\frac{\alpha^2}{\mu}\Big|_{x=\varphi}\right)\right) + \frac{\mu}{\alpha^2}\frac{d\alpha(\varphi)}{dt}\frac{d}{dt}\left(\frac{\alpha}{\mu}\Big|_{x=\varphi}\right)\right.\\
&\left. - \frac{\mu}{\rho^2}\frac{d\rho(\varphi)}{dt}\frac{d}{dt}\left(\frac{\rho}{\mu}\Big|_{x=\varphi}\right)\right\},
\end{aligned}$$

$$\begin{aligned}
g = {}& \lambda^{3/2}\frac{d}{dt}\left\{\frac{\dot{\varphi}}{\lambda^{3/2}}\left(\frac{\eta\sinh^2\beta}{\kappa\beta} - 2\frac{\kappa_1}{\rho}u_{0x}\Big|_{x=\varphi}\right)\right\}\\
&+ 2\beta\dot{\varphi}^2\frac{d}{dt}\left(\frac{\eta\dot{\varphi}}{\lambda}\right) + \dot{\varphi}\rho u_{0x}\Big|_{x=\varphi}\left\{2\frac{\kappa\kappa_1}{\lambda}\dot{\lambda} - \lambda\left(\frac{\rho}{\alpha}\right)^2\frac{d}{dt}\left(\frac{\alpha}{\rho}\Big|_{x=\varphi}\right)\right\}\\
&+ 2\frac{\dot{\beta}\kappa^2}{\sinh^2\beta} + \lambda\left(1 - \frac{\sinh^2\beta}{\beta^2}\right)\left\{2\frac{\rho\dot{\varphi}}{\alpha}\rho u_{0x}\Big|_{x=\varphi}\sinh\beta + \frac{1}{2}\eta^2\right\},
\end{aligned}$$

$$\kappa = \beta\cot\beta - 1, \quad \kappa_1 = \frac{1}{\kappa}\left(\frac{\sinh^2\beta}{\beta} - \sinh 2\beta + \beta\right),$$

and $E > 0$ is the same constant as in (3.47).

We conclude this section with two possible extensions of the above results.

First, we point out that a limiting problem similar to (3.53), (3.54) and an asymptotic solution of the form (3.55), (3.56) can be also obtained in the case where the coefficients α, ρ, μ are smooth functions of x and t. Then the trajectory $x = \varphi(t)$ of the motion of discontinuity of the limiting solution will be described by the model equations

$$\left(\frac{d\varphi}{dt}\right)^2 = \frac{\alpha(\varphi,t)\rho(\varphi,t)}{\mu(\varphi,t)}\frac{\sinh^2\beta}{\beta^2},$$
$$\frac{d}{dt}\left\{(\sinh 2\beta - 2\beta)\frac{\alpha(\varphi,t)}{\rho(\varphi,t)}\right\} - 2\left\{\frac{\sinh^2\beta}{\beta} - \beta\right\}\frac{\partial}{\partial t}\left(\frac{\alpha(x,t)}{\rho(x,t)}\right)\Big|_{x=\varphi}$$
$$+\left\{2\frac{\sinh^2\beta}{\beta} - \sinh 2\beta\right\}\frac{\alpha}{\mu\rho^3}\frac{\partial}{\partial t}(\mu(x,t)\rho^2(x,t))\Big|_{x=\varphi} = 0.$$

Further, when deriving the limiting problem (3.53), (3.54) we assumed that the background perturbation u_0 is of order $O(1)$. As a consequence, we obtained that the velocity of motion of the smoothed shock wave depends on u_0 only in the higher order term of magnitude $O(\varepsilon)$. We remark now that, nevertheless, formally the situation is possible where the velocity of the motion of the wave front depends on the background, namely, in the case of abnormally large deviations from the equilibrium state, i.e., when $q_k \sim 1$.

To see this, make the change of variables

$$U(x,t,\varepsilon) = u(x+\varepsilon,t,\varepsilon) - u(x,t,\varepsilon). \tag{3.94}$$

Then, using the notation (3.44), (3.45), the system (3.40) reduces to the pseudo-differential equation

$$\varepsilon^2\frac{\partial^2 U}{\partial t^2} = 4\sinh\left(\frac{\varepsilon}{2}\frac{\partial}{\partial x}\right)\mu^{-1}\sinh\left(\frac{\varepsilon}{2}\frac{\partial}{\partial x}\right)\{\alpha(e^{\rho U}-1)\}, \tag{3.95}$$

where the function U to be determined is of magnitude $O(1)$.

Omitting all calculations, we present the leading term of the asymptotic solution to (3.95):

$$U = U_0(x,t) + Y\left(\frac{x-\varphi}{\varepsilon} + \varphi_1, t\right) + O(\varepsilon),$$

where

$$Y(\tau,t) = \frac{1}{\rho(\varphi)}\log\left\{\frac{\cosh\beta(\tau+1)\cosh\beta(\tau-1)}{\cosh^2\beta\tau}\right\}. \tag{3.96}$$

Here the background U_0 is the solution of the nonlinear wave equation

$$\frac{\partial^2 U_0}{\partial t^2} = \frac{\partial}{\partial x}\frac{1}{\mu(x)}\frac{\partial}{\partial x}\left(\alpha(x)(e^{\rho(x)U_0} - 1)\right),$$

and we assume that U_0 is a smooth function.

The functions $\varphi = \varphi(t)$ and $\beta = \beta(t)$ are obtained as solutions of the model equations

$$\begin{aligned}&\left(\frac{d\varphi}{dt}\right)^2 = \lambda(\varphi,t)\frac{\sinh^2\beta}{\beta^2},\\ &\frac{d}{dt}\{\mu(\varphi)\lambda(\varphi,t)(\sinh 2\beta - 2\beta)\} + 2\frac{(\beta^2 - \sinh^2\beta)}{\beta}\frac{\mu}{\rho^2}\frac{\partial\lambda(x,t)}{\partial t}\Big|_{t=\varphi} = 0,\end{aligned} \tag{3.97}$$

where

$$\lambda(x,t) = \frac{\alpha(x)\rho(x)}{\mu(x)} \exp\big(\rho(x)U_0(x,t)\big),$$

and the shift $\varphi_1 = \varphi_1(t)$ is found from an equation similar to (3.93).

It is not hard to see that the second term in the right-hand side of (3.96) is a distorted soliton whose velocity is determined by the Hugoniot type conditions (3.97), and the amplitude A is given by the formula

$$A = \frac{2}{\rho(\varphi)} \log(\cosh\beta).$$

In the limit (as $\varepsilon \to 0$) the function Y/ε is Dirac's δ-function, i.e., as $\varepsilon \to 0$,

$$\frac{1}{\varepsilon} Y\Big(\frac{x-\varphi}{\varepsilon} + \varphi_1, t\Big) \to a(t)\delta(x-\varphi(t)),$$

where $a = \int_{-\infty}^{\infty} Y(\tau,t)\,d\tau$.

3.3. Soliton-type asymptotics for a Langmuir chain

The study of the spectrum of Langmuir oscillations in a plasma leads to the dicrete dynamical system

$$M_k \frac{dN_k}{dt} + N_k(N_{k+1} - N_{k-1}) = 0, \quad k = 0, \pm 1, \pm 2, \dots.$$

For constant M_k this system can be integrated by the method of the inverse scattering problem [**266**].

We will assume that the $M_k > 0$ are small functions of t slowly varying in k. Denoting by ε a small parameter characterizing the magnitude of M_k, we introduce a smooth function $\mu(x,t) > 0$ such that

$$M_k(t) = 2\varepsilon\mu(x,t)\big|_{x=k\varepsilon}, \quad k = 0, \pm 1, \pm 2, \dots.$$

It is not hard to see that in the case of smooth dependence of the N_k on k averaging of (3.98), i.e., the passage to the limit as $\varepsilon \to 0$, leads to the Hopf equation. Our goal is averaging (3.98) in the case where the N_k form a discrete analog of a soliton-type function.

In order to state the main result, we need some auxiliary functions. Let $w_0 = w_0(x,t) > 0$ be a smooth solution of the equation

$$\frac{\partial w_0}{\partial t} + \frac{w_0}{\mu}\frac{\partial}{\partial x} w_0 = 0, \tag{3.99}$$

and let the functions $\varphi = \varphi(t)$ and $\beta = \beta(t)$ satisfy the system of equations

$$\frac{d\varphi}{dt} = \frac{\sinh 2\beta}{2\mu\beta} w_0\Big|_{x=\varphi}, \tag{3.100}$$

$$\frac{d}{dt}\Big\{(\sinh 2\beta - 2\beta)\mu w_0\big|_{x=\varphi}\Big\} = (\sinh 2\beta - 2\beta)\Big(\frac{\partial}{\partial t}\mu w_0\Big)\Big|_{x=\varphi}. \tag{3.101}$$

Then the leading term of the asymptotic solution to (3.98) has the form

$$N_k = Y\Big(\frac{x-\varphi(t)}{\varepsilon} + \varphi_1(t), t\Big) w_0(x,t)\Big|_{x=k\varepsilon} + O(\varepsilon), \tag{3.102}$$

where

$$Y(\tau,t) = \frac{\cosh(\beta(\tau-2))\cosh(\beta(\tau+1))}{\cosh(\beta(\tau-1))\cosh\beta\tau}. \tag{3.103}$$

In some particular cases the equation (3.101) can be integrated. Namely, if $w_0 = 1$ and $\mu = \mu(x)$, we obtain from (3.101)

$$\mu(\varphi)(\sinh 2\beta - 2\beta) = E_1, \tag{3.104}$$

and for $\mu = \text{const}$ we obtain

$$\sqrt{w_0(\varphi,t)}\cosh\beta = E_2.$$

Here E_1, E_2 are positive constants. It is easily seen that if $w_0 = 1$ and $\mu = \text{const}$, then (3.100), (3.101) imply that $\beta = \text{const}$ and φ is a linear function of t. In this case the first term in (3.102) is an exact solution to the Langmuir chain with constant coefficients.

With the aid of the explicit formula (3.103), it is not hard to see that the asymptotic formula (3.102) describes a distorted solitary wave with amplitude $A = w_0(\varphi,t)(\cosh 2\beta - 1)$ and velocity $\dot\varphi$ determined by (3.100), (3.101) which moves over the smooth background $w_0(x,t)$. Thus in the limit (as $\varepsilon \to 0$) averaging of (3.98) leads to the Hopf equation (3.99) describing the limit of N_k at the points other than $x = \varphi(t)$ and to equations (3.100), (3.101) describing the motion of the singular point. This singularity has the form of an isolated peak, i.e., it equals zero everywhere except at the point $x = \varphi(t)$, where it is equal to $A(t) > 0$. The same singularity results from the pointwise passage to the limit for the soliton-type asymptotic solution to the Korteweg–de Vries equation with small dispersion. In this respect it is instructive to compare the first Hugoniot type condition (3.100) for the Langmuir chain with the Hugoniot condition

$$\dot\psi = \frac{1}{\mu}\Big(U_0(\psi,t) + \frac{a(t)}{3}\Big) \tag{3.105}$$

for the soliton-type solution

$$v = U_0(x,t) + a(t)\cosh^{-2}\Big(\beta_1\Big(\frac{x-\psi(t)}{\varepsilon} + \psi_1\Big)\Big) \tag{3.106}$$

of the Korteweg–de Vries equation

$$\frac{\partial v}{\partial t} + \frac{1}{2\mu}\frac{\partial v^2}{\partial x} = -\varepsilon^2\rho\frac{\partial^3 v}{\partial x^3} \tag{3.107}$$

(see §1.3). Here the background U_0 is a smooth solution to the Hopf equation (3.99), and $\rho = \rho(x,t) \neq 0$ is an arbitrary smooth function. Note that the relation (3.105) does not depend on the dispersion coefficient ρ. This is related to the fact that, in the weak sense, the dispersion term in the right-hand side of (3.107) vanishes as $\varepsilon \to 0$.

Now we rewrite the solution (3.102) in an additive form of type (3.106) and rewrite the condition (3.100) in the form similar to (3.105):

$$\dot{\varphi} = \frac{1}{\mu}\Big(w_0(\varphi, t) + \frac{A(t)}{3} q\Big), \tag{3.108}$$

where

$$q = \frac{3}{4}\,\frac{\sinh 2\beta - 2\beta}{\beta \sinh^2 \beta}.$$

It is not hard to calculate that $q = 1 + O(\beta^2)$ for small β. However, for β finite q differs from one. The distinction between conditions (3.105) and (3.108) can be easily explained by the "non-divergent" form of equations (3.98). Indeed, pass from (3.98) to the differential-difference equation

$$\mu \frac{\partial u(x,t)}{\partial t} + \frac{1}{2\varepsilon} u(x,t)\big(u(x+\varepsilon, t) - u(x-\varepsilon, t)\big) = 0. \tag{3.109}$$

Next, neglecting the derivatives of the fifth order (which, strictly speaking, is incorrect), rewrite (3.109) as

$$\mu \frac{\partial u}{\partial t} + u \frac{\partial u}{\partial x} + \frac{1}{6}\varepsilon^2 u \frac{\partial^3 u}{\partial x^3} = 0. \tag{3.110}$$

Now consider the equation (3.110) in the weak sense:

$$\int_0^\infty \int_{-\infty}^\infty \Big\{(\mu f)_t u + \frac{1}{2} f_x u^2 + \frac{\varepsilon^2}{12} f_{xxx} u^2 - \frac{1}{4} f_x (\varepsilon u_x)^2\Big\}\, dx\, dt = 0, \tag{3.111}$$

where $f = f(x,t)$ is an arbitrary smooth compactly supported function such that $f(x,0) = 0$.

We see that the third term in (3.111) vanishes as $\varepsilon \to 0$, whereas for soliton-type functions ($u_x \sim \varepsilon^{-1}$) the last term has the same magnitude as the first two.

It can be easily seen that a similar treatment of equation (3.107) leads to the equality

$$\int_0^\infty \int_{-\infty}^\infty \Big\{(\mu f)_t v + \frac{1}{2} f_x v^2 + \varepsilon^2 (\mu \rho f)_{xxx} v\Big\}\, dx\, dt = 0,$$

in which only two terms remain as $\varepsilon \to 0$.

In the light of the distinction mentioned above the Langmuir chain (3.107) is a good model example for the theory of finite-difference schemes. Indeed, if we regard (3.98) as a differential-difference scheme, we can see the consequence of non-divergent approximation

$$u u_x \sim \frac{1}{2\varepsilon} N_k (N_{k+1} - N_{k-1})$$

of the quadratic term. It is easily seen that the divergent approximation, e.g.,

$$u u_x \sim \frac{1}{4\varepsilon}(N_{k+1}^2 - N_{k-1}^2),$$

does not give rise to any additional terms when taking the weak limit.

Construction of an asymptotic solution of (3.102) is carried through along the same lines as in previous examples. We only point out here that it is technically convenient to pass from (3.98) to the pseudodifferential equation

$$\varepsilon\mu\frac{\partial u}{\partial t}+\sinh\left(\varepsilon\frac{\partial}{\partial x}\right)e^{u}=0,\qquad N_k=\exp u(x,t)\big|_{x=k\varepsilon},$$

and that for $w_0=1$, $\mu=\mu(x)$ the equation for the phase shift φ_1 takes a relatively simple form:

$$\frac{d}{dt}\left\{\alpha\frac{d\widetilde{\varphi}_1}{dt}-\frac{E_1\mu_x}{4\mu^2}\bigg|_{x=\varphi}\widetilde{\varphi}_1\right\}+g=0.$$

Here $\widetilde{\varphi}_1=(2\varphi_1-1)\mu(\varphi)$,

$$\begin{aligned}\alpha&=\frac{\mu\beta}{2\sinh 2\beta(1-2\beta\sinh 2\beta)}\left\{\frac{E_1}{\mu}+I\beta\coth\beta\right.\\&\qquad\left.+\sinh 2\beta(1-2\beta\coth 2\beta)\right\}\bigg|_{x=\varphi},\\ g&=\frac{d}{dt}\left\{\frac{2\beta\coth\beta-1}{2\beta\coth 2\beta-1}\frac{dr}{dt}\right\}-\mu\frac{\sinh 2\beta}{E_1}\frac{d^2r}{dt^2}\\&\qquad+\frac{1}{\mu\beta E_1}\left(\frac{dr}{dt}\right)^2+\frac{1}{4}E_1\dot{\varphi}\frac{\partial u^-}{\partial x}\bigg|_{x=\varphi},\\ I&=\int_{-\infty}^{\infty}(Y(\tau,t)-1)\ln Y(\tau,t)\,d\tau,\end{aligned}$$

where $r=\mu^2(\varphi)\beta^2$, the constant E_1 and the function Y are defined by (3.103), (3.104), and the function $u^-=u^-(x,t)$ is found as the solution to the problem

$$\begin{aligned}&\mu\frac{\partial u^-}{\partial t}+\frac{\partial u^-}{\partial x}=0,\qquad x<\varphi(t),\quad t>0,\\&u^-\big|_{x=\varphi}=4\big(1-\dot{\varphi}\mu(\varphi)\big)\frac{d}{dt}(\beta\mu(\varphi)),\end{aligned}$$

completed with an initial condition for $t=0$, $x<\varphi(0)$. Clearly, the meaning of u^- is the tail perturbation arising after the passage of the distorted solitary wave.

CHAPTER 4

Problems with a Free Boundary

Traditionally, a problem with free boundary means a problem of dynamical description of a medium which may exist in one of two phases (e.g., water and ice) when the boundary (interface) between the domains occupied by these phases moves with time and its position is unknown *a priori*. Such problems arise in many areas, not only in physics (see, e.g., [**159, 171**]), but even in financial mathematics [**230**].

The first mathematical model for problems with a free boundary was proposed by Stefan. It consists in solving the heat equation in each of the domains,

$$\frac{\partial \theta^\pm}{\partial t} = k^\pm \frac{\partial^2 \theta^\pm}{\partial x^2}, \qquad x \in \Omega_t^\pm, \quad t > 0, \tag{4.1}$$

completed with conditions on the boundary $x = \varphi(t)$ between the phases

$$\theta^+\big|_{x=\varphi} = \theta^-\big|_{x=\varphi} = 0, \tag{4.2}$$

$$\left[k\frac{\partial \theta}{\partial x}\right] = -2V, \tag{4.3}$$

as well as by natural initial and boundary conditions (the latter on the outer boundary). Here $\Omega_t^\pm$ are the subintervals of $(a, b) \subset \mathbb{R}^1$ separated by the point $x = \varphi(t)$, i.e., $\Omega_t^- = \{x \in (a, \varphi(t))\}$ and $\Omega_t^+ = \{x \in (\varphi(t), b)\}$, the functions $\theta^\pm = \theta^\pm(x, t)$ mean the temperatures of the phases occupying the intervals $\Omega_t^\pm$ respectively, V is the velocity of the boundary motion, and the square brackets denote, as usual, the amplitude of the jump of the function on the boundary $x = \varphi(t)$, i.e.,

$$\left[k\frac{\partial \theta}{\partial x}\right] = k^+\frac{\partial \theta^+}{\partial x}\bigg|_{x=\varphi(t)+0} - k^-\frac{\partial \theta^-}{\partial x}\bigg|_{x=\varphi(t)-0},$$

where $k^\pm > 0$ are the coefficients of heat conductivity, which are assumed to be constant.

It is easily seen that the problem (4.1)–(4.3) is nonlinear and involves three unknown functions: the temperatures $\theta^\pm$ and the function $\varphi(t)$ describing the boundary position.

Subsequently the problem (4.1)–(4.3), which was called the classical Stefan problem, as well as its natural multidimensional extension, were intensively studied by both physicists and mathematicians (see, e.g., the book [**171**] and references therein). It was established, in particular, that condition (4.2), implying a fixed temperature on the free boundary, is too coarse because in actual media the temperatures at various points of the interface surface Γ_t may differ slightly, and their values $\theta^{\pm}|_{\Gamma_t}$ depend on the geometry of the surface Γ_t. To remedy this drawback, physicists proposed a more adequate model, which was called the modified Stefan problem or the Stefan problem with surface tension. In the multidimensional case this model has the form

$$\frac{\partial \theta^{\pm}}{\partial t} = k^{\pm} \Delta \theta^{\pm}, \qquad x \in \Omega_t^{\pm}, \quad t > 0, \tag{4.4}$$

$$[\theta] = 0, \tag{4.5}$$

$$\theta^{\pm}\big|_{\Gamma_t} = \kappa_1 \mathcal{K} + \kappa_2 V_\nu, \tag{4.6}$$

$$\left[k \frac{\partial \theta}{\partial \nu}\right] = -\, 2V_\nu, \tag{4.7}$$

where ν is the normal to Γ_t, V_ν is the normal velocity of Γ_t, and $\mathcal{K}$ is the mean curvature of Γ_t. We see that the problem (4.4)–(4.7), like the classical Stefan problem, contains the condition of continuity of the temperature on the boundary (for methodological reasons we have written down this condition in the explicit form (4.5)), as well as the so-called Stefan condition (4.7) which relates the magnitude of the jump of the normal flow $\kappa \partial\theta/\partial\nu$ to the normal (to Γ_t) velocity V_ν of the boundary motion. Moreover, (4.2) is replaced by the so-called Gibbs–Thomson condition (4.6) which relates the temperature on the interface to the mean curvature $\mathcal{K}$ of the surface Γ_t and the normal velocity of its motion. Finally, κ_1, κ_2 in (4.6) are constants whose physical meaning is the surface tension.

However, it turned out that neither the classical nor the modified Stefan model is universal: there are a number of physically meaningful situations where the motion of the free boundary is described by other laws (the Mullins–Sekerka, Hele–Shaw, and other models). Moreover, detailed physical research revealed that actually there is no strict boundary between different phases, but rather a thin boundary layer containing the substance in an intermediate state. The microscopic description of processes in the boundary layer is given by the Ginzburg–Landau theory of phase transitions (see, e.g., [**245**]). Based on this theory, detailed and very complicated mathematical models, as well as simplified equationss that preserve physical adequacy, were obtained, see, e.g., [**4–6, 29, 30, 34–36, 80, 208**] and references therein.

This brings up the physically and mathematically meaningful question of averaging these microscopic models, i.e., of deriving limiting macroscopic equations describing the motion of the free boundary. It is clear that the solution of this global problem would enable us to determine the form of the limiting problems corresponding to the various particular physical settings. There is an extensive literature devoted to the study of this problem. Without aiming at completeness, we refer the reader to the papers [**31–33, 49, 50, 60, 61, 113, 130–132, 185, 186, 196–200, 211, 216, 234, 235, 237**] and references therein.

In this chapter we treat two microscopic models by Caginalp to demonstrate that the method of constructing asymptotic solutions developed in this book enables us to reduce the derivation of the limiting equations to very simple calculations. The material of this chapter is based on the results of [**49, 50, 196–201**].

4.1. Dynamics of the free boundary in the phase field model

As the basic model providing a microscopic description of processes inside and outside the transition zone, we will use the phase field equations

$$
\begin{aligned}
&\frac{\partial \theta}{\partial t} + \frac{l}{2}\frac{\partial \psi}{\partial t} = \langle \nabla, k(\psi, x)\nabla\theta \rangle, \\
&\tau_0 \frac{\partial \psi}{\partial t} = \xi^2 \Delta\psi + \frac{1}{2a}(\psi - \psi^3) + g(\psi)\theta.
\end{aligned}
\tag{4.8}
$$

Here θ is a normalized function with the meaning of temperature, and ψ is the so-called phase function (also referred to as order function), which is normalized so that the values $\psi = \pm 1$ correspond to the main phases of the medium (we will call them solid for $\psi = 1$ and liquid for $\psi = -1$). The meaning of the function $k(\psi, x) > 0$ is the coefficient of heat conductivity, and

$$
k(\psi, x) = \frac{k^+ + k^-}{2} + \frac{k^+ - k^-}{2}\eta(\psi), \tag{4.9}
$$

where $k^\pm = k^\pm(x) > 0$ are the coefficients of heat conductivity in the solid and liquid phases, and $\eta(\psi)$ is a smooth function such that $\eta \to 1$ as $\psi \to 1$ and $\eta \to -1$ as $\psi \to -1$. The constant coefficients l, τ_0, ξ, a characterize the substance and mean physically the latent heat of melting, the relaxation time, the surface tension, and the width of the transition zone, respectively. Finally, $g(\psi)$ is a smooth function. According to the physical meaning of equations (4.8) we assume that

$$
\tau_0 \ll 1, \quad \xi \ll 1, \quad a \ll 1,
$$

i.e., these coefficients may be treated as small parameters.

Under the constancy assumption on the coefficient of heat conductivity, $k^+ = k^- = \text{const}$, this model was proposed by Caginalp [**29**] and was afterwards studied by many authors, see [**40, 50, 196, 200, 211, 234**] and references therein. In particular, it was established in the case $k^+ = k^- = \text{const}$, $g = \text{const}$ that for different relations between the parameters τ_0, ξ, a, and l the equation (4.8) gives rise to different limiting problems, including the classical and modified Stefan problems.

We will obtain a similar result for variable coefficient of heat conductivity $k = k(\psi, x)$, which may be different in the solid and liquid phases, and for the function g localized in a neighborhood of the free boundary,

$$
g(\psi) = \kappa(\psi^2 - 1), \quad \kappa = \text{const}.
$$

The latter condition enables us to use the model (4.8) outside the transition zone.

Note that originally the asymptotic solution of the phase field equations was constructed by the matching method, which allowed one to obtain only its local description (see, e.g., [**29, 32**]). The first asymptotic formulas applicable in the entire range of spatial variables were obtained in [**49, 196**] using the method set forth in this book.

We will restrict ourselves to the question of correspondence between the microscopic model (4.8) and the macroscopic Stefan problem. From the physical viewpoint this corresponds to the following relationship between the parameters l, τ_0, ξ, and a:

$$l = \text{const}, \qquad \frac{\tau_0}{a} = \text{const}, \qquad \frac{\sqrt{a}}{\xi} = \text{const}, \qquad a \ll 1.$$

To simplify the notation, we will assume that $\varepsilon = 2a$ is a small parameter and $\tau_0 = \varepsilon$, $\xi^2 = \varepsilon$, $l = 2$. Then the original microscopic model becomes

$$\frac{\partial \theta}{\partial t} + \frac{\partial \psi}{\partial t} = \langle \nabla, k(\psi, x)\nabla\theta \rangle, \qquad x \in \Omega, \quad t \in (0, T), \tag{4.10}$$

$$\varepsilon^2 \frac{\partial \psi}{\partial t} = \varepsilon^2 \Delta\psi + \psi - \psi^3 + \varepsilon\kappa(\psi^2 - 1)\theta, \tag{4.11}$$

$$\theta\big|_{t=0} = \theta^0(x, \varepsilon), \quad \psi\big|_{t=0} = \psi^0(x, \varepsilon),$$

$$\theta\big|_{\Sigma} = \theta_b, \quad \psi\big|_{\Sigma} = 1.$$

Here Ω is a smooth domain in $\mathbb{R}^n$, $n = 2, 3$, with smooth boundary $\partial\Omega$, $\Sigma = \partial\Omega \times (0, T]$, $T > 0$, and $\theta_b = \theta_b(x)\big|_{\Sigma}$ is a given smooth function; for definiteness, we assume that in the boundary condition $\psi\big|_{\Sigma} = 1$ the domain adjacent to the outer boundary is occupied by the solid phase. Moreover, we assume that θ^0, ψ^0 for $\varepsilon = \text{const}$ are smooth functions. Furthermore, a smooth surface Γ_0 of codimension 1 is given which is disjoint from the outer boundary and divides Ω into two parts $\Omega_0^{\pm}$ such that

$$\psi^0(x, \varepsilon) \to \pm 1, \quad \theta^0(x, \varepsilon) \to \theta^0_{\pm}(x) \qquad \text{as} \quad \varepsilon \to 0, \ x \in \Omega_0^{\pm},$$

with $\theta^0_+\big|_{\Gamma_0} = \theta^0_-\big|_{\Gamma_0}$.

From the limiting point of view (as $\varepsilon \to 0$), Γ_0 is the initial position of the boundary between the phases, and $\theta^0_{\pm}$ are their temperatures which appear in the macroscopic description, while the prelimiting initial data $\psi^0(x, \varepsilon), \theta^0(x, \varepsilon)$ are certain regularizations of the Heaviside function and $\theta^0_{\pm}$. More detailed assumptions on θ^0 and ψ^0 will be formulated later on.

We will derive the limiting problem for (4.11) as $\varepsilon \to 0$ by constructing its asymptotic solution. First of all, note that by the physical meaning of the problem (4.10), (4.11) as described above, the prelimiting phase function $\psi(x, t, \varepsilon)$ must be a regularization of the Heaviside function $H(x, t)$ which takes the values ± 1 in the domains $\Omega_t^{\pm}$ occupied at time t by the solid and liquid phases. Hence we can use the same tools in constructing the asymptotics of ψ as in constructing a solution of smoothed shock wave type. Further, as follows from the conditions on the free boundary (4.5), (4.7), the limiting temperature is continuous, but its derivatives are discontinuous on the surface Γ_t. The model description of such behavior in the theory of distributions is given by the function τ_+, where $\tau_+ = \tau$ for $\tau > 0$ and $\tau_+ = 0$ for $\tau < 0$. Thus the prelimiting temperature $\theta = \theta(x, t, \varepsilon)$ must be a regularization of a function of type τ_+.

These preliminary arguments enable us to choose the form of the tentative asymptotic solution. Namely, we will seek the asymptotic solution to the problem (4.10), (4.11) in the form

$$\begin{aligned} \psi &= \chi(\tau, x) + \varepsilon\{\psi_1(x, t) + W_1(\tau, x, t)\} + \dots, \\ \theta &= \theta_0(x, t) + \rho(x, t)V(\tau, x, t) + \varepsilon\{\theta_1(x, t) + U_1(\tau, x, t)\} + \dots, \end{aligned} \tag{4.12}$$

where the fast variable has the form

$$\tau = \frac{t - \varphi(x)}{\varepsilon}, \tag{4.13}$$

$\varphi(x), \psi_i(x,t), \theta_i(x,t), \rho(x,t)$ are smooth functions,

$$\Gamma_0 = \{x,\ \varphi(x) = 0\}, \quad \rho\big|_{t=\varphi(x)} = 0, \tag{4.14}$$

and χ, V, W_i, U_i are smooth functions of τ, x, t stabilizing exponentially fast as $\tau \to \pm\infty$. We assume here that

$$\chi \to \pm 1 \quad \text{as} \quad \tau \to \pm\infty, \qquad V, W_i, U_i \to 0 \quad \text{as} \quad \tau \to \infty. \tag{4.15}$$

Let us discuss the form of the asymptotic expansion (4.12) in comparison with the similar expansion for a smoothed shock wave (see §2.5). The choice of the fast variable in the form (4.13) means the assumption that the free boundary Γ_t, like the front of the shock wave, moves monotonically (i.e., its velocity has a nonzero component normal to Γ_t), being a smooth surface of codimension 1. It is equivalent to the assumption that the position of Γ_t at each time moment t can be described by a smooth function $\varphi(x)$, i.e., $\Gamma_t = \{x \in \Omega,\ \varphi(x) = t\}$. The function φ is not known in advance, but the position of its zero level surface is known, which is expressed by the first equality in (4.14). Moreover, unlike the problem on the smoothed shock wave, we know in advance the limiting values of the phase function outside the free boundary. For this reason we impose the first condition (4.15) and can choose the function χ to be independent of t.

The leading term of the expansion for temperature in (4.12) contains no term similar to χ, since the limit of θ as $\varepsilon \to 0$ must be continuous. However, as compared to the smoothed shock wave, it contains an additional term ρV. According to the second condition (4.14), this term vanishes on Γ_t, so it does not destroy continuity of the limiting temperature. At the same time, the condition

$$\frac{\partial \rho}{\partial \nu}\bigg|_{\Gamma_t} \neq 0, \tag{4.16}$$

which will be shown to hold, implies that the limiting value of the normal derivative of the temperature is discontinuous on Γ_t. Thus the term ρV models a smoothed function of type τ_+. The remaining terms in (4.12) have the same meaning as in the smoothed shock wave case, the second condition in (4.15) being simply a normalization condition.

Now we construct the terms of the asymptotic expansion (4.12). Following our standard scheme, we obtain first the equations for the regular terms of (4.12), i.e., the background functions $\psi_i(x,t)$, $i = 1, 2, \dots$, and $\theta_j(x,t)$, $j = 0, 1, \dots$, and the tail functions $W_i^-(x,t)$, $U_i^-(x,t)$, $i = 1, 2, \dots$, and $\rho V^-(x,t)$. As before, we denote

$$f^-(x,t) = \lim_{\tau \to -\infty} f(\tau, x, t).$$

For definiteness, let the function φ decrease when passing from Ω_0^- into Ω_0^+. (This means that we consider the case of freezing, i.e., the free boundary moves in the direction of the liquid phase. This assumption is not important for constructing the asymptotics, since the case of melting is quite similar.) Then the domains $\Omega_t^- = \{x,\ t-\varphi(x)<0\}$ and $\Omega_t^+ = \{x,\ t-\varphi(x)>0\}$ for all $t\ge 0$ correspond to the ones occupied by the liquid and solid phases, respectively. Accordingly, as $\tau\to+\infty$ we eventually get into the solid phase, and by (4.15)

$$\psi = 1+\sum_{i=1}^{N}\varepsilon^i\psi_i(x,t),\quad \theta=\sum_{j=0}^{N}\varepsilon^j\theta_j(x,t), \tag{4.17}$$

where $N\ge 1$ is the number of terms in the asymptotic expansion (4.12). As $\tau\to-\infty$ we get into the liquid phase, and

$$\begin{aligned}\psi &= -1+\sum_{i=1}^{N}\varepsilon^i(\psi_i(x,t)+W_i^-(x,t)),\\ \theta &= \theta_0(x,t)+\rho_0V_0^-(x,t)+\sum_{i=1}^{N}\varepsilon^i(\theta_i(x,t)+U_i^-(x,t)).\end{aligned} \tag{4.18}$$

Clearly in Ω_t^- the terms ψ_i and W_i^-, as well as $\theta_i,\rho_0V_0^-,U_i^-$, for each i are unidentifiable, because the solution contains their sums. Hence, instead of (4.18), it is natural to use the expansions

$$\psi=-1+\sum_{i=1}^{N}\varepsilon^i\psi_i^-(x,t),\quad \theta=\sum_{j=0}^{N}\varepsilon^j\theta_j^-(x,t), \tag{4.19}$$

where $\psi_i^-=\psi_i+W_i^-$, $\theta_0^-=\theta_0+\rho_0V_0^-$, $\theta_j^-=\theta_j+U_j^-$, and $x\in\Omega_t^-$. For symmetry of notation we rewrite (4.17) as

$$\psi=1+\sum_{i=1}^{N}\varepsilon^i\psi_i^+(x,t),\quad \theta=\sum_{j=0}^{N}\varepsilon^j\theta_j^+(x,t), \tag{4.20}$$

where $\psi_i^+=\psi_i$, $\theta_j^+=\theta_j$, and $x\in\Omega_t^+$.

Substitute the expansion (4.12) into equations (4.10), (4.11), pass to the limit as $\tau\to\pm\infty$, and use the notation (4.19), (4.20). Then, equating to zero the terms of order ε^0, we see that (4.11) holds $\operatorname{mod} O(\varepsilon)$, while (4.10) implies the equations

$$\frac{\partial\theta_0^\pm}{\partial t}=\langle\nabla,k^\pm(x)\nabla\theta_0^\pm\rangle,\qquad x\in\Omega_t^\pm,\quad t>0. \tag{4.21}$$

Obviously, we have obtained an analog of the heat equations (4.4) from the modified Stefan problem.

Next, consider the regular terms of order ε in (4.10), (4.11) to see that (4.10) yields a homogeneous linear equation for $\theta_1^\pm$ similar to (4.21), while (4.11) within $O(\varepsilon^2)$ implies that $\psi_1^\pm$ are equal to zero since $\chi^\pm=\pm1$. By a similar treatment of the higher order terms we again obtain from (4.10) equations of the form (4.21), i.e.,

$$\frac{\partial\theta_j^\pm}{\partial t}=\langle\nabla,k^\pm(x)\nabla\theta_j^\pm\rangle,\qquad x\in\Omega_t^\pm,\quad t>0, \tag{4.22}$$

while (4.11) implies that $\psi_j^\pm = 0$, because all the previous $\psi_k^\pm$ were equal to zero and $\eta(\chi) = \text{const}$ outside a small neighborhood of Γ_t. Clearly, the equations $\psi_j^\pm = 0$ imply that $\psi_j = 0$ and $W_j^- = 0$, $j = 1, \dots, N$, which, in its turn, means that the $W_j(\tau, x, t)$ must be soliton-type functions. This enables us to simplify the form of the asymptotic expansion (4.12). Namely, we set $\psi_j = 0$ and $W_j = W_j(\tau, x)$ for all $j = 1, \dots, N$ in (4.12).

Now we determine rapidly varying components of the asymptotic expansion for the solution. Substituting (4.12) into (4.11) and equating to zero the terms of order ε^0, we obtain the equation

$$|\nabla\varphi|^2 \frac{\partial^2 \chi}{\partial \tau^2} + \chi - \chi^3 = 0. \tag{4.23}$$

Using the first condition (4.15), we immediately find the function χ from this equation:

$$\chi = \tanh\Big(\frac{\beta}{\sqrt{2}}(\tau + \varphi_1)\Big),$$

where $\beta = 1/|\nabla\varphi|$, and $\varphi_1 = \varphi_1(x)$ is the "constant" of integration.

In the treatment of equation (4.10) we have to take the following identity into account:

$$\begin{aligned}
&\Big\langle \nabla_x, k\Big(\psi\Big(\frac{t-\varphi(x)}{\varepsilon}, x, \varepsilon\Big), x\Big) \nabla_x \Big(\theta_0(x,t) + \rho(x,t) V\Big(\frac{t-\varphi(x)}{\varepsilon}, x, t\Big)\Big)\Big\rangle \\
&= \Big\{ \frac{|\nabla\varphi|^2}{\varepsilon^2} \rho(x,t) \frac{\partial}{\partial\tau} \big(k(\psi(\tau,x,\varepsilon)) V(\tau,x,t)\big) \\
&\quad - \frac{\rho(x,t)}{\varepsilon} \frac{\partial}{\partial\tau} \Big(2\big\langle \nabla\varphi, \nabla_x (k(\psi(\tau,x,\varepsilon),x) V(\tau,x,t))\big\rangle + kV\Delta\varphi \Big) \\
&\quad - \frac{1}{\varepsilon} \Big(\langle \nabla\varphi, \nabla\theta_0\rangle \frac{\partial}{\partial\tau} k(\psi(\tau,x,\varepsilon),x) \\
&\quad + \langle\nabla\varphi, \nabla\rho\rangle \Big(2k \frac{\partial V(\tau,x,t)}{\partial\tau} + V \frac{\partial k(\psi(\tau,x,\varepsilon),x)}{\partial\tau}\Big)\Big) \\
&\quad + \langle \nabla_x, k(\psi(\tau,x,\varepsilon),x) \nabla_x (\theta_0 + \rho V(\tau,x,t))\rangle \Big\}\Big|_{\tau=(t-\varphi)/\varepsilon}.
\end{aligned} \tag{4.24}$$

Observe that the first term in the right-hand side of (4.24) has the form $\varepsilon^{-2}\rho v$, where v is a function exponentially decreasing as $\tau \to \pm\infty$. This enables us to expand ρ by the Taylor formula about the point $t = \varphi(x)$ and to see, by the second condition (4.14), that this term is of order $O(\varepsilon^{-1})$.

These transformations allow us to rewrite the right-hand side of (4.24) as

$$\begin{aligned}
&\frac{1}{\varepsilon}\Big\{ |\nabla\varphi|^2 \frac{\partial\rho}{\partial t} \tau \frac{\partial}{\partial\tau}\Big(k\frac{\partial V}{\partial\tau}\Big) - \langle\nabla\varphi, \nabla\theta_0\rangle \frac{\partial k}{\partial\tau} \\
&\quad - \langle\nabla\varphi, \nabla\rho\rangle \Big(2k\frac{\partial V}{\partial\tau} + V\frac{\partial k}{\partial\tau}\Big)\Big\}\Big|_{t=\varphi,\, \tau=(t-\varphi)/\varepsilon} \\
&+ \Big\{ |\nabla\varphi|^2 \tau^2 \Big(\frac{\partial\rho}{\partial t}\frac{\partial}{\partial\tau}\Big(k\frac{\partial V}{\partial t}\Big) + \frac{1}{2}\frac{\partial^2\rho}{\partial t^2}\frac{\partial}{\partial\tau}\Big(k\frac{\partial V}{\partial\tau}\Big)\Big)
\end{aligned} \tag{4.25}$$

$$
\begin{aligned}
&-\tau\frac{\partial\rho}{\partial t}\frac{\partial}{\partial\tau}\big(2\langle\nabla\varphi,\nabla_x(kV)\rangle+kV\Delta\varphi\big)\\
&-\tau\frac{\partial}{\partial t}\Big(\langle\nabla\varphi,\nabla\rho\rangle\Big(2k\frac{\partial V}{\partial\tau}+V\frac{\partial k}{\partial\tau}\Big)\Big)\Big\}\Big|_{t=\varphi,\,\tau=(t-\varphi)/\varepsilon}\\
&+\big\langle\nabla_x,k\nabla_x(\theta_0+\rho V)\big\rangle\big|_{\tau=(t-\varphi)/\varepsilon}.
\end{aligned}
$$

We assume here that V and ψ are functions of τ, x, t, which are treated as independent variables when differentiating.

Thus we see that when substituting (4.10) into the right-hand side of equation (4.12) the resulting expression is of order $O(\varepsilon^{-1})$ rather than $O(\varepsilon^{-2})$, which would be the case for a solution of shock wave type.

Treating similarly the other terms resulting from substitution of (4.12) into (4.10) and equating to zero all the terms of order ε^{-1}, we obtain the equality

$$
\begin{aligned}
\frac{\partial\chi}{\partial\tau}=\Big\{&|\nabla\varphi|^2\frac{\partial\rho}{\partial t}\tau\frac{\partial}{\partial\tau}\Big(k_0\frac{\partial V}{\partial\tau}\Big)-\langle\nabla\varphi,\nabla\theta_0\rangle\frac{\partial k_0}{\partial\tau}\\
&-\langle\nabla\varphi,\nabla\rho\rangle\Big(2k_0\frac{\partial V}{\partial\tau}+V\frac{\partial k_0}{\partial\tau}\Big)+|\nabla\varphi|^2\frac{\partial}{\partial\tau}\Big(k_0\frac{\partial}{\partial\tau}U_1\Big)\Big\}\Big|_{\Gamma_t},
\end{aligned}
\tag{4.26}
$$

where $k_0 = k(\chi, x)$.

Obviously, the following condition is necessary for (4.26) to hold:

$$
\begin{aligned}
\int_{-\infty}^{\infty}\Big\{|\nabla\varphi|^2\frac{\partial\rho}{\partial t}\tau\frac{\partial}{\partial\tau}(k_0V)-\langle\nabla\varphi,\nabla\rho_0\rangle\Big(2k_0\frac{\partial V}{\partial\tau}+V\frac{\partial k_0}{\partial\tau}\Big)\Big\}\,d\tau\Big|_{\Gamma_t}&\\
=2+\langle\nabla\varphi,\nabla\theta_0\rangle(k^+-k^-)\big|_{\Gamma_t}&.
\end{aligned}
\tag{4.27}
$$

Observe that for any function ρ vanishing on the surface Γ_t we have the equality

$$
\nabla\rho\big|_{\Gamma_t}=-\frac{\partial\rho}{\partial t}\nabla\varphi\Big|_{\Gamma_t}.
\tag{4.28}
$$

Using this relation, we can easily evaluate the integral in (4.27) to obtain

$$
|\nabla\varphi|^2\frac{\partial\rho}{\partial t}k^-V^-\Big|_{\Gamma_t}=-2-\langle\nabla\varphi,\nabla\theta_0\rangle(k^+-k^-)\big|_{\Gamma_t}.
\tag{4.29}
$$

Combined with (4.28), this implies (4.16) because generically the right-hand side of (4.29) does not vanish.

According to the above assumptions, define the unit normal vector to the surface Γ_t and the velocity of motion of Γ_t along this normal as

$$
\nu=-\frac{\nabla\varphi}{|\nabla\varphi|},\qquad V_\nu=-\frac{1}{|\nabla\varphi|}.
\tag{4.30}
$$

Then, using once again (4.28) and the condition $\rho_0|_{\Gamma_t} = 0$, the relation (4.29) can be transformed to

$$
\big\langle\nu,\{k^+\nabla\theta_0-k^-\nabla(\theta_0+\rho V^-)\}\big\rangle\big|_{\Gamma_t}=-2V_\nu.
\tag{4.31}
$$

It remains to observe that the expression in braces is the difference of flows from the right and from the left of Γ_t. Thus (4.31) is nothing but the Stefan condition (4.7).

Using the relations obtained, we integrate the equality (4.26) and write it as

$$\begin{aligned} |\nabla\varphi|^2 k_0 \frac{\partial \check{U}_1}{\partial \tau} &= -|\nabla\varphi|^2 \frac{\partial \check{\rho}}{\partial t} k_0 \left(\tau \frac{\partial \check{V}}{\partial \tau} + \check{V}\right) \\ &\quad + \langle \nabla\varphi_0, \nabla\check{\theta}_0\rangle\big|_{t=\varphi}(k_0 - k^+) + \chi - 1. \end{aligned} \tag{4.32}$$

It is easily seen that we have obtained a relation which involves two free functions $\check{U}_1$ and $\check{V}$, and that the boundary condition (4.31) does not depend on the local behavior of V_0 for finite τ. Using these properties, we can arbitrarily choose the function $\check{V}$ involved in the leading term of the asymptotics, and compensate for this arbitrariness by the choice of the function $\check{U}_1$ which enters into the first correction in the asymptotics (4.12). Note that the Stefan condition (4.31) also does not depend on the function $\eta(\psi)$ involved in the definition of the coefficient $k(\psi, x)$ of heat conductivity. This means that we can choose an arbitrary way of smoothing the discontinuous limiting coefficient $k^\pm$ of heat conductivity.

So, to simplify the formulas, we take

$$V = \frac{1}{2}(1 - \chi)$$

and, accordingly, $V^- = 1$. Then evaluation of $\check{U}_1$ reduces to elementary integration of (4.32) and choosing the constant of integration by the second condition (4.15).

Now consider the terms of order ε arising when (4.12) is substituted into (4.11). After some calculations we obtain the equation

$$LW_1 = F_1(\tau, x), \tag{4.33}$$

where

$$\begin{aligned} L &= |\nabla\varphi|^2 \frac{\partial^2}{\partial \tau^2} + 1 - 3\chi^2, \\ F_1 &= (2\langle\nabla\varphi, \nabla\rangle + \Delta\varphi + 1)\frac{\partial \chi}{\partial \tau} - \kappa(\chi^2 - 1)\check{\theta}_0. \end{aligned}$$

Obviously, L is the variation of the nonlinear operator in the left-hand side of (4.23). Therefore one of the solutions of $Lw = 0$ is the function $\partial\chi/\partial\tau$. It is not hard to check that the second solution to this equation is a function growing exponentially in τ. Hence for solvability of (4.33) in the class of functions stabilizing at infinity it is necessary and sufficient that

$$\int_{-\infty}^{\infty} F_1 \frac{\partial \chi}{\partial \tau}\, d\tau = 0. \tag{4.34}$$

Moreover, since F_1 is a soliton function, W_1 will be one too under the condition (4.34).

After some algebra, (4.34) can be rewritten as

$$\left\langle \nabla, \int_{-\infty}^{\infty} \left(\frac{\partial \chi}{\partial \tau}\right)^2 d\tau \nabla\varphi \right\rangle + \left(1 + \frac{\sqrt{2}}{\beta}\kappa\check{\theta}_0\right) \int_{-\infty}^{\infty} \left(\frac{\partial \chi}{\partial \tau}\right)^2 d\tau = 0.$$

On integrating, this equality becomes

$$\langle \nabla, \beta \nabla \varphi \rangle + \sqrt{2}\kappa \check{\theta}_0 + \beta = 0. \tag{4.35}$$

To see that (4.35) is the Gibbs–Thomson condition (4.6) with $\kappa_2 = -\kappa_1 = 1/\sqrt{2}\kappa$, it remains to recall that $\beta = 1/|\nabla \varphi|$ and that the mean curvature of Γ_t is given by the formula

$$\mathcal{K} = \operatorname{div}\Big(\frac{\nabla \varphi}{|\nabla \varphi|}\Big). \tag{4.36}$$

To complete the construction of the leading term of the asymptotics, we have to determine the function $\rho(x,t)$ outside the surface Γ_t. To this end observe that, according to the above computations, the following relations hold outside an $O(\sqrt{\varepsilon})$-neighborhood of Γ_t:

$$\begin{aligned} &\theta = \theta_0(x,t) + \rho(x,t) + O(\varepsilon), \quad x \in \Omega_t^-, \quad t > 0, \\ &\theta = \theta_0(x,t) + O(\varepsilon), \quad x \in \Omega_t^+, \quad t > 0. \end{aligned} \tag{4.37}$$

Let us use the notation (4.19), i.e., define

$$\begin{aligned} &\theta_0^+(x,t) = \theta_0(x,t) \qquad \text{for} \quad x \in \Omega_t^-, \quad t > 0, \\ &\theta_0^-(x,t) = \theta_0(x,t) + \rho(x,t) \qquad \text{for} \quad x \in \Omega_t^+, \quad t > 0. \end{aligned} \tag{4.38}$$

Then (4.21), (4.31), (4.35), and the condition $\rho|_{\Gamma_t} = 0$ lead us to a natural analog of the problem (4.4)–(4.7):

$$\begin{aligned} &\frac{\partial \theta_0^\pm}{\partial t} = \langle \nabla, k^\pm(x) \nabla \theta_0^\pm \rangle, \qquad x \in \Omega_t^\pm, \quad t > 0, \\ &[\theta_0^\pm] = 0, \quad \Big[k^\pm(x) \frac{\partial \theta^\pm}{\partial \nu} \Big] = -2V_\nu, \\ &\theta^\pm \big|_{\Gamma_t} = \kappa'(V_\nu - \mathcal{K}), \end{aligned} \tag{4.39}$$

where $\kappa' = 1/\sqrt{2}\kappa$ and the quantities ν, V_ν, and $\mathcal{K}_t$ are defined by (4.30), (4.36). We complete (4.39) with the initial and boundary conditions

$$\theta^+\big|_\Sigma = \theta_b, \qquad \theta_0^\pm\big|_{t=0} = \theta_\pm^0(x) \quad \text{for} \quad x \in \Omega_0^\pm, \tag{4.40}$$

with follow in a natural way from the corresponding conditions (4.10), (4.11).

The classical solvability of the modified Stefan problem (4.39), (4.40) (provided that the initial and boundary conditions conform to each other and Γ_0 is sufficiently smooth) is well known, see [**215**]. Therefore $\theta_0^\pm$ are smooth functions defined in the domains $D^\pm = \bigcup_{t=0}^T \Omega_t^\pm$. Thus for ε sufficiently small we can determine a smooth extension $\theta_c^\pm$ in such a way that $\theta_c^\pm = \theta_0^\pm$ for $x \in \overline{\Omega_t^\pm}$, $\theta_c^\pm$ satisfy the same heat equations for $x \in \Omega_t^\mp$ and $\operatorname{dist}(x, \Gamma_t) \le \sqrt{\varepsilon}$, and $\theta_c^\pm$ are arbitrary smooth functions for $x \in \Omega_t^\mp$ and $\operatorname{dist}(x,t) > \sqrt{\varepsilon}$.

Now we set

$$\rho(x,t) = \frac{1}{2}\big(\theta_c^+(x,t) - \theta_c^-(x,t)\big), \quad \theta_0(x,t) = \theta_c^+(x,t).$$

It is clear that the functions ρ, θ_0 so defined satisfy all the requirements imposed above and ensure the validity of (4.37), (4.38). It remains to observe that the function $\chi = \tanh(\beta\tau/\sqrt{2})$ satisfies the boundary condition $\psi|_\Sigma = 1$ within $O(\varepsilon^\infty)$ (i.e., within $O(\varepsilon^N)$ for any $N > 0$) provided that the distance between $\partial\Omega$ and Γ_t is positive. For $t = 0$ this condition is fulfilled, and, since the boundary Γ_t moves inwards, it will also be fulfilled for all $t \in [0, T]$ for which there exists a smooth solution to the problem (4.39), (4.40). This completes our construction of the leading term of the asymptotics.

We will outline the scheme of constructing the higher order terms of the expansion (4.12). Following our standard procedure, for the determination of W_j and $\check{U}_{j+1} = U_{j+1}|_{t=\varphi}$ we obtain the equations

$$|\nabla\varphi|^2 \frac{\partial}{\partial\tau}\left(k_0 \frac{\partial}{\partial\tau}\check{U}_{j+1}\right) = \mathcal{F}_{j+1}(\tau, x), \tag{4.41}$$

$$LW_j = F_j(\tau, x), \tag{4.42}$$

similar to (4.26) and (4.33). The following conditions are necessary and sufficient for solvability of these equations in the class of functions stabilizing at infinity and satisfying (4.15):

$$\mathcal{F}_{j+1} \to 0 \quad \text{as } \tau \to \pm\infty, \qquad F_j \to 0 \quad \text{as } \tau \to \infty, \tag{4.43}$$

$$\int_{-\infty}^{\infty} \mathcal{F}_{j+1}\, d\tau = 0, \tag{4.44}$$

$$\int_{-\infty}^{\infty} F_j \frac{\partial\chi}{\partial\tau}\, d\tau = 0. \tag{4.45}$$

We ensured the fulfillment of conditions (4.43) when constructing the regular terms of the asymptotic expansion (4.12); moreover, their fulfillment was provided for by the special procedure of extending the $\check{U}_i$, $i = 1, \ldots, j$, outside the surface Γ_t. Due to the specific properties of equation (4.11), these conditions also imply that

$$F_j \to 0 \quad \text{as} \quad \tau \to -\infty.$$

This ensures that the W_j are soliton-type functions. Next, it is not hard to see that condition (4.44), similarly to (4.27), implies the Stefan condition on the jump of the normal derivative of the next correction to the temperature. In its turn, condition (4.45) entails the nonhomogeneous linearized Gibbs–Thomson condition. These conditions are imposed on the surface Γ_t, which is already known from the solution to the problem (4.39), (4.40). However, it should be noted that the functions W_j are not determined uniquely by (4.41), but include arbitrary "constants" of integration,

$$W_j = \omega_j(\tau, x) + \varphi_{j+1}(x)\chi_\tau, \quad j = 1, 2, \ldots, \tag{4.46}$$

where ω_j is a particular solution of (4.42), and $\varphi_{j+1}\chi_\tau$ is a general solution of the homogeneous equation in the class of functions stabilizing at $\pm\infty$. The presence of arbitrariness in determining the φ_{j+1}, and the function φ_1 which arises when integrating (4.23) makes the linearized modified Stefan problems for higher order corrections solvable. Finally, by appropriately defining the extension of $\check{U}_{j+1}$

outside Γ_t we provide for the possibility of performing this procedure at the next step.

This scheme was realized, for example, in [**200**], where the first correction for the system (4.8), (4.10), (4.11) with $k = 1$ and $g = 1$ was computed; in particular, the linearized Gibbs–Thomson condition

$$\theta_1^{\pm}\big|_{\Gamma_t} = \kappa'(\widehat{V}_\nu' - \widehat{\mathcal{K}}')\varphi_1 + \frac{\varphi_1}{4}\left(\frac{\partial\theta^+}{\partial t}\bigg|_{\Gamma_t+0} + \frac{\partial\theta^-}{\partial t}\bigg|_{\Gamma_t-0}\right) - \frac{|\nabla\varphi|}{2\sqrt{2}}$$

was obtained. Here $(\widehat{V}_\nu' - \widehat{\mathcal{K}}')\varphi_1$ is the variation of $V_\nu - \mathcal{K}$, which is computed by the formula

$$(\widehat{V}_\nu' - \widehat{\mathcal{K}}')\varphi_1 = \langle\nu, \nu_1\rangle - \langle\nabla, (\nu_1 - \langle\nu, \nu_1\rangle\nu)\rangle,$$

where $\nu_1 = \nabla\varphi_1/|\nabla\varphi|$.

Let us return to the question of an admissible choice of the initial data $\psi^0(x,\varepsilon)$ and $\theta^0(x,\varepsilon)$ in the problem (4.10), (4.11). It follows from the explicit form of the leading term of the asymptotics and the remark we made when solving equation (4.32) that in the leading term there must be a specific smoothing of the phase function $\overline{\psi^0} = \pm 1$, viz., $\psi^0(x,\varepsilon) = \tanh \beta\tau/\sqrt{2} + O(\varepsilon)$, whereas the limiting temperature $\overline{\theta^0} = \theta_{\pm}^0(x)$ may be smoothed arbitrarily. We also pointed out that in the leading term the way of smoothing the discontinuous coefficient $k^{\pm}$ of heat conductivity is immaterial. In the terms of higher order in ε the function ψ^0 must also have a fixed form up to the arbitrariness indicated in (4.46), since $\varphi_{j+1}|_{\Gamma_0}$ can take any value. It is easily seen that the higher order terms of θ^0 can be chosen arbitrarily outside a small neighborhood of Γ_0, but they must have a specific form near Γ_0. Thus, as is usually the case when constructing self-similar soliton-type asymptotics, all restrictions on the choice of initial data concern only their form in a small neighborhood of the initial surface Γ_0.

It remains to remark that the choice of a certain way of smoothing the functions $k^{\pm}$ and $\theta_{\pm}^0$ will affect the right-hand sides which appear in the linearized conditions on the free boundary in the modified Stefan problem for the first correction.

4.2. Initial stage of solidification in binary alloys

In this and the next section we discuss the processes connected with solidification in binary alloys. By binary alloys we mean the ones which may separate into two constituent parts (phases), and we refer to the process of separation as solidification. For example, this may be an alloy which separates into pure metal and slag. However, we will not go into physical details of this question.

The simplest mathematical model that describes the solidification in binary alloys taking the temperature into account is the so-called conserved phase field system (or the non-isothermic Cahn–Hilliard system)

$$\begin{aligned} &\frac{\partial}{\partial t}\Big(\theta + \frac{l}{2}\psi\Big) = k\Delta\theta + f(x), \\ &-\tau_0\frac{\partial\psi}{\partial t} = \xi^2\Delta\Big(\xi^2\Delta\psi + \frac{1}{2a}(\psi - \psi^3) + \kappa_1\theta\Big). \end{aligned} \tag{4.47}$$

All the notation here has a similar meaning as in (4.8), but the phase function ψ is usually treated as a normalized concentration; $f(x)$ is a given smooth function corresponding to a permanent source of heat. To simplify the subsequent calculations, we restrict ourselves to the case of a constant coefficient of heat conductivity, $k = \text{const}$, equal for both phases, and set the function $g(\psi)$, which characterizes the influence of temperature on the variation of the phase function, equal to a constant κ_1.

The system (4.47) is called conserved because for zero boundary conditions on the outer boundary of the domain $\Omega \subset \mathbb{R}^n$ occupied by the alloy, integration of the second equation (4.47) obviously leads to the conservation law

$$\frac{d}{dt}\int_\Omega \psi\,dx = 0. \tag{4.48}$$

It is not hard to see that the phase field system (4.8) does not possess this property.

For $\kappa_1 = 0$ the second equation in (4.47) splits off and becomes the so-called Cahn–Hilliard equation (which explains the other name of the system). This equation was derived by Cahn and Hilliard [**35–36**] to describe the solidification process under a constant temperature.

The system (4.47), which allows for non-isothermic solidification processes, was proposed by Caginalp [**30**] and was studied in [**31, 197–199**]. There is also an extensive literature on the Cahn–Hilliard equation, see, e.g., [**185, 186, 237**]. It was established that, similarly to (4.8), averaging of the microscopic model (4.47) leads to different limiting problems for different relationships between the parameters τ_0, ξ, a, and l. In [**198, 199**] an asymptotic solution to the system (4.47) was constructed by the method discussed in this book. By the same method it was established for the first time that the system (4.47) has a soliton-type solution [**197**].

Here we restrict our consideration to the case of the following relationship between the parameters of equation (4.47):

$$\frac{\sqrt{a}}{\xi} = \text{const}, \qquad \tau_0 = \text{const}, \qquad l = \text{const}, \qquad a \ll 1.$$

We let $\varepsilon = 2a$ and regard ε as a small parameter. Moreover, to simplify the formulas we set

$$\xi^2 = \varepsilon, \qquad \tau_0 = \kappa = \text{const}, \qquad l = 2, \qquad k = 1.$$

Then we can conveniently rewrite (4.47) as

$$\frac{\partial}{\partial t}(\theta + \psi) = \Delta\theta + f(x), \qquad x \in \Omega, \quad t > 0, \tag{4.49}$$

$$-\kappa\frac{\partial \psi}{\partial t} = \Delta(\varepsilon^2\Delta\psi + \psi - \psi^3 + \varepsilon\kappa_1\theta). \tag{4.50}$$

We complete these equations with the natural boundary and initial conditions

$$\left.\frac{\partial\theta}{\partial n_\Omega}\right|_\Sigma = 0, \qquad \left.\frac{\partial\psi}{\partial n_\Omega}\right|_\Sigma = 0, \qquad \left.\frac{\partial}{\partial n_\Omega}\Delta\psi\right|_\Sigma = 0, \tag{4.51}$$

$$\theta\big|_{t=0} = \theta^0(x,\varepsilon), \quad \psi\big|_{t=0} = \psi^0(x,\varepsilon), \tag{4.52}$$

where $\Sigma = \partial\Omega \times (0,T)$, the boundary $\partial\Omega$ of the domain $\Omega \subset \mathbb{R}^n$ is assumed to be smooth, n_Ω is the outer normal to $\partial\Omega$, and θ^0, ψ^0 are smooth functions for $\varepsilon = \text{const}$, subject to conditions to be stated more precisely below.

The full treatment of the solidification problem in a binary alloy must begin with an almost homogeneous state of the alloy, $\psi^0 \approx 0$ (recall that the zero phase function means a completely homogeneous state of the alloy), and end with the state of completely separated phases when the domains $\Omega^\pm$ have formed, each occupied by a single phase, $\psi = \pm 1$ for $x \in \Omega^\pm$. However, already a qualitative analysis of equation (4.50) reveals that this process splits into a number of intermediate processes of essentially different physical nature, which require different mathematical descriptions.

To explain this statement, consider the model equation

$$\kappa \frac{\partial \widetilde{\psi}}{\partial t} = \Delta\big((3\psi_0^2 - 1)\widetilde{\psi}\big), \tag{4.53}$$

which arises in the investigation for stability of the regular state $\psi_0 = \psi_0(x,t)$ with respect to small, regular in x and t, perturbations $\widetilde{\psi} = \widetilde{\psi}(x,t)$. It is clear from (4.53) that for

$$|\psi_0| < 1/\sqrt{3} \tag{4.54}$$

such a state is very unstable because the amplitude of perturbations increases at the same rate as the solution of the inverse heat equation. On the other hand, if

$$1/\sqrt{3} < |\psi_0| < 1, \tag{4.55}$$

then this state is stable. At the same time the function ψ_0 itself cannot stay for any long time in the intervals $(-1, -1/\sqrt{3})$ and $(1/\sqrt{3}, 1)$, because it must satisfy the equation

$$\kappa \frac{\partial \psi_0}{\partial t} = \Delta(c(\psi_0)\psi_0), \tag{4.56}$$

where $c = \psi_0^2 - 1$ is a sort of heat conductivity coefficient. Moreover, (4.56) immediately implies that the only stable states for ψ_0 are ± 1.

Of course, the above arguments are very rough because they do not take into account the fast variation of ψ in both x and t. Nevertheless our conclusions coincide with the generally recognized conception about the existence of unstable (for (4.54)) and metastable (for (4.55)) stages of the solidification process. In its turn, the metastable stage can be divided into two parts: the main part when the domains $\Omega_t^\pm$ occupied by pure phases have already formed, but their position changes with time, and the preliminary one, which we call the soliton stage. This is related to the fact that at this preliminary stage there exist soliton-type solutions of the conserved phase field system. As will be seen from the formulas to be derived, the soliton-type solutions describe the appearance of a kind of nuclei which give rise to the domains $\Omega_t^\pm$.

Before proceeding to the construction of asymptotic solutions to (4.49), (4.50), we explain why soliton-type solutions arise by the following qualitative arguments. Consider an overcooled alloy where the phase function ψ^0 satisfies the inequalities

$$\frac{1}{\sqrt{3}} < \psi^0(x) < 1, \tag{4.57}$$

and its "mass" is

$$m(\psi^0) = \int_\Omega \psi^0(x)\,dx < |\Omega|. \tag{4.58}$$

Then, as we explained above, the function $\psi_0(x,t)$ will increase with time and tend to its stable state $+1$. However, ψ_0 cannot be equal to or close to one at each point $x \in \Omega$, because then we would have $m(\psi_0) \approx |\Omega|$, which contradicts (4.58) and the law of conservation of "mass" (4.48). Hence there must be domains Ω_ε^i of small volume (of magnitude $O(\varepsilon)$) where ψ_0 takes negative values so as to compensate for convergence of ψ_0 to 1 outside Ω_ε^i. It is also clear that in general there must be sufficiently many such domains, i.e., $i = 1, \ldots, M$ and $M \sim 1/\varepsilon$.

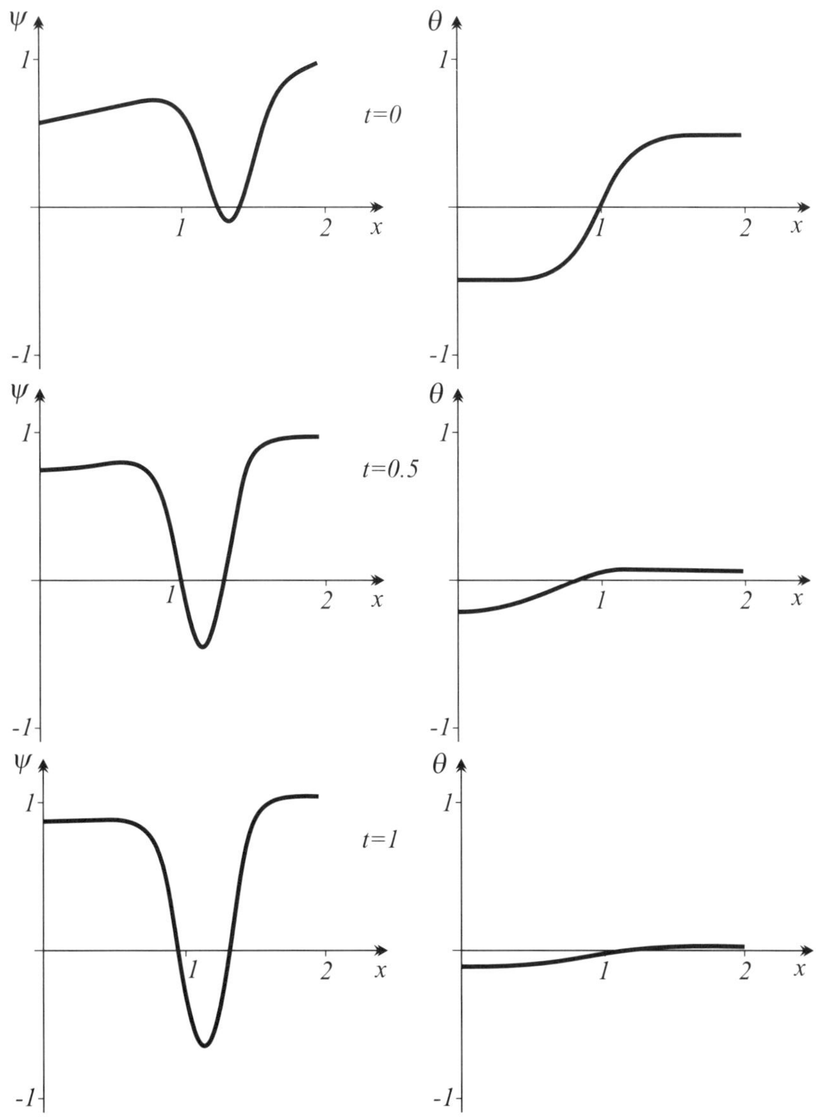

FIGURE 4.1. Spatial profile of the initial data.

In Figure 4.1 we demonstrate the results of numerical solution of the system (4.49), (4.50) in the one-dimensional case ($x \in (0,2)$). We let $\varepsilon = 0.1$, $\kappa_1 = 1$, $f = 0$ and take the initial data

$$\theta\big|_{t=0} = \tanh\Big(\frac{x-1}{2}\Big), \quad \psi\big|_{t=0} = \psi_0^0(x) + \chi\Big(1.48\frac{x-1.3}{\varepsilon}\Big),$$

which correspond to a single-soliton perturbation χ (see (4.73)) of the smooth background ψ_0^0,

$$\psi_0^0(x) = \frac{1}{\sqrt{3}} + \Big(1 - \frac{1}{\sqrt{3}}\Big)\frac{x}{2}, \quad \chi(\xi) = -\frac{9.52}{e^{\xi} + 2.16e^{-\xi} + 6.8}.$$

The plots in Figure 4.1 clearly show the dynamics of the soliton-type perturbation. We see that, along with increasing background, the amplitude of the soliton grows, which eventually gives rise to the domain Ω_ε^1. We point out that the numerical results are in excellent agreement with the asymptotic formulas we will obtain.

Now we proceed to the specific problem of this section, that is, the construction of an asymptotic solution which would model the appearance of "nuclei" Ω_ε^i. We begin by formally constructing a self-similar soliton-type solution to equations (4.49), (4.50). Following the standard procedure, we write it in the form of the expansion

$$\begin{aligned} \theta(x,t,\varepsilon) &= \theta^M(x,t,\varepsilon) + \varepsilon V^M\Big(\frac{S(x,t)}{\varepsilon}, \frac{z}{\varepsilon}, x, t, \varepsilon\Big) + O(\varepsilon^{M+1}), \\ \psi(x,t,\varepsilon) &= \Psi^M(x,t,\varepsilon) + W^M\Big(\frac{S(x,t)}{\varepsilon}, \frac{z}{\varepsilon}, x, t, \varepsilon\Big) + O(\varepsilon^{M+1}). \end{aligned} \tag{4.59}$$

Here θ^M, Ψ^M are the regular part of the solution,

$$\theta^M(x,t,\varepsilon) = \sum_{j=0}^{M} \varepsilon^j \theta_j(x,t), \quad \Psi^M(x,t,\varepsilon) = \sum_{j=0}^{M} \varepsilon^j \psi_j(x,t),$$

with functions θ_j, ψ_j assumed to be smooth; the functions V^M, W^M form the rapidly varying part of the solution:

$$\begin{aligned} V^M(\tau,\zeta,x,t,\varepsilon) &= \sum_{j=1}^{M} \varepsilon^{j-1}\{U_j(\tau,x,t) + Y_j(\zeta,x',t)\}, \\ W^M(\tau,\zeta,x,t,\varepsilon) &= \chi(\tau,x) + \sum_{j=1}^{M} \varepsilon^j\{W_j(\tau,x,t) + Z_j(\zeta,x',t)\}, \end{aligned} \tag{4.60}$$

where χ, U_j, W_j are standard functions describing a soliton-type solution, i.e.,

$$\chi \to 0 \quad \text{as} \quad \tau \to \pm\infty, \qquad U_j, W_j \to 0 \quad \text{as} \quad \tau \to -\infty, \tag{4.61}$$

and U_j, W_j stabilize as $\tau \to +\infty$. Moreover, in (4.59) z is the distance from $\partial\Omega$ along the inner normal to the boundary, $x' \in \partial\Omega$, and the Y_j, Z_j are functions of boundary layer type, i.e.,

$$Y_j(\zeta,x',t), \ Z_j(\zeta,x',t) \to 0 \qquad \text{as} \quad \zeta \to \infty. \tag{4.62}$$

These functions appear because in this problem, unlike the phase field system (4.8), we cannot satisfy the boundary conditions (4.51) with the functions θ_j, φ_j, U_j, W_j only.

We have incorporated into (4.59) the property that, for soliton-type solutions of (4.49), (4.50), the leading terms of the temperature cannot depend on the fast variable. Moreover, assuming that the front of the initial soliton is a smooth surface Γ_0 of codimension one, we set

$$S = t - \varphi(x),$$

where φ is a smooth function vanishing on Γ_0, with Γ_0 and the outer boundary $\partial\Omega$ being bounded away from each other by a finite distance. This implies, in particular, the condition that Γ_0 is a closed surface.

To determine the functions involved in (4.59), substitute this expansion into equations (4.49), (4.50) and collect the resulting terms as follows:

$$\tag{4.63} \begin{aligned} &\frac{\partial}{\partial t}(\theta^M + \Psi^M) - \Delta\theta^M - f \\ &\quad = \frac{1}{\varepsilon}\Big\{\mathcal{L}_2 V^M - \frac{\partial}{\partial\tau}W^M\Big\} - \Big\{\Big(\frac{\partial}{\partial\tau} + \mathcal{L}_1\Big)V^M + \frac{\partial}{\partial t}W^M\Big\} \\ &\qquad + \varepsilon\Big(\Delta - \frac{\partial}{\partial t}\Big)V^M + O(\varepsilon^{M+1}), \end{aligned}$$

$$\tag{4.64} \begin{aligned} &\frac{1}{\varepsilon^2}\mathcal{L}_2\{\mathcal{L}_2 W^M + W^M - (\Psi^M + W^M)^3\} \\ &+ \frac{1}{\varepsilon}\Big\{\Big(\kappa\frac{\partial}{\partial\tau} - \mathcal{L}_2\mathcal{L}_1\Big)W^M - \mathcal{L}_1\big(\mathcal{L}_2 W^M + W^M - (\Psi^M + W^M)^3\big)\Big\} \\ &+ \Big\{(\mathcal{L}_2\Delta + \mathcal{L}_1^2)W^M + \kappa_1\mathcal{L}_2 V^M + \kappa\frac{\partial}{\partial t}(\Psi^M + W^M) \\ &\qquad + \Delta\big(\mathcal{L}_2 W^M + \Psi^M + W^M - (\Psi^M + W^M)^3\big)\Big\} \\ &- \varepsilon\{\mathcal{L}_1(\Delta W^M + \kappa_1 W^M) + \Delta(\mathcal{L}_1 W^M - \kappa_1\theta^M)\} \\ &+ \varepsilon^2\Delta\{\Delta(\Psi^M + W^M) + \kappa_1 V^M\} \\ &\quad = O(\varepsilon^{M+1}). \end{aligned}$$

Here

$$\begin{aligned} \mathcal{L}_2 &= |\nabla\varphi|^2\frac{\partial^2}{\partial\tau^2} + |\nabla z|^2\frac{\partial^2}{\partial\zeta^2}, \quad \mathcal{L}_1 = \frac{\partial}{\partial\tau}\widehat{\Pi} - \frac{\partial}{\partial\zeta}\widehat{\pi}_b, \\ \widehat{\Pi} &= 2\langle\nabla\varphi, \nabla\rangle + \Delta\varphi, \quad \widehat{\pi}_b = 2\langle\nabla z, \nabla\rangle + \Delta z. \end{aligned}$$

First, we determine the regular terms of (4.59). Passing to the limit as $\tau \to \pm\infty$, $\zeta \to \infty$, we obtain from (4.63), (4.64) the relations

$$\tag{4.65} \begin{aligned} &\frac{\partial}{\partial t}(\theta^M + \Psi^M) - \Delta\theta^M - f = -\frac{\partial}{\partial t}W^{M\pm} - \varepsilon\Big(\frac{\partial}{\partial t} - \Delta\Big)V^{M\pm} + O(\varepsilon^{M+1}), \\ &\kappa\frac{\partial}{\partial t}(\Psi^M + W^{M\pm}) + \Delta(\Psi^M + W^{M\pm} - (\Psi^M + W^{M\pm})^3) \\ &\qquad + \varepsilon\kappa_1\Delta\theta^M + \varepsilon^2\Delta(\Delta(\Psi^M + W^{M\pm}) + \kappa_1 V^{M\pm}) = O(\varepsilon^{M+1}), \end{aligned}$$

where $V^{M\pm}, W^{M\pm}$ denote the limiting values of V^M, W^M as $\tau \to \pm\infty$.

Denote

$$\theta_j^\pm = \theta_j + U_j^\pm, \quad \Psi_j^\pm = \psi_j + W_j^\pm, \tag{4.66}$$

where $j = 1, 2, \ldots, M$ and $U_j^- = W_j^- = 0$. Then, equating the terms of equal order in ε, we obtain from (4.65) the equations

$$\kappa \frac{\partial \psi_0}{\partial t} = \Delta(\psi_0^3 - \psi_0), \qquad x \in \Omega, \quad t > 0, \tag{4.67}$$

$$\frac{\partial \theta_0}{\partial t} = \Delta \theta_0 + f(x) - \frac{\partial \psi_0}{\partial t}, \qquad x \in \Omega, \quad t > 0, \tag{4.68}$$

for the leading regular terms of the asymptotics, and the equations

$$\Big(\frac{\partial}{\partial t} - \Delta\Big)\theta_k^\pm = f_{k,\theta}^\pm(x,t), \quad x \in \Omega_t^\pm, \quad t > 0, \tag{4.69}$$

$$\kappa \frac{\partial \Psi_k^\pm}{\partial t} - \Delta((3\psi_0^2 - 1)\Psi_k^\pm) = f_{k,\psi}^\pm(x,t), \quad x \in \Omega_t^\pm, \quad t > 0, \tag{4.70}$$

for higher order terms.

Here $\Omega_t^\pm$ are subsets of Ω such that

$$\Omega_t^+ = \{x \in \Omega,\ \varphi(x) < t\}, \quad \Omega_t^- = \{x \in \Omega,\ \varphi(x) > t\}, \quad \Omega = \Omega_t^+ \cup \Omega_t^- \cup \Gamma_t,$$

where $\Gamma_t = \{x \in \Omega,\ \varphi(x) = t\}$ and we assume that $\operatorname{dist}(\Gamma_t, \partial\Omega) \geq \text{const} > 0$. In (4.69), (4.70) $f_{k,\theta}^\pm, f_{k,\psi}^\pm$ denote polynomials in previous terms of the expansion and their derivatives. In particular,

$$f_{1,\theta}^\pm = -\frac{\partial \Psi_1^\pm}{\partial t}, \quad f_{1,\psi}^\pm = \kappa_1 \Delta \theta_0.$$

Notice that although the solution of (4.67) increases in t (for small $|\psi_n|$), we have nevertheless $\psi_0 \to \pm 1$. We will derive an asymptotic solution in the metastable zone, i.e., we assume the condition (4.55) to hold. To simplify the formulas, we will only consider a positive background ψ_0. Under the assumption

$$\frac{1}{\sqrt{3}} < \psi_0(x, 0) < 1, \tag{4.71}$$

(4.67) immediately implies that $\psi_0(x,t)$ will possess the same property, and $\psi_0 \to 1$ at an exponential rate. Clearly, this implies boundedness of the solution to (4.70).

Now we observe that the supports of fast variation of the boundary layer functions Y_j, Z_j on one hand and of the functions χ, V_j, W_j rapidly varying in a neighborhood of Γ_t on the other hand are disjoint within $O(\varepsilon^\infty)$. This allows us to construct these functions separately.

Consider a neighborhood of the surface Γ_t, which corresponds to the treatment of (4.63), (4.64) as $\zeta \to \infty$. It is easily seen that by (4.65) all the terms in (4.63), (4.64) will tend to zero as $\tau \to \pm\infty$. This enables us to use the Taylor expansion with respect to t about $t = \varphi(x)$ and to employ the standard techniques for obtaining the rapidly varying functions χ, V_j, W_j.

Consider the terms of order ε^{-2} in (4.64),

$$\frac{\partial^2}{\partial\tau^2}\left\{|\nabla\varphi|^2\frac{\partial^2\chi}{\partial\tau^2}+\chi-(\check{\psi}_0+\chi)^3\right\}=0,$$

where, as usual, $\check{\psi}_0=\psi_0(x,t)|_{t=\varphi(x)}$. After integrating twice and taking (4.61) into account, we obtain

$$|\nabla\varphi|^2\frac{\partial^2\chi}{\partial\tau^2}+\chi-(\check{\psi}_0+\chi)^3=-\check{\psi}_0^3. \tag{4.72}$$

We solve this equation to find the soliton-type function

$$\chi(\tau,x)=-\frac{8Q^2}{e^{\xi}+8be^{-\xi}+8\check{\psi}_0}, \tag{4.73}$$

where

$$\xi=\beta(\tau-\varphi_1),\quad \beta=\frac{Q}{|\nabla\varphi|},\quad Q=\sqrt{3\check{\psi}_0^2-1},\quad b=1-\check{\psi}_0^2, \tag{4.74}$$

and $\varphi_1=\varphi_1(x)$ denotes the "constant" of integration. It is seen from (4.74) that (4.55) is a necessary condition for existence of a soliton-type solution of (4.73).

Next, equating to zero the terms of order ε^{-2+k}, we obtain from (4.63), (4.64), and (4.61) the problems

$$\frac{\partial^2\check{U}_k}{\partial\tau^2}=\check{F}_k^{\theta},\qquad \check{U}_k\to 0\quad\text{as}\quad \tau\to-\infty, \tag{4.75}$$

$$\frac{\partial^2}{\partial\tau^2}L\check{W}_k=\check{F}_k^{\psi},\qquad \check{W}_k\to 0\quad\text{as}\quad \tau\to-\infty. \tag{4.76}$$

Here

$$L=|\nabla\varphi|^2\frac{\partial^2}{\partial\tau^2}+1-3(\check{\psi}_0+\chi)^2,$$

and the $\check{F}_k^{\theta}$, $\check{F}_k^{\psi}$ are polynomials in τ and $\theta_0,\psi_0,\dots,U_{k-1},W_{k-1}$ and their derivatives taken at the point $t=\varphi$. In particular,

$$\begin{aligned}\check{F}_1^{\psi}=&\frac{\partial^2}{\partial\tau^2}\left\{\widehat{\Pi}\frac{\partial\chi}{\partial\tau}+3(\check{\psi}_0+\chi)^2\check{\psi}_1\right\}\\&+3\frac{\partial\psi_0}{\partial t}\bigg|_{t=\varphi}\left(2+\tau\frac{\partial}{\partial\tau}\right)\frac{\partial}{\partial\tau}(\check{\psi}_0+\chi)^2-\frac{\kappa}{|\nabla\varphi|^2}\frac{\partial\chi}{\partial\tau},\end{aligned} \tag{4.77}$$

$$\check{F}_1^{\theta}=|\nabla\varphi|^{-2}\frac{\partial\chi}{\partial\tau}. \tag{4.78}$$

It is not hard to establish that the problems (4.75), (4.76) have solutions stabilizing as $\tau\to\pm\infty$ if and only if the following conditions are fulfilled:

$$\check{F}_k^{\theta},\check{F}_k^{\psi}\to 0\qquad\text{as}\quad\tau\to\pm\infty, \tag{4.79}$$

$$\int_{-\infty}^{\infty}\check{F}_k^{\theta}\,d\tau=0,\qquad\int_{-\infty}^{\infty}\check{F}_k^{\psi}\,d\tau=0, \tag{4.80}$$

$$\int_{-\infty}^{\infty}f_k^{\psi}\frac{\partial\chi}{\partial\tau}\,d\tau=0, \tag{4.81}$$

where

$$f_k^\psi = \int_{-\infty}^{\tau} \int_{-\infty}^{\tau'} \check{F}_k^\psi(\tau'', x)\, d\tau''\, d\tau'.$$

Recall once again that we ensured the fulfillment of conditions (4.79) for all k when constructing the regular terms of the asymptotics.

Consider the other conditions for $k = 1$. Since condition (4.80) is fulfilled by (4.77), (4.78), we only have to consider (4.81), where f_1^ψ is given by the formula

$$f_1^\psi = \widehat{\Pi}\frac{\partial \chi}{\partial \tau} + 3(\check{\psi}_1 + \xi \check{\psi}_{0t})(2\check{\psi}_0 \chi + \chi^2) + \sqrt{2}\kappa\beta \log(J^2 R)$$

with

$$R = \frac{e^\xi + 4\check{\psi}_0 - 2\sqrt{2}Q}{e^\xi + 4\check{\psi}_0 + 2\sqrt{2}Q}, \quad J = \frac{1}{\sqrt{b}}(Q + \sqrt{2}\check{\psi}_0), \tag{4.82}$$

and ξ, β, Q, b are the same as in (4.74).

On integrating in the left-hand side of (4.81), this condition for $k = 1$ becomes

$$\kappa V_\nu = \frac{1}{3}\mathcal{K}(1 + G) + G\frac{\partial}{\partial \nu} \log \check{\psi}_0, \tag{4.83}$$

where $V_\nu = 1/|\nabla\varphi|$ is the velocity of motion of Γ_t along the normal $\nu = \nabla\varphi/|\nabla\varphi|$, $\mathcal{K} = \operatorname{div} \nu$ is the mean curvature of Γ_t, $\partial/\partial\nu = \langle \nu, \nabla \rangle$, and

$$G = \frac{IQ^2}{2(Q - I)}, \quad I = \sqrt{2}\check{\psi}_0 \log J. \tag{4.84}$$

Thus the determination of the leading term of the self-similar asymptotics (4.59), (4.60) reduces to a typical setup for soliton-type solutions: the background functions θ_0, ψ_0 are obtained from equations (4.67), (4.68) defined in the entire domain Ω, while the front of the soliton-type function χ as in (4.73) is found from the separate equation (4.83).

Now we return to the determination of the first correction. Under the condition (4.83), we obtain from (4.75), (4.76) that

$$\check{U}_1 = \check{U}_1^+(x)\eta(\tau, x), \quad \check{W} = \check{W}_1^+(x)\eta(\tau, x) + \omega_1(\tau, x), \tag{4.85}$$

where

$$\begin{gathered} \eta = \frac{\sqrt{2}}{a} \log(J^2 R), \quad a = 2\frac{I}{\check{\psi}_0}, \\ \check{U}_1^+ = -aV_\nu, \quad \check{W}_1^+ = -\frac{a\kappa}{Q^2}V_\nu, \end{gathered} \tag{4.86}$$

$\omega_1(\tau, x)$ is a function defined up to an additive term $\varphi_2(x)\partial\chi/\partial\tau$ and decreasing exponentially fast as $\tau \to \pm\infty$, and η is a function such that

$$\eta \to 1 \quad \text{as} \quad \tau \to \infty, \qquad \eta \to 0 \quad \text{as} \quad \tau \to -\infty$$

at an exponential rate.

Define a smooth extension of the functions $\check{U}_1, \check{W}_1$ in the form

$$U_1(\tau, x, t) = u_1(x,t)\eta(\tau, x),$$
$$W_1(\tau, x, t) = w_1(x,t)\eta(\tau, x) + \omega_1(\tau, x), \tag{4.87}$$

where

$$u_1 = \theta^+_{1c} - \theta^-_{1c}, \quad w_1 = \Psi^+_{1c} - \Psi^-_{1c}, \tag{4.88}$$

and $\theta^{\pm}_{1c}, \Psi^{\pm}_{1c}$ are smooth extensions of $\theta^{\pm}_1, \Psi^{\pm}_1$. Recall that for the determination of $\theta^{\pm}_1, \Psi^{\pm}_1$ we have obtained so far only equations (4.69), (4.70) defined in different domains Ω^+_t and Ω^-_t. However, comparing (4.75) and (4.76), we see that equalities (4.86) and (4.88) require the conditions

$$[\theta^{\pm}_1] = aV_\nu, \quad [\Psi^{\pm}_1] = \frac{\kappa a}{Q^2} V_\nu, \tag{4.89}$$

i.e., the functions $\theta^{\pm}_1$ and $\Psi^{\pm}_1$ must have jumps on the surface Γ_t proportional to the velocity of its motion.

Consider now the equations (4.75), (4.76) and conditions (4.80), (4.81), respectively, for $k = 2$. It is not hard to show after some algebra that conditions (4.80) for $k = 2$ yield the equalities

$$\begin{aligned} \left[\frac{\partial \theta^{\pm}_1}{\partial \nu}\right] &= -aV_\nu \Big\{ \Big(1 + \frac{\kappa}{Q^2}\Big) V_\nu + \mathcal{K} - \frac{4}{abQ}\frac{\partial \check{\psi}_0}{\partial \nu} \Big\}, \\ \left[\frac{\partial}{\partial \nu}(3\check{\psi}_0^2 - 1)\Psi^{\pm}_1\right] &= -12|\nabla\varphi|\check{\psi}_{0t}\Big\{ Q\mathcal{K} - V_\nu \operatorname{div}(Q\nabla\varphi) + \frac{a}{2}\frac{\partial\check{\psi}_0}{\partial\nu}\Big\} \\ &\quad + V_\nu^2 \frac{\kappa a}{2Q^2}(\kappa + Q^2\Delta\varphi). \end{aligned} \tag{4.90}$$

Moreover, it can be shown after very tedious calculations that condition (4.81) for $k = 2$ reduces to the nonhomogeneous linearized equation (4.83), i.e.,

$$\kappa \frac{\partial \varphi_1}{\partial \nu} = G_1 \mathcal{K}' \varphi_1 + G_2 \varphi_1 + f_\varphi, \tag{4.91}$$

where G_1, G_2, f_φ are functions which depend on $x, \varphi, \theta^{\pm}_1, \Psi^{\pm}_1$, and $\mathcal{K}'$ is the variation of the operator $\mathcal{K}$ as in (4.83).

It is clear that conditions (4.90) complement (4.89) to a complete system of conditions to be imposed on the surface Γ_t, and the separate equation (4.91) may be used to correct (in terms of order ε) the position of the soliton front. Thus, neglecting the simple question about setting the initial and boundary conditions (on $\partial\Omega$) for equations (4.69), (4.70) for $k = 1$, we have finished construction of the first correction in the expansion (4.59), (4.60).

Subsequent terms of the asymptotics near Γ_t are constructed similarly to the above. Namely, for each $k \geq 1$ we have the following recurrence procedure:

- Compute $\check{U}^+_k, \check{W}^+_k$ as limits of the solution to (4.75), (4.76) as $\tau \to \infty$ to obtain the conditions on the jump of $\theta^{\pm}_k, \Psi^{\pm}_k$ on the surface Γ_t.

- Using formulas similar to (4.87), extend U_k, W_k outside Γ_t. Since $\theta_k^\pm, \Psi_k^\pm$ solve the equations (4.68), (4.69), the conditions (4.79) are fulfilled.
- By means of (4.80), conditions on the jumps of the normal derivatives $\partial\theta_k^\pm/\partial\nu$, $\partial\Psi_k^\pm/\partial\nu$ on Γ_t are obtained.
- Condition (4.81) leads to an equation similar to (4.91) for the "constant" of integration $\varphi_k(x)$.

Now we turn to boundary conditions on the outer boundary. Since (4.67), (4.68) is a parabolic system, (4.51) immediately implies the boundary conditions

$$\frac{\partial\theta_0}{\partial n_\Omega}\bigg|_\Sigma = 0, \qquad \frac{\partial\psi_0}{\partial n_\Omega}\bigg|_\Sigma = 0, \tag{4.92}$$

which make this problem well posed. For definiteness, let the domain Ω_t^- be adjacent to the outer boundary $\partial\Omega$. Then we have to use the fact that if dist $(\Gamma_t, \partial\Omega) \geq$ const > 0, then for ε small enough the functions χ, U_j, W_j, rapidly varying in a neighborhood of Γ_t, differ in a neighborhood of $\partial\Omega$ from their limiting values (as $\tau \to \infty$) by $O(\varepsilon^\infty)$. Thus the discrepancy in the second boundary condition (4.51) arises only from the regular part of the solution, because in the general case

$$\frac{\partial}{\partial n_\Omega}\Delta\psi_0\big|_\Sigma \neq 0.$$

We set $Z_j = 0$ for $j = 1, 2$, since

$$\frac{\partial}{\partial n_\Omega}\Delta_x\varepsilon^3 Z_3\Big(\frac{z(x)}{\varepsilon}, x', t\Big)\bigg|_{z=0} = O(1).$$

Then, using the standard construction of a solution of boundary layer type, we derive from (4.64) the equation for $Z_3 = Z_3(\zeta, x')$:

$$\frac{\partial^2 Z_3}{\partial\zeta^2} - qZ_3 = 0, \qquad Z_3 \to 0 \quad \text{as} \quad \zeta \to \infty, \tag{4.93}$$

where $q = 3\psi_0^2|_\Sigma - 1$. Obviously,

$$Z_3 = c_3(x', t)e^{-\sqrt{q}\zeta}.$$

It is now easily seen that the boundary condition

$$\frac{\partial^3}{\partial\zeta^3}Z_3\bigg|_{\zeta=0} = \frac{\partial}{\partial n_\Omega}\Delta\psi_0\bigg|_\Sigma \tag{4.94}$$

leads to the formula

$$c_3 = -q^{-3/2}\frac{\partial}{\partial n_\Omega}\Delta\psi_0\bigg|_\Sigma.$$

Next we observe that the appearance of the boundary layer function $\varepsilon^3 Z_3$ necessarily entails a correction of the Neumann conditions in the terms of order $O(\varepsilon^2)$. Therefore the Neumann condition for Ψ_2^- becomes

$$\frac{\partial\Psi_2^-}{\partial n_\Omega}\bigg|_\Sigma = \frac{\partial Z_3}{\partial\zeta}\bigg|_{\zeta=0} = \frac{1}{q}\frac{\partial}{\partial n_\Omega}\Delta\psi_0\bigg|_\Sigma. \tag{4.95}$$

Moreover, the appearance of the boundary layer functions Z_k gives rise to the appearance of the boundary layer terms Y_j in the asymptotic expansion for the temperature. Namely, having in mind that the position of the boundary $\partial\Omega$ does not change with time, we can set $Y_j = 0$ for $j = 1, \ldots, 4$. Then we immediately derive from (4.63) the equation

$$\frac{\partial^2 Y_5}{\partial \zeta^5} = \frac{\partial}{\partial t} Z_3(\zeta, x', t), \qquad Y_5 \to 0 \quad \text{as} \quad \zeta \to \infty. \tag{4.96}$$

Thus we obtain from (4.96) that

$$Y_5 = -q^{-5/2} \frac{\partial^2}{\partial n_\Omega \partial t} \Delta \psi_0 \Big|_\Sigma e^{-\sqrt{q}\zeta},$$

and this function in general does not satisfy conditions of the form (4.51). To compensate for the discrepancy, we have to appropriately correct the Neumann condition for temperature in the term of order ε^4. After some simple calculations we obtain

$$\frac{\partial \theta_4^-}{\partial n_\Omega}\Big|_\Sigma = \frac{\partial Y_5}{\partial \zeta}\Big|_{\zeta=0} = \frac{1}{q^2} \frac{\partial^2}{\partial n_\Omega \partial t} \Delta \psi_0 \Big|_\Sigma . \tag{4.97}$$

Clearly θ_4^- is the first function in the asymptotic solution for which the Neumann condition contains nonzero data.

So, the asymptotic expansion in a small neighborhood of the outer boundary $\partial\Omega$ has the form

$$\theta = \theta_0(x,t) + \sum_{j=1}^{4} \varepsilon^j \theta_j^-(x,t) + \sum_{j=5}^{M} \varepsilon^j \big(\theta_j^-(x,t) + Y_j(\zeta, x', t)\big) + O(\varepsilon^{M+1}),$$

$$\psi = \psi_0(x,t) + \sum_{j=1}^{2} \varepsilon^j \Psi_j^-(x,t) + \sum_{j=3}^{M} \varepsilon^j \big(\Psi_j^-(x,t) + Z_j(\zeta, x', t)\big) + O(\varepsilon^{M+1}).$$

Here Z_3, Y_5 are the functions constructed above, and the boundary layer functions Z_k for $k \geq 4$ and Y_j for $j \geq 6$ are found from linear nonhomogeneous equations of the form (4.93), (4.96).

Accordingly, we obtain the following boundary conditions for equations (4.69), (4.70):

1. The conditions (4.92) for θ_0, ψ_0.
2. The conditions

$$\frac{\partial \theta_j^-}{\partial n_\Omega}\Big|_\Sigma = 0, \qquad \frac{\partial \Psi_1^-}{\partial n_\Omega}\Big|_\Sigma = 0$$

for θ_j^-, $j = 1, 2, 3$, and Ψ_1^-.

3. The conditions (4.95), (4.97) for Ψ_2^- and θ_4^-.
4. The conditions

$$\frac{\partial \theta_j^-}{\partial n_\Omega}\Big|_\Sigma = \frac{\partial Y_{j+1}}{\partial \zeta}\Big|_{\zeta=0}, \qquad \frac{\partial \Psi_k^-}{\partial n_\Omega}\Big|_\Sigma = \frac{\partial Z_{k+1}}{\partial \zeta}\Big|_{\zeta=0}$$

for θ_j^-, $j \geq 5$, and Ψ_k^-, $k \geq 3$.

It remains to discuss the possible initial conditions for equations (4.67)–(4.70), as well as the admissible freedom in the choice of the initial data (4.52). The answer to the first question is obvious: $\theta_0|_{t=0}$, $\psi_0|_{t=0}$, $\theta_k^\pm|_{t=0}$, and $\Psi_k^\pm|_{t=0}$ may be any smooth functions satisfying natural conditions of mutual conformity between the initial and boundary data (including the data for $\theta_k^\pm, \Psi_k^\pm$ which are set on Γ_0). Moreover, it is assumed that $\psi_0|_{t=0}$ satisfies (4.57) (or some alternative inequalities that follow from the metastability condition (4.55), in which case the construction of the asymptotics must be altered in a natural way). Conversely, since the soliton-type function χ and the functions of the shock wave type U_j, W_j have a fixed form (up to terms of the form $\varphi_j \partial\chi/\partial\tau$), we obtain the conditions on the form of $\theta^0(x,\varepsilon)$ and $\psi^0(x,\varepsilon)$ in a neighborhood of Γ_0. The same holds for the behavior of initial data in a small neighborhood of the outer boundary, except for specific initial data $\psi_0|_{t=0}$, $\psi_k^-|_{t=0}$ satisfying the two boundary conditions (4.51). In this case the boundary layer terms of the asymptotics appear only for $t > 0$.

Finally, the initial front Γ_0 of the distorted soliton may be any closed surface of codimension one bounded off from the outer boundary by a finite distance. It is natural to take the normal distance to Γ_0 for the argument of the initial function χ as in (4.73).

For clarity, we formulate once again the limiting problems which arise in construction of the leading term of the asymptotic expansion:

$$\begin{aligned} \theta(x,t,\varepsilon) &= \theta_0(x,t) + O(\varepsilon), \\ \psi(x,t,\varepsilon) &= \psi_0(x,t) + \chi\Big(\frac{t-\varphi(x)}{\varepsilon} - \varphi_1(x), x\Big) + O(\varepsilon). \end{aligned} \tag{4.98}$$

So, for construction of the smooth background θ_0, ψ_0 we have the problems (4.67), (4.68), (4.92), i.e.,

$$\begin{aligned} \kappa\frac{\partial\psi_0}{\partial t} &= \Delta(\psi_0^3 - \psi_0), \qquad x \in \Omega, \quad t > 0, \\ \frac{\partial\psi_0}{\partial n_\Omega}\Big|_\Sigma &= 0, \quad \psi_0\big|_{t=0} = \bar{\psi}^0(x), \end{aligned} \tag{4.99}$$

$$\begin{aligned} \frac{\partial\theta_0}{\partial t} &= \Delta\theta_0 + f(x) - \frac{\partial\psi_0}{\partial t}, \qquad x \in \Omega, \quad t > 0, \\ \frac{\partial\theta_0}{\partial n_\Omega}\Big|_\Sigma &= 0, \quad \theta_0\big|_{t=0} = \bar{\theta}^0(x). \end{aligned} \tag{4.100}$$

The soliton front χ in the leading term is the surface $\Gamma_t = \{x \in \Omega,\ \varphi(x) = t\}$, where the function φ is found from the problem

$$\kappa V_\nu = \alpha(x,\psi_0,\varphi)\mathcal{K} + F(x,\psi_0,\varphi), \quad \varphi\big|_{\Gamma_0} = 0. \tag{4.101}$$

Here $\alpha = \frac{1}{3}(1+G)$, $F = G\,\partial\log\psi_0(x,\varphi)/\partial\nu$, V_ν and $\mathcal{K}$ are the normal velocity and the curvature of Γ_t, and the notations (4.74) and (4.84) have been used. It is not hard to establish that this is a quasilinear parabolic equation (where the distance along the normal ν to Γ_t plays the role of time); therefore the problem (4.101) is well posed. Note also that the equation (4.101) can be viewed as a version of the Gibbs–Thomson condition on the free boundary. However, unlike the modified Stefan problem (4.39), the equation (4.101) splits off from the problem (4.99), (4.101).

The soliton part χ of the phase function is computed by formula (4.73), which may be rewritten as

$$\chi(\tau', x) = -\frac{8Q^2}{e^{\beta\tau'} + 8be^{-\beta\tau'} + 8\psi_0(x, \varphi)}, \tag{4.102}$$

where $\beta = Q/|\nabla\varphi|$, $Q = \sqrt{3\psi_0^2(x, \varphi) - 1}$, $b = 1 - \psi_0^2(x, \varphi)$, $\tau' = \tau - \varphi_1$.

The higher order corrections to the leading term (4.98) of the asymptotic expansion have the form of a smoothed shock wave, and are computed by solving the limiting equations (4.69), (4.70) with Hugoniot-type conditions (4.89), (4.90) on the surface of discontinuity.

It is important to note, however, that the magnitude of the first correction φ_1 to the phase (see (4.74), (4.91)) increases when approaching the boundary of validity of soliton-type asymptotics, i.e., as $\psi_0 \to 1$.

Let us discuss the leading term of the asymptotics. For $\tau' = 0$ the function χ has the amplitude

$$A = \frac{3\psi_0^2(x, \varphi) - 1}{\frac{9}{8} + \psi_0(x, \varphi) - \psi_0^2(x, \varphi)}.$$

One can see by simple calculations that $A \to 0$ as $\psi_0 \to 1/\sqrt{3}$, in which case $\alpha \to \frac{1}{3}$, $F \to 0$, so that (4.101) becomes the Gibbs–Thomson condition with zero temperature,

$$3\kappa V_\nu = \mathcal{K}. \tag{4.103}$$

Thus for $\psi_0 = 1/\sqrt{3} + \delta$, $\delta \ll 1$, the function (4.102) describes a hollow of depth $O(\delta)$, which moves with a finite velocity given, within $O(\delta)$, by condition (4.103).

Next, it is not hard to calculate that A is a monotonically increasing function of ψ_0, $\partial A/\partial\psi_0 > 0$. Hence there exists a value ψ_0^* such that $A < \psi_0(x, \varphi)$ for $\check{\psi}_0 \in (1/\sqrt{3}, \psi_0^*)$ and $A > \psi_0(x, \varphi)$ for $\check{\psi}_0 \in (\psi_0^*, 1)$. Therefore, moving into the domain with $\check{\psi}_0 \in (\psi_0^*, 1)$, the soliton solution describes the appearance of the domain $\Omega_{t,\varepsilon}^- = \{x \in \Omega,\ \psi_0 + \chi < 0\}$ with a negative value of the phase function.

Let us discuss the behavior of the solution as $\psi_0 \to 1$. For sufficiently small $\delta > 0$, let $\psi_0 = 1 - \widetilde{\psi}(x, t)\exp(-1/\delta)$. Then it is not hard to calculate that $G = -1 + O(\delta)$, $F = O(\delta^\infty)$, and we obtain from (4.101) that $V_\nu \sim \delta$, i.e., the soliton slows down as $\psi_0 \to 1$. In this case the domain $\Omega_{t,\varepsilon}^-$ with negative ψ grows in volume, since

$$b \sim \exp\Big(-\frac{1}{\delta}\Big) \quad \text{and} \quad |\Omega_{t,\varepsilon}^-| \sim |\Omega_{t,\varepsilon/\delta}^-| \sim \frac{\varepsilon}{\delta}$$

with $\delta \ll 1$. As δ decreases further, the volume of $\Omega_{t,\varepsilon}^-$ continues to grow, and the maximal value of $|\chi|$ tends to 2 (taking the growth of the correction φ_1 into account). From the physical viewpoint this means the appearance of a domain occupied by the other phase. From the mathematical viewpoint this means the beginning of the bifurcation stage when the soliton-type solution turns into a pair of kinks. The behavior of the solution at the bifurcation stage, i.e., when $\psi_0+\chi\sim-1$ and $|\Omega^-_{t,\varepsilon/\delta}|\sim$ const, cannot be described any longer by a soliton-type asymptotics, because all our constructions are valid only in the case $|\Omega^-_{t,\varepsilon/\delta}|\to 0$ as $\varepsilon\to 0$.

Nevertheless we can point out an interesting property of the formula for the distorted soliton χ [**195**]. If $\psi_0-1\sim-\exp(-1/\delta)$, $\delta\ll 1$, then the exponents in the denominator in (4.102) are

$$+\frac{\delta}{\varepsilon}(t-\varphi)\quad\text{and}\quad-\Big\{\frac{1}{\varepsilon}+\frac{\delta}{\varepsilon}(t-\varphi)\Big\},$$

where $\varphi=\varphi_0+\varepsilon\varphi_1$. Accordingly, the soliton is concentrated in a neighborhood of the front $\Gamma_t=\{\varphi=t\}$ of width $\sim(\varepsilon/\delta)^{1-\mu}$, $\mu>0$, in the direction of the normal $\nu^-=-\nabla\varphi/|\nabla\varphi|$ and of width $\sim(\varepsilon/\delta^2)^{1-\mu}$ in the direction of the normal $\nu^+=\nabla\varphi/|\nabla\varphi|$. Hence one can easily derive that, as ψ_0 approaches 1 with growing t (by (4.99)) and, accordingly, δ decreases, the "left" border of the soliton (on the side of ν^+) moves faster than the "right" (on the side of ν^-). Surprisingly, this is precisely the kind of behavior indicated by direct numerical computations. Figures 4.2 and 4.3 show the results of numerical modeling of of the Cahn–Hilliard system (4.49), (4.50) with initial data

$$\psi\big|_{t=0}=\psi_0+\chi\big|_{t=0},\quad\theta\big|_{t=0}=\frac{1}{2}\tanh\big(5(x-1)/\sqrt{2}\big),$$

where $x\in[0,2]$, $\psi_0\approx 1$, $\chi\big|_{t=0}$ is defined by (4.102), and $\varepsilon=0.1$.

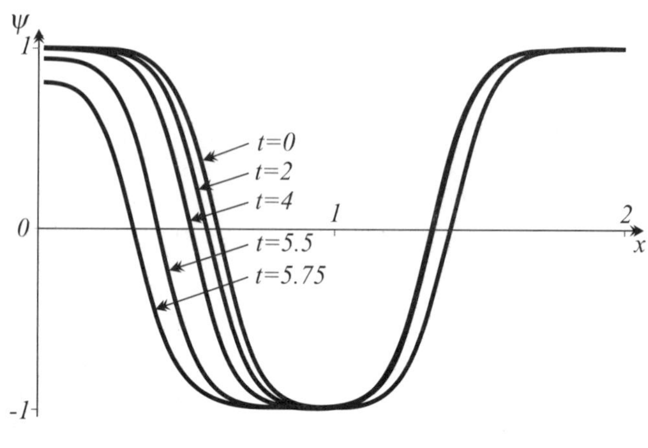

FIGURE 4.2. Dynamics of the phase function in a bifurcation soliton/two kinks.

The plots in Figure 4.2 show that the motion of the right border of the soliton slows down, and for $t>2$ it practically stops. On the contrary, the left border moves with a finite velocity, growing with t. The plots in Figure 4.3 demonstrate the corresponding dynamics of the temperature. It is clearly seen that the temperature dissipates rather fast, being an odd function with respect to the point $x=1$ (at least, visually). Hence the asymmetric behavior of the left and right borders of the soliton cannot be caused by temperature.

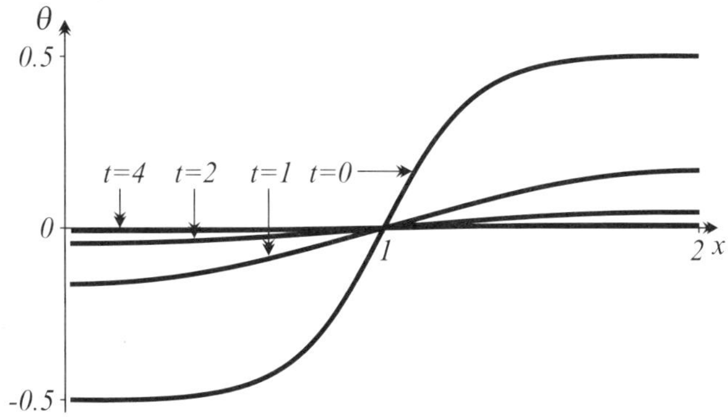

FIGURE 4.3. Dynamics of temperature.

4.3. Motion of the free boundary for large times

Our aim in this section is to describe the motion of the interface between the domains occupied by the different phases after they have already formed. It can be shown [**199**] that for equations (4.49), (4.50) a solution similar to the one constructed in §4.1 can exist only for times $t \sim 1/\varepsilon$. This suggests that, although by equations (4.67) the smooth background ψ_0 converges to its stable states ± 1 very rapidly, the transformation of the "nuclei" $\Omega^{\pm}_{t,\varepsilon/\delta}$ into the domains $\Omega^{\pm}_t$ of full measure (as $\varepsilon \to 0$) takes a long time, of order $1/\varepsilon$.

Let us make the normalizing change of variable $t = t'/\varepsilon$. Then, dropping the prime of the new variable t' to simplify the formulas, we rewrite the system (4.49), (4.50) as

$$\begin{aligned} \varepsilon\frac{\partial}{\partial t}(\theta + \psi) &= \Delta\theta + f(x), \\ -\kappa\varepsilon\frac{\partial\psi}{\partial t} &= \Delta(\varepsilon^2\Delta\psi + \psi - \psi^3 + \varepsilon\kappa_1\theta). \end{aligned} \tag{4.104}$$

As before, we set the boundary conditions for (4.104) in the form (4.51), and concerning the initial data

$$\theta\big|_{t=0} = \theta^0(x,\varepsilon), \quad \psi\big|_{t=0} = \psi^0(x,\varepsilon)$$

we will assume that we are given subsets $\Omega_0^{\pm} \subset \Omega$ separated by a smooth surface Γ_0 of codimension 1 such that $\Omega = \Omega_0^+ \cup \Omega_0^- \cup \Gamma_0$, and that

$$\lim_{\varepsilon\to 0}\varphi^0 = 1, \quad x \in \Omega_0^+, \qquad \lim_{\varepsilon\to 0}\varphi^0 = -1, \quad x \in \Omega_0^-. \tag{4.105}$$

We assume that dist $(\Gamma_0, \partial\Omega) \geq$ const > 0 and, for definiteness, that it is the domain Ω_0^- which is adjacent to the outer boundary $\partial\Omega$.

By analogy with the phase field system, as well as with the conserved phase field system for small relaxation times [**198**], one can naturally expect that the first derivative of the temperature $\theta^0(x,\varepsilon)$ has a discontinuity on Γ_0 as $\varepsilon \to 0$. However, this is not the case. It can be shown [**199**] that the system (4.104) has a solution with discontinuous limiting phase function which we are interested in only if θ^0, up to $O(\varepsilon)$, is a smooth function for all $\varepsilon \in [0,1]$. This result agrees with the behavior of soliton-type asymptotics as $\psi_0 \to \pm 1$. Indeed, the leading term of the asymptotic expansion for temperature constructed in §4.1 is a smooth function, while the amplitude of the jump of the first correction remains bounded as $\psi_0 \to \pm 1$. No wonder, therefore, that the leading term of the temperature remains smooth throughout the bifurcation period, in spite of the fact that during this time the phase function turns from a soliton-type function into a function of Heaviside type.

Thus we will assume that $\varphi^0(x,\varepsilon)$ is a smooth (for $\varepsilon = \text{const}$) regularization of the Heaviside function, while $\theta^0(x,\varepsilon) = \theta_0^0(x) + \varepsilon\theta_1^0(x,\varepsilon)$ and the smooth function θ_0^0 does not depend on ε. However, we will show that the first correction θ_1^0 as $\varepsilon \to 0$ has discontinuous first derivatives on Γ_0. Moreover, we will show that for the existence of the solution of interest it is necessary that the heat source f have a zero mean, i.e.,

$$\int_\Omega f(x)\,dx = 0. \tag{4.106}$$

In order to formulate the main result of this section, we will introduce auxiliary functions. Denote by $\theta_0(x)$ the solution to the Neumann problem

$$\Delta\theta_0 = -f(x), \quad x \in \Omega, \qquad \left.\frac{\partial\theta_0}{\partial n_\Omega}\right|_{\partial\Omega} = 0, \tag{4.107}$$

and let the functions $\Psi_1^\pm(x,t), \varphi(x)$ solve the following model problems:

$$\begin{gathered}\Delta\Psi_1^\pm = -\frac{\kappa_1}{2} f(x), \qquad x \in \Omega_t^\pm, \quad t > 0,\\ \left.\frac{\partial\Psi_1^-}{\partial n_\Omega}\right|_\Sigma = 0, \quad [\Psi_1^\pm] = 0, \quad \left[\frac{\partial\Psi_1^\pm}{\partial\nu}\right] = -\kappa V_\nu,\end{gathered} \tag{4.108}$$

$$\mathcal{K} = \kappa_2\Psi_1^\pm(x,\varphi), \quad \varphi|_{\Gamma_0} = 0, \tag{4.109}$$

where $\Omega_t^\pm$ are subsets of Ω separated by the surface $\Gamma_t = \{x \in \Omega,\ \varphi(x) = t\}$, $\nu = \nabla\varphi/|\nabla\varphi|$, $V_\nu = 1/|\nabla\varphi|$, $\mathcal{K}_t = \operatorname{div}\nu$, $\kappa_2 = 3\sqrt{2}$.

Further, define the functions $\theta_1^\pm(x,t)$ as a solution to the problem

$$\begin{gathered}\Delta\theta_1^\pm = 0, \qquad x \in \Omega_t^\pm, \quad t > 0,\\ \left.\frac{\partial\theta_1^-}{\partial n_\Omega}\right|_\Sigma = 0, \quad [\theta_1^\pm] = 0, \quad \left[\frac{\partial\theta_1^\pm}{\partial\nu}\right] = -2V_\nu.\end{gathered} \tag{4.110}$$

For uniqueness of the solution to (4.107)–(4.110) we require the normalizing conditions

$$\int_\Omega \theta_0\,dx = -K_0, \quad \int_{\Omega_t^+} \Psi_1^+\,dx + \int_{\Omega_t^-} \Psi_1^-\,dx = K_1,$$
$$\int_{\Omega_t^+} \theta_1^+\,dx + \int_{\Omega_t^-} \theta_1^-\,dx = -K_1.$$

Here K_0 and K_1 are the coefficients in the expansion

$$\int_\Omega \psi^0(x,\varepsilon)\,dx = \sum_{i\geq 0} \varepsilon^i K_i.$$

Now we can state the formulas for the first two terms of the asymptotic solution to equations (4.104):

$$\begin{aligned}
\theta(x,t,\varepsilon) &= \theta_0(x) + \varepsilon\Big\{\frac{1}{2}(\theta^+_{1,c}(x,t) + \theta^-_{1,c}(x,t)) \\
&\qquad + \frac{1}{2}(\theta^+_{1,c}(x,t) - \theta^-_{1,c}(x,t))\chi(\tau,x)\Big\} + O(\varepsilon^2), \\
\psi(x,t,\varepsilon) &= \chi(\tau,x) + \varepsilon\Big\{\frac{1}{2}(\Psi^+_{1,c}(x,t) + \Psi^-_{1,c}(x,t)) \\
&\qquad + \frac{1}{2}(\Psi^+_{1,c}(x,t) - \Psi^-_{1,c}(x,t))\chi(\tau,x) + \omega_1(\tau,x)\Big\} + O(\varepsilon^2),
\end{aligned} \tag{4.111}$$

where

$$\chi = \tanh\beta\tau, \qquad \tau = \frac{t-\varphi(x)}{\varepsilon} + \varphi_1(x), \qquad \beta = \frac{1}{\sqrt{2}|\nabla\varphi|},$$

the $\theta^\pm_{1,c}$, $\Psi^\pm_{1,c}$ are smooth extensions of $\theta^\pm_1$, $\Psi^\pm_1$ into layers of width $O(\sqrt{\varepsilon})$ lying in $\Omega^\mp_t$, and the function ω_1 decreases exponentially fast as $\tau \to \pm\infty$. The construction of ω_1 and φ_1 will be described later on.

These formulas clearly show that the leading term of the temperature asymptotics corresponds to the smooth stationary solution of (4.104), i.e., to Neumann's problem for the Poisson equations (4.107), while the variations of temperature related to the motion of the free boundary Γ_t occur only in terms of order $O(\varepsilon)$ and are given by the solution of the problem (4.110), where a Stefan condition is set on the surface Γ_t, which is already known. This surface is determined by the solution to the "stationary" version (4.108), (4.109) of the modified Stefan problem, which involves, however, the auxiliary function $\Psi^\pm_1$, rather than the temperature, in the right-hand side of the "Gibbs–Thomson condition" (4.109). It follows from (4.111) that $\Psi^\pm_1$ is the first correction to the smoothed Heaviside function χ. Moreover, it should be pointed out that in (4.110) the value of $\theta^\pm_1$ on the free boundary Γ_t is not specified, and the temperature is only required to be continuous. This is what makes the problem (4.110) solvable, because Γ_t is already specified.

In Figure 4.4 we demonstrate the results of numerical modeling of the system (4.111) in the one-dimensional case ($x \in (0,2)$). We take $\varepsilon = 0.1$, $\kappa_1 = 1$, $f = 3x^2 - 4x$, and set the initial data:

$$\theta\big|_{t=0} = \tanh\Big(\frac{x-1}{2}\Big), \quad \psi\big|_{t=0} = \tanh\Big(\frac{x-1}{\sqrt{2\varepsilon}}\Big).$$

The computations were carried out for time $t \in [0,\varepsilon]$, which corresponds to $t \sim 1$ if we discard the normalization made when passing to (4.104). The graphs in Figure 4.4 show that the temperature approaches the steady state rather fast. It is also seen that at the same time the free boundary moves with finite velocity (the zero point of the phase function shifts by $O(\varepsilon)$ during the time interval $\delta t \sim \varepsilon$), which also agrees with the asymptotic description of the solution given above.

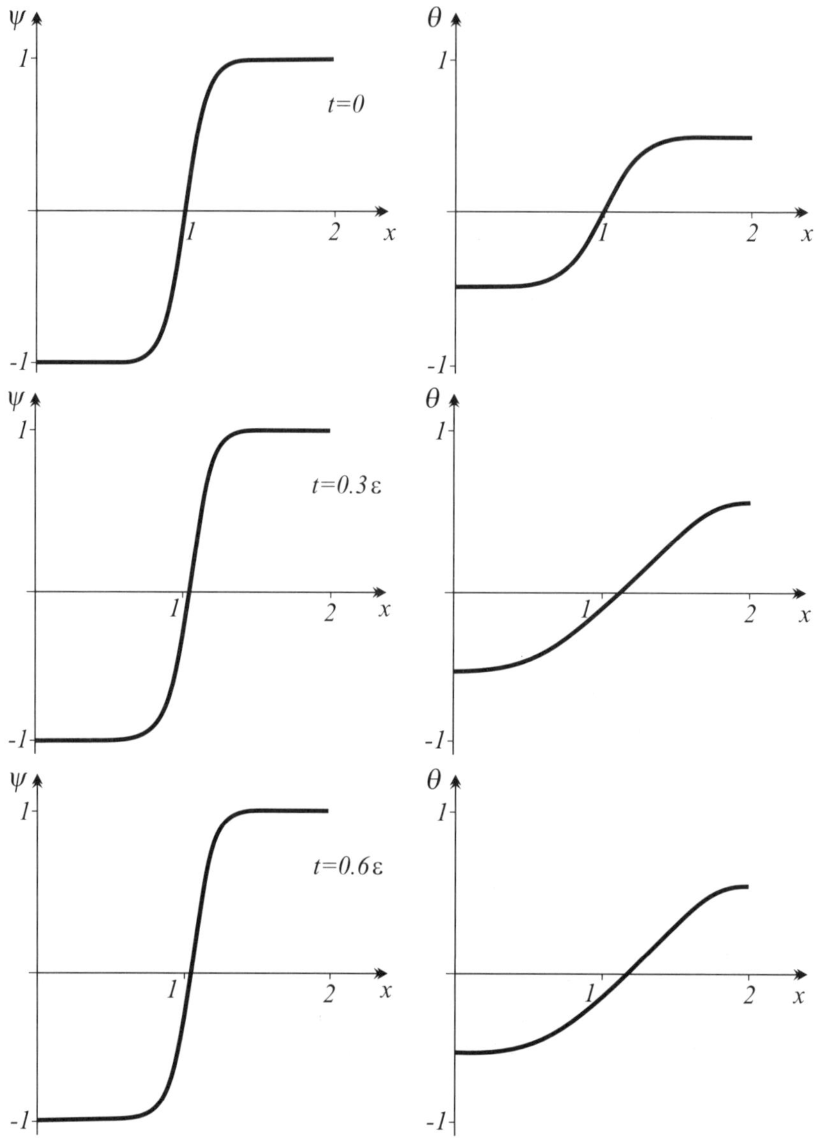

FIGURE 4.4. Spatial profile of the initial data.

The asymptotic solution to (4.111) is constructed as in §4.1. For completeness we present the main steps.

We write the asymptotic solution to (4.104) as

$$\begin{aligned}\theta(x,t,\varepsilon) &= \theta^M(x,t,\varepsilon) + \varepsilon V^M\Big(\frac{S}{\varepsilon},\frac{z}{\varepsilon},x,t,\varepsilon\Big) + O(\varepsilon^{M+1}),\\ \psi(x,t,\varepsilon) &= \varepsilon\Psi^M(x,t,\varepsilon) + W^M\Big(\frac{S}{\varepsilon},\frac{z}{\varepsilon},x,t,\varepsilon\Big) + O(\varepsilon^{M+1}).\end{aligned} \tag{4.112}$$

Here

$$
\begin{aligned}
\theta^M &= \sum_{j=0}^{M} \varepsilon^j \theta_j(x,t), \quad \Psi^M = \sum_{j=1}^{M} \varepsilon^{j-1} \psi_j(x,t), \\
V^M &= \rho(x,t)V(\tau,x) + U_1(\tau,x,t) + Y_1(\zeta,x',t) \\
&\qquad + \sum_{j=2}^{M} \varepsilon^{j-1}\{U_j(\tau,x,t) + Y_j(\zeta,x',t)\}, \\
W^M &= \chi(\tau,x) + \varepsilon\big\{\gamma(x,t)V(\tau,x) + W_1(\tau,x,t) + Z_1(\zeta,x',t)\big\} \\
&\qquad + \sum_{j=2}^{M} \varepsilon^{j}\{W_j(\tau,x,t) + Z_j(\zeta,x',t)\},
\end{aligned} \tag{4.113}
$$

where the same notation as in (4.59), (4.60) has been used. Moreover, we assume that

$$
\chi \to \pm 1 \quad \text{as} \quad \tau \to \pm\infty, \qquad \rho\big|_{\Gamma_t} = 0, \qquad \gamma\big|_{\Gamma_t} = 0,
$$

and that, according to the results of §4.1,

$$
V \to 1 \quad \text{as} \quad \tau \to \infty, \qquad V \to 0 \quad \text{as} \quad \tau \to -\infty.
$$

Substituting (4.112) into (4.104) and passing to the limit as $\tau \to \infty$ and $\zeta \to 0$, we obtain the following relations for the regular part of the asymptotics:

$$
\begin{aligned}
&\Delta\theta^M + f = \varepsilon\Big\{\frac{\partial}{\partial t}(W^{M\pm} - \theta^M) - \Delta V^{M\pm}\Big\} \\
&\qquad + \varepsilon^2 \frac{\partial}{\partial t}(V^{M\pm} + \Psi^M) + O(\varepsilon^{M+1}), \\
&\Delta\big(W^{M\pm} - (W^{M\pm})^3\big) \\
&\qquad + \varepsilon\Big\{\kappa\frac{\partial}{\partial t}W^{M\pm} - \Delta\big((3(W^{M\pm})^2 - 1)\Psi^n + \kappa_1\theta^M\big)\Big\} \\
&\qquad + \varepsilon^2\Big\{\Delta\big((\Delta - 3(\Psi^M)^2)W^{M\pm} + \kappa_1 V^{M\pm}\big) + \kappa\frac{\partial}{\partial t}\Psi^M\Big\} \\
&\qquad + \varepsilon^3\Delta\big(\Delta\Psi^M - (\Psi^M)^3\big) = O(\varepsilon^{M+1}).
\end{aligned} \tag{4.114}
$$

Equating to zero the terms of order ε^0 in (4.114), we immediately obtain the Poisson equation (4.107) and the relation

$$
\Delta\big(\chi^\pm - (\chi^\pm)^3\big) = 0,
$$

which conforms to our assumption $\chi^\pm = \pm 1$. Next, we introduce the notation

$$
\begin{aligned}
\theta_1^\pm &= \theta_1 + \rho V^\pm + U_1^\pm, \qquad & \Psi_1^\pm &= \psi_1 + \gamma V^\pm + W_1^\pm, \\
\theta_k^\pm &= \theta_k + U_k^\pm, & \Psi_k^\pm &= \varphi_k + W_k^\pm, \quad k \geq 2,
\end{aligned}
$$

and consider the terms of order ε^j in (4.114) to obtain (4.108), (4.110), and the similar equations

$$
\Delta\theta_k^\pm = f_{k,\theta}^\pm(x,t), \quad \Delta\Psi_k^\pm = f_{k,\psi}^\pm(x,t)
$$

for the higher order terms.

Let us discuss the construction of the asymptotic solution in a neighborhood of Γ_t. Carrying out the same constructions as in §§4.1, 4.2 and equating to zero the terms of order ε^{-2} in the relation similar to (4.64), we arrive at the equation

$$\frac{\partial^2}{\partial\tau^2}\left\{|\nabla\varphi|^2\frac{\partial^2\chi}{\partial\tau^2}+\chi-\chi^3\right\}=0.$$

Integrating this equation, we immediately obtain that $\chi=\tanh(\beta(\tau+\varphi_1))$, where $\beta=1/\sqrt{2}|\nabla\varphi|$ and $\varphi_1=\varphi_1(x)$ is the "constant" of integration.

Next, equating the higher order terms to zero, we obtain (4.75) and (4.76), where

$$L=|\nabla\varphi|^2\frac{\partial^2}{\partial\tau^2}+1-3\chi^2,$$

and F_k^θ, F_k^ψ are functions of τ and of previous terms of the asymptotic expansion taken at the point $t=\varphi$. In particular,

$$F_1^\theta=0,\quad F_1^\psi=\frac{\partial^2}{\partial\tau^2}\left\{\widehat{\Pi}\frac{\partial\chi}{\partial\tau}+3\chi^2\psi_1(x,\varphi)\right\},$$

$$F_2^\theta=2\beta^2\left\{2\langle\nabla\varphi,\nabla\rho\rangle\frac{\partial V}{\partial\tau}+\frac{\partial\chi}{\partial\tau}-\frac{\tau}{2\beta^2}\frac{\partial\rho}{\partial t}\frac{\partial^2V}{\partial\tau^2}\right\}\bigg|_{t=\varphi}.$$

Obviously, $\check{U}_1=0$, and we have the same conditions (4.79)–(4.81) for solvability of the remaining equations.

We have ensured the fulfillment of (4.79) when constructing θ_0, $\theta_k^\pm$, $\Psi_k^\pm$. Conditions (4.80) for $k=1$ are fulfilled automatically. Further, since in our case

$$f_1^\psi=\widehat{\Pi}\frac{\partial\chi}{\partial\tau}+3\check{\psi}_1(\chi^2-1),$$

one can show by simple calculations that condition (4.81) for $k=1$ implies

$$\operatorname{div}\left(\frac{\nabla\varphi}{|\nabla\varphi|}\right)=3\sqrt{2}\psi_1(x,\varphi). \tag{4.115}$$

Taking (4.115) into account, equation (4.76) for $k=1$ can easily be solved explicitly. Then we obtain that

$$\check{W}_1(\tau,x)\to 0\qquad\text{as}\quad\tau\to\pm\infty.$$

Hence $\check{W}_1^\pm=0$, and (4.115) can be rewritten in the form (4.109) since $\gamma|_{\Gamma_t}=0$. Now we define the function γ outside the surface Γ_t by

$$\gamma=\Psi_{1,c}^+(x,t)-\Psi_{1,c}^-(x,t),$$

where $\Psi_{1,c}^\pm$ is a smooth extension of $\Psi_1^\pm$ such that in the layer of width $\sqrt{\varepsilon}$ adjacent to Γ_t and lying in $\Omega_t^\mp$ the Poisson equation holds, while $\Psi_{1,c}^\pm$ are arbitrary smooth functions in $\Omega_t^\mp$ at larger distances from Γ_t. Setting $\psi_1=\Psi_{1,c}^-$, $V=\frac{1}{2}(\chi-1)$, we define the first correction to ψ as

$$\psi_1+\gamma V+W_1=\Psi_{1,c}^-(x,t)+\big(\Psi_{1,c}^+(x,t)-\Psi_{1,c}^-(x,t)\big)V(\tau,x)+\check{W}_1(\tau,x).$$

Now let us discuss the conditions (4.80), (4.81) for $k = 2$. After some algebra we obtain from (4.80) for $k = 2$ the equalities

$$\left\{2\langle\nabla\varphi, \nabla\rho\rangle + |\nabla\varphi|^2\frac{\partial\rho}{\partial t}\right\}\bigg|_{\Gamma_t} = -2,$$

$$\left\{2\langle\nabla\varphi, \nabla\gamma\rangle + |\nabla\varphi|^2\frac{\partial\gamma}{\partial t}\right\}\bigg|_{\Gamma_t} = -2\kappa.$$

Using (4.58), these equalities can be easily reduced to the Stefan conditions in (4.108), (4.110).

The subsequent constructions are carried through in the same manner as in §§4.1, 4.2. It is clear that condition (4.81) for $k = 2$ yields a linearized equation (4.109) for the function φ_1.

It remains to discuss the boundary conditions on the outer boundary and the boundary layer terms of the asymptotics. The leading term χ of the asymptotic expansion for the phase function satisfies both boundary conditions within $O(\varepsilon^\infty)$. Therefore, a discrepancy in fulfillment of the second boundary condition for ψ arises only in the terms of order $O(\varepsilon)$. Accordingly, letting $Z_j = 0$ for $j = 1, 2, 3$, we obtain for Z_4 the equation

$$\frac{\partial^2 Z_4}{\partial\zeta^2} - 2Z_4 = 0, \qquad Z_4 \to 0 \quad \text{as} \quad \zeta \to \infty.$$

Obviously,

$$Z_4 = -2^{-3/2}\frac{\partial}{\partial n_\Omega}\Psi_1^-\bigg|_\Sigma e^{-\sqrt{2}\zeta}.$$

Further, we observe that the appearance of the boundary layer function $\varepsilon^4 Z_4$ makes a nontrivial contribution of order $O(\varepsilon^3)$ to the boundary value of the normal derivative ψ. To compensate for this addition, we have to impose the following boundary condition on Ψ_3^-:

$$\frac{\partial\Psi_3^-}{\partial n_\Omega}\bigg|_\Sigma = -\frac{\partial Z_4}{\partial\zeta}\bigg|_{\zeta=0} = \frac{1}{2}\frac{\partial}{\partial n_\Omega}\Delta\Psi_1^-\bigg|_\Sigma.$$

Moreover, the appearance of the boundary layer functions Z_k, $k \geq 4$, yields similar terms in the asymptotic expansions for the temperature. It is not hard to calculate that $Y_j = 0$ for $j = 1, \dots, 6$, and we have the following equation for the first nontrivial boundary layer term:

$$\frac{\partial^2 Y_7}{\partial\zeta^2} = \frac{\partial}{\partial t}Z_4(\zeta, x', t), \qquad Y_7 \to 0 \quad \text{as} \quad \zeta \to \infty.$$

Therefore,

$$Y_7 = -2^{-5/2}\frac{\partial^2}{\partial n_\Omega\partial t}\Delta\Psi_1^-\bigg|_\Sigma e^{-\sqrt{2}\zeta}.$$

In its turn, the appearance of the boundary layer function $\varepsilon^7 Y_7$ gives rise to an additional quantity of order $O(\varepsilon^6)$ in the normal derivative of θ on the boundary. To compensate for this addition, we have to impose the following condition for θ_6^-:

$$\frac{\partial\theta_6^-}{\partial n_\Omega}\bigg|_\Sigma = \frac{\partial Y_7}{\partial\zeta}\bigg|_{\zeta=0} = \frac{1}{4}\frac{\partial^2}{\partial n_\Omega\partial t}\Delta\Psi_1^-\bigg|_\Sigma.$$

The boundary conditions in the higher order terms are treated similarly.

It remains to observe that nontrivial terms in the right-hand side of the Neumann conditions begin only with Ψ_3^- and θ_6^-. This fact and (4.51) imply the boundary condition in (4.107), as well as

$$\left.\frac{\partial \Psi_i^-}{\partial n_\Omega}\right|_\Sigma = 0, \quad i = 1, 2, \qquad \text{and} \qquad \left.\frac{\partial \theta_i^-}{\partial n_\Omega}\right|_\Sigma = 0, \quad i = 1, \dots, 6.$$

Obviously, (4.106) is the classical condition for solvability of the Neumann problem for the Poisson equation.

CHAPTER 5

Multi-phase Asymptotic Solutions

In the previous chapters we considered only single-phase asymptotic solutions, i.e., solutions representable as

$$u(x,t,\varepsilon) = U\left(\frac{S(x,t)}{\varepsilon}, x, t, \varepsilon\right), \tag{5.1}$$

where S is a scalar function, and $U(\tau, x, t, \varepsilon)$ regularly depends on the parameter ε. The class of such solutions is rather narrow: dealing with solutions of the form (5.1) only excludes problems of interaction. Of much more interest is the problem of constructing multi-phase asymptotics, i.e., solutions representable as (5.1) with a vector function $S = (S_1, \dots, S_m)$. Methodologically the problem on constructing multi-phase asymptotics with localized fast variation is treated along the same lines as the single-phase case. However, this problem is much more complicated technically, because construction of multi-phase asymptotics leads to model partial differential equations, the first of which is nonlinear. For this reason, multi-phase asymptotics have been effectively constructed for a very narrow class of problems. For the most part, these are problems that can be reduced in some way to those integrable by the method of the inverse scattering problem (see, e.g., [**53, 68, 103–105, 120, 150, 169**]), with some exceptions to be indicated below. A nontrivial development of the theory of multi-phase asymptotics was given recently in [**224–228**], where a two-phase soliton-type asymptotic solution to the Navier–Stokes equations was constructed (see Chapter 6).

5.1. Interaction of a distorted soliton with a rapidly oscillating background

The simplest setup in which a two-phase asymptotics can be constructed is related to the study of the nonlinear Schrödinger equation with a "wrong" sign:

$$i\varepsilon\frac{\partial\psi}{\partial t} + \varepsilon^2\Delta\psi - \alpha(x,t)|\psi|^2\psi = 0. \tag{5.2}$$

Here and later, $i = \sqrt{-1}$ is the imaginary unit, ψ is a complex-valued scalar function, Δ is the Laplace operator in $\mathbb{R}^n_x$, $n \geq 1$, and $\alpha(x,t)$ is a smooth function, which will be assumed to be positive,

$$\alpha(x,t) \geq \text{const} > 0. \tag{5.3}$$

The nonlinear Schrödinger equation, which is one of the fundamental equations of mathematical physics, arises in many applied problems [**94, 97, 124, 183, 222, 162**]. In particular, equation (5.2) with positive coefficient α is used to describe the phenomena of self-focusing and self-compression of wave trains in plasma [**94**]. In the one-dimensional case ($n = 1$) for constant α this equation (with an arbitrary sign of α) can be integrated by the method of the inverse scattering problem [**268**]. In particular, equation (5.2) has the following exact solution:

$$\psi = \exp\Big\{i\Big(\frac{x - V_1 t}{\varepsilon} + \sigma\Big(\frac{x - V_2 t}{\varepsilon}\Big)\Big)\Big\}\sqrt{a^2 - A\cosh^{-2}\Big(\beta\frac{x - V_2 t}{\varepsilon}\Big)}, \tag{5.4}$$

where $\sigma(\tau) = \kappa \arctan(\gamma \tanh(\beta\tau))$ and $V_j, \kappa_j, \gamma, a > 0$, $A > 0$, $\beta \in \mathbb{R}$ are constants.

It is not hard to establish that outside a small neighborhood of the soliton front Γ_t, i.e., the line $x = V_2 t$, (5.4) is a rapidly oscillating planar wave with amplitude a. Inside the neighborhood of Γ_t a modulation of the fast oscillations occurs, so that on the soliton front the amplitude reduces to $a' = \sqrt{a^2 - A}$. This leads to the natural condition

$$A < a^2. \tag{5.5}$$

Moreover, in the neighborhood of Γ_t the function σ varies from $-\kappa \arctan\gamma$ to $\kappa \arctan\gamma$. Thus σ can be naturally interpreted as the phase shift of oscillations which occurs after the passage of the soliton.

Of course, the wave function ψ can be described as a planar wave with constant amplitude and phase shift only in the simplest one-dimensional setup, when, in addition, the coefficient α in the potential $\alpha|\psi|^2$ does not depend on x and t. A natural attempt to extend the solution (5.4) to the case of a variable coefficient $\alpha = \alpha(x,t)$ was made in [**77**]. However, a complete solution to this problem in the one-dimensional case was obtained in [**238**]. In this section we construct an asymptotic solution describing the interaction between a rapidly oscillating background and a soliton in the multidimensional ($n \geq 1$) case.

Before we proceed to the very cumbersome calculations required to construct the asymptotics, we formulate our main result. For that we need some auxiliary functions. Denote by $\Phi = \Phi(x,t)$ and $a_0 = a_0(x,t)$ real-valued scalar functions satisfying the equations

$$\frac{\partial\Phi}{\partial t} + |\nabla\Phi|^2 + \alpha a_0^2 = 0, \tag{5.6}$$

$$\frac{\partial a_0}{\partial t} + 2\langle\nabla\Phi, \nabla a_0\rangle + a_0\Delta\Phi = 0, \tag{5.7}$$

and retaining their smoothness for $t \in [0,T]$. The meaning of Φ and a_0 is the phase and amplitude of the rapidly oscillating background. Let us remark that the equations (5.6), (5.7) can be interpreted as equations of ideal gas dynamics,

$$\begin{aligned} \rho\Big(\frac{\partial u}{\partial t} + \langle u, \nabla\rangle u\Big) + \nabla p = 0, \\ \frac{\partial\rho}{\partial t} + \langle\nabla, \rho u\rangle = 0, \end{aligned} \tag{5.8}$$

if we introduce the "density" and "velocity" by the equalities

$$\rho = a_0^2, \quad u = 2\nabla\Phi, \tag{5.9}$$

and define the "pressure" p by the equation of state $dp = 2\rho d(\alpha\rho)$. As we will see later, this interpretation is by no means accidental.

As before, we will describe the front of the distorted soliton Γ_t as a level surface of an auxiliary function $\varphi(x)$, i.e., $\Gamma_t = \{x,\ \varphi(x) = t\}$. Denote by ν a unit normal to Γ_t, and let

$$\nu = \frac{\nabla\varphi}{|\nabla\varphi|}, \quad V_\nu = \frac{1}{|\nabla\varphi|}, \quad u_\nu = \langle \nu, u\rangle, \tag{5.10}$$

where V_ν is the normal velocity of Γ_t, and u_ν is the projection of the vector u on the normal ν. We will also use the notation

$$\kappa = V_\nu - \check{u}_\nu, \quad A = \rho - \frac{\kappa^2}{2\alpha}\Big|_{\Gamma_t}, \tag{5.11}$$

where κ is the relative velocity of the front motion along ν, and A is the amplitude of the distorted soliton. As before, the hat denotes the trace of the function on the surface Γ_t, i.e., $\check{f} = f(x, \varphi(x))$.

Now we write the main model equation,

$$\langle \nabla, A^{3/2}\nu\rangle + \frac{3}{2}\frac{\sqrt{A}}{\kappa}\langle \nabla, A\check{u}\rangle - \frac{1}{2}\frac{A^{3/2}}{\alpha}\left\{3\frac{\alpha}{V_\nu\kappa}\frac{\partial u_\nu}{\partial t} - \frac{\partial\alpha}{\partial\nu}\right\}\Big|_{\Gamma_t} = 0, \tag{5.12}$$

which describes the dynamics of the soliton front. Here and in the sequel, $\partial f/\partial\nu = \langle \nu, \nabla f\rangle$ denotes the normal derivative of the function f.

We assume that in the domain $\{x,\ 0 \le \varphi(x) \le T\}$ the solution $\varphi(x)$ of (5.12) remains smooth and satisfies the condition

$$\kappa^2 < 2\alpha\rho\big|_{\Gamma_t}, \tag{5.13}$$

which is similar to (5.5).

To describe the phase shift resulting from the interaction between the soliton and the rapidly oscillating background, we introduce the functions $\sigma^\pm(x,t)$ and the auxiliary functions $\rho_1^\pm(x,t)$ as a solution to the problem

$$\frac{\partial\sigma^\pm}{\partial t} + \langle u, \nabla\sigma^\pm\rangle + 2\alpha\rho_1^\pm = 0, \qquad x \in \Omega_t^\pm, \quad t > 0, \tag{5.14}$$

$$\frac{\partial\rho_1^\pm}{\partial t} + \langle \nabla, \rho_1^\pm u + \rho\nabla\sigma^\pm\rangle = 0, \tag{5.15}$$

$$[\sigma^\pm] = 2\arctan\gamma, \quad [\rho_1^\pm] - \frac{\check{\rho}}{\kappa}\left[\frac{\partial\sigma^\pm}{\partial\nu}\right] = \delta. \tag{5.16}$$

Here $\Omega_t^+ = \{x,\ t - \varphi(x) > 0\}$ and $\Omega_t^- = \{x,\ t - \varphi(x) < 0\}$ are the domains separated by the surface Γ_t; the square brackets denote the magnitude of the jump on Γ_t,

$$\gamma = \sqrt{2\frac{\alpha\rho}{\kappa^2} - 1}\,\Big|_{\Gamma_t}, \quad \delta = \sqrt{2}\frac{V_\nu}{\kappa}\left\langle \nabla, \sqrt{\frac{A}{\check{\alpha}}}(\nu + V_\nu^{-1}\check{u}^{\parallel})\right\rangle, \tag{5.17}$$

where $u^{\|} = u - u_\nu \nu$ denotes the components of the vector (5.9) that are tangent to Γ_t.

Now we can write the formula for the leading term of the asymptotic solution to equation (5.2):

$$\psi = e^{i((\Phi(x,t)/\varepsilon)+\sigma(x,t,\tau))}\sqrt{a_0^2(x,t) - \frac{A(x)}{\cosh^2(\beta\tau)}} + O(\varepsilon), \tag{5.18}$$

where

$$\begin{gathered}\sigma = \frac{\sigma_c^+(x,t) + \sigma_c^-(x,t)}{2} + \frac{\sigma_c^+(x,t) - \sigma_c^-(x,t)}{2}\eta(\tau,x),\\ \eta = \frac{\arctan(\gamma\tanh(\beta\tau))}{\arctan\gamma}, \quad \tau = \frac{t-\varphi(x)}{\varepsilon} + \varphi_1(x), \quad \beta = \sqrt{\frac{\breve{\alpha}A}{2|\nabla\varphi|^2}},\end{gathered} \tag{5.19}$$

$\sigma_c^\pm$ are smooth extensions of the functions $\sigma^\pm$, and φ_1 is the correction to the position of the soliton front. The method of evaluating it will be given below. As is easily seen, outside a small neighborhood of Γ_t the function η equals ± 1. Hence the full phase of the wave function ψ is equal to $\Phi(x,t)/\varepsilon + \sigma^\pm(x,t)$ in the domains $\Omega_t^\pm$ respectively.

The equation (5.12) is very complicated. This is clearly seen from its geometric interpretation

$$A\mathcal{K} - \frac{3}{2\alpha}a = f, \tag{5.20}$$

where $\mathcal{K}$ is the mean curvature of the surface Γ_t, the meaning of $a = \langle V_\nu \nu + u^{\|}, \nabla\kappa\rangle$ is acceleration, and f is a function which depends on the first derivatives of φ. It is easily seen that the dynamics of the surface Γ_t is substantially affected by the "outer flow" $u^{\|}$ parallel to Γ_t.

Consider some particular cases where (5.12) can be considerably simplified.

1. Let $n = 1$ and $\alpha = \text{const}$. Assuming for definiteness that φ is an increasing function of x, we can easily reduce (5.12) to the following equation:

$$\left.\frac{dA}{dx} + \frac{A}{V_\nu}\frac{\partial u}{\partial x}\right|_{t=\varphi} = 0. \tag{5.21}$$

Taking into account the dependence of κ on φ, we derive from the relation $A = \breve{\rho} - \kappa^2/2\breve{\alpha}$ the second equation

$$\frac{d\varphi}{dx} = \left(u + \sqrt{2\alpha(\rho - A)}\right)^{-1}\Big|_{t=\varphi}. \tag{5.22}$$

Invoking the equality $V_\nu = (d\varphi/dx)^{-1}$, we see that equations (5.21), (5.22) form a closed system.

In case $\rho = \text{const}$, $u = \text{const}$ (5.21), (5.22) obviously imply that $A = \text{const}$ and φ is a linear function of x. In this case the leading term of the asymptotic solution (5.18) is an exact solution of (5.2).

2. Let $n \geq 2$ and let α, ρ, u be constants. Then (5.12) can be readily reduced to a simplified version of (5.20):

$$A\mathcal{K} - \langle V_\nu \nu + u^{\|}, \nabla \kappa \rangle = 0.$$

Hence it is clearly seen that in the multidimensional case the surface Γ_t has very complicated dynamics even in the simplest setting of constant α, ρ, u.

3. Let $n = 2$, $\alpha = \text{const}$, and let Γ_0 be a circle of radius $R \gg 1$. Moreover, let the "outer flow" ρ and u be functions of t and the radius r. Then, obviously, $A = A(r)$ and $\varphi = \varphi(r)$. Consider the behavior of the solution for large r. Letting $\rho \sim r^{-2/3}$, $u \sim r^{-1/3}$, we obtain from (5.12)

$$\frac{dA}{dr} + \frac{2}{3}\frac{A}{r} + r^{-1/3}\frac{d\varphi}{dr}\frac{\partial}{\partial r}\Big(r^{1/3}\frac{\partial u}{\partial r}\Big)\bigg|_{t=\varphi(r)} = 0.$$

Hence we find that the surface Γ_t moves with velocity $V_\nu \sim r^{-1/3}$ and the amplitude decays as $r^{-2/3}$. Note that we have obtained the same dependence of the amplitude of the soliton on the radius as for the amplitude of a solitary wave moving in a basin of constant depth (see §2.4). Apparently there must be deep physical analogies.

Let us proceed to construction of the self-similar asymptotic solution to equation (5.2). We represent it by an expansion

$$(5.23)\qquad \begin{gathered} \psi = e^{i\Phi(x,t)/\varepsilon}\mathcal{V}\Big(\frac{t-\varphi(x)}{\varepsilon}, x, t, \varepsilon\Big), \\ \mathcal{V}(\tau, x, t) = V_0(\tau, x, t) + \varepsilon V_1(\tau, x, t) + \dots, \end{gathered}$$

where Φ, φ are smooth real-valued functions, and the V_j are smooth complex-valued functions such that

$$V_j(\tau, x, t) \to V_j^{\pm}(x, t) \qquad \text{as} \quad \tau \to \pm\infty$$

for all $j \geq 0$, and

$$(5.24)\qquad |V_0^+| = |V_0^-|.$$

First we determine the rapidly oscillating background. To this end we substitute (5.23) into (5.2) and pass to the limit as $\tau \to \pm\infty$. Denoting

$$(5.25)\qquad \mathcal{V}^{\pm} = V_0^{\pm}(x, t) + \varepsilon V_1^{\pm}(x, t) + \dots,$$

we obtain the equation

$$(5.26)\qquad -\Big(\frac{\partial \Phi}{\partial t} + |\nabla\Phi|^2 + \alpha|\mathcal{V}^{\pm}|^2\Big)\mathcal{V}^{\pm} + i\varepsilon\widehat{\Pi}\mathcal{V}^{\pm} + \varepsilon^2\Delta\mathcal{V}^{\pm} = 0,$$

where

$$\widehat{\Pi} = \frac{\partial}{\partial t} + 2\langle\nabla\Phi, \nabla\rangle + \Delta\Phi$$

is a translation operator.

Using the standard technique of the nonlinear WKB method, we obtain from (5.26) the equations

$$\frac{\partial \Phi}{\partial t} + |\nabla \Phi|^2 + \alpha |V_0^\pm|^2 = 0, \tag{5.27}$$

$$\left(\frac{\partial \Phi}{\partial t} + |\nabla \Phi|^2 + 2\alpha |V_0^\pm|^2\right) V_k^\pm + \alpha (V_0^\pm)^2 \overline{V_k^\pm} = f_k^\pm(x,t), \tag{5.28}$$

for determination of the phase Φ and the coefficients of the expansion (5.25). Here $k = 1, 2, \dots$, the bar means the complex conjugate, and the right-hand sides f_k are polynomials in previous terms of the expansion and their derivatives. We will need the explicit formulas for the first two functions $f_k^\pm$:

$$f_1^\pm = i\widehat{\Pi} V_0^\pm,$$
$$f_2^\pm = i\widehat{\Pi} V_1^\pm - \alpha\big(\overline{V_0^\pm}(V_1^\pm)^2 + 2|V_1^\pm|^2 V_0^\pm\big) + \Delta V_0^\pm.$$

Using the assumption (5.24) and denoting $a_0 = |V_0^\pm|$, we immediately obtain equation (5.6) from (5.27). Next, taking (5.27) into account and dividing each equation (5.28) by $V_0^\pm$, we obtain the equalities

$$\alpha\{\overline{V_0^\pm} V_k^\pm + V_0^\pm \overline{V_k^\pm}\} = (V_0^\pm)^{-1} f_k^\pm. \tag{5.29}$$

Let us introduce the notation

$$\rho_k^\pm = \operatorname{Re}\big(\overline{V_0^\pm} V_k^\pm\big), \quad w_k = \operatorname{Im}\big(\overline{V_0^\pm} V_k^\pm\big), \quad \sigma^\pm = \arg\big(V_0^\pm\big), \tag{5.30}$$

so that $V_0^\pm = a_0 \exp(i\sigma^\pm)$, $\rho_0^+ = \rho_0^- = a_0^2$. Consider (5.29) for $k = 1$. Using the explicit form of $f_1^\pm$, we obtain

$$2\alpha\rho_1^\pm + \frac{d\sigma^\pm}{dt} - \frac{i}{a_0^2}\left(\frac{1}{2}\frac{da_0^2}{dt} + \Delta\Phi a_0^2\right) = 0, \tag{5.31}$$

where

$$\frac{d}{dt} = \frac{\partial}{\partial t} + 2\langle \nabla\Phi, \nabla\rangle \tag{5.32}$$

is the total derivative along characteristics of equation (5.6) treated as a Hamilton–Jacobi equation with "known" right-hand side αa_0^2.

Now, equating to zero the imaginary and real parts of (5.31), we obtain the equations (5.7), (5.14), respectively.

We see that we have obtained a closed system (5.6), (5.7) for determination of Φ and a_0; however, equation (5.14) contains an additional unknown function $\rho_1^\pm$. To determine this function, consider equation (5.29) for $k = 2$:

$$2\alpha\rho_2^\pm = \big(V_0^\pm\big)^{-1}\{i\widehat{\Pi} V_1^\pm + \Delta V_0^\pm\} - \frac{\alpha}{\rho}\{(\overline{V_0^\pm} V_1^\pm)^2 + 2\rho|V_1^\pm|^2\}, \tag{5.33}$$

where we have used the notation (5.9). Equating the real and imaginary parts of (5.33), we obtain an equation similar to (5.14) that relates $\sigma_1^\pm = \arg(V_1^\pm)$ to $\rho_2^\pm$, as well as the equality

$$\operatorname{Im}\big\{\overline{V_0^\pm}\big(i\widehat{\Pi} V_1^\pm + \Delta V_0^\pm\big) - \alpha\big(\overline{V_0^\pm} V_1^\pm\big)^2\big\} = 0. \tag{5.34}$$

To transform (5.34), observe that equation (5.29) for $k = 1$ can be written as

$$\frac{dV_0^\pm}{dt} + \Delta\Phi V_0^\pm + i2\alpha\rho_1^\pm V_0^\pm = 0.$$

Using this equality, we can easily derive the formula

$$\overline{V_0^\pm}\frac{dV_1^\pm}{dt} = \frac{d\rho_1^\pm}{dt} + \big(\Delta\Phi + 2\alpha w_1^\pm\big)\rho_1^\pm + i\Big(\frac{dw_1^\pm}{dt} + \Delta\Phi w_1^\pm - 2\alpha(\rho_1^\pm)^2\Big).$$

Therefore

$$\operatorname{Re}\big(\overline{V_0^\pm}\widehat{\Pi}V_1^\pm\big) - \alpha \operatorname{Im}\big(\overline{V_0^\pm}V_1^\pm\big)^2 = \frac{d\rho_1^\pm}{dt} + 2\Delta\Phi\rho_1^\pm. \tag{5.35}$$

It is also not hard to calculate that

$$\operatorname{Im}\big(\overline{V_0^\pm}\Delta V_0^\pm\big) = a_0\big(2\langle\nabla a_0, \nabla\sigma^\pm\rangle + a_0\Delta\sigma^\pm\big).$$

This equality combined with (5.34) and (5.35) implies (5.15).

A similar procedure is used to construct subsequent terms of (5.25). It is clear that we derive the equation for $\sigma_k^\pm = \arg(V_k^\pm)$ and $\rho_{k+1}^\pm$ from the real part of the kth equality (5.29), while the missing equation similar to (5.15) is obtained by equating to zero the imaginary part of the $(k+1)$st equality (5.29).

Now we proceed to constructing the "amplitude" $\mathcal{V}(\tau, x, t, \varepsilon)$ of the solution (5.23). First of all, rewrite the outcome of substituting (5.23) into (5.2) in the more convenient form

$$i\varepsilon\frac{d\mathcal{V}}{dt} + \varepsilon^2\Delta\mathcal{V} - \alpha\big(|\mathcal{V}|^2 - \rho\big)\mathcal{V} + i\varepsilon\Delta\Phi\mathcal{V} = 0, \tag{5.36}$$

where $\mathcal{V} = \mathcal{V}\big((t - \varphi(x))/\varepsilon, x, t, \varepsilon\big)$, and we have used (5.6) and the notation (5.9), (5.32). The passage from (5.2) to (5.36), though trivial computationally, is conceptually important. The meaning of this transformation consists in reducing the problem of constructing the two-phase asymptotics $\psi = \psi(\Phi/\varepsilon, S/\varepsilon, x, t, \varepsilon)$ to the much simpler problem of constructing a single-phase solution $\mathcal{V} = \mathcal{V}(S/\varepsilon, x, t, \varepsilon)$. Of course, such a reduction is only possible owing to the specific form of the Schrödinger equation, viz., its linearity with respect to the argument of the wave function. Since construction of the rapidly oscillating background outside a small neighborhood of Γ_t has been almost completed (we have not derived the Hugoniot conditions on the front Γ_t, and the position of the front itself is not yet known), we can use all the techniques developed above with minor modifications to analyze (5.36).

Namely, we pass to "independent" variables τ, x, t in (5.36) and use the expansion (5.23) for $\mathcal{V}$. Then after some algebra we can rewrite (5.36) as

$$\begin{aligned} &i\kappa_1\frac{\partial V_0}{\partial\tau} + \frac{\partial^2 V_0}{\partial\tau^2} - \alpha_1(|V_0|^2 - \rho)V_0 + \sum_{k\geq 1}\varepsilon^k\Big\{i\kappa_1\frac{\partial V_k}{\partial\tau} \\ &\quad + \frac{\partial^2 V_k}{\partial\tau^2} - \alpha_1(2|V_0|^2 - \rho)V_k - \alpha_1 V_0^2\overline{V_k} + \mathcal{F}_k\Big\}\Big|_{\tau=(t-\varphi)/\varepsilon} = 0. \end{aligned} \tag{5.37}$$

Here $\kappa_1 = \kappa/|\nabla\varphi|$, $\alpha_1 = \alpha/|\nabla\varphi|^2$, and the $\mathcal{F}_k$ are polynomials in $V_0, \dots, V_{k-1}$ and their derivatives. In particular,

$$\mathcal{F}_1 = \frac{1}{|\nabla\varphi|^2}\left\{i\left(\frac{dV_0}{dt} + \Delta\Phi V_0\right) - \widehat{\pi}\frac{\partial V_0}{\partial\tau}\right\}, \quad \widehat{\pi} = 2\langle\nabla\varphi, \nabla\rangle + \Delta\varphi.$$

Following our method, we will consider consecutively the terms of order ε^j, $j = 0, 1, \dots$, in (5.37). Under the condition (5.24) (which is yet to be justified) all the terms of magnitude ε^0 in (5.37) tend to zero as $\tau \to \pm\infty$. Hence, writing down the Taylor expansion about the point $t = \varphi(x)$ and denoting $v_j = V_j(\tau, x, \varphi)$, we obtain from (5.37) the equation

$$\frac{\partial^2 v_0}{\partial\tau^2} + i\kappa_1\frac{\partial v_0}{\partial\tau} - \check{\alpha}_1\big(|v_0|^2 - \check{\rho}\big)v_0 = 0. \tag{5.38}$$

We make the change of variables

$$v_0 = f\exp(i\check{\sigma}), \tag{5.39}$$

where $f = f(\tau, x)$ and $\check{\sigma} = \check{\sigma}(\tau, x)$ are real-valued functions stabilizing as $\tau \to \pm\infty$. Then, substituting (5.39) into (5.38) and equating to zero the real and imaginary parts, we get the equations

$$\frac{\partial^2 f}{\partial\tau^2} - f\left(\frac{\partial\check{\sigma}}{\partial\tau}\right)^2 - \kappa_1 f\frac{\partial\check{\sigma}}{\partial\tau} - \check{\alpha}_1(f^2 - \check{\rho})f = 0, \tag{5.40}$$

$$f\frac{\partial^2\check{\sigma}}{\partial\tau^2} + 2\frac{\partial\check{\sigma}}{\partial\tau}\frac{\partial f}{\partial\tau} + \kappa_1\frac{\partial f}{\partial\tau} = 0. \tag{5.41}$$

Integrating (5.41) once more, we obtain the equality

$$f^2\frac{\partial\check{\sigma}}{\partial\tau} = \frac{\kappa_1}{2}(\check{\rho} - f^2). \tag{5.42}$$

We exclude $\check{\sigma}$ from (5.40) and integrate to transform this equation into

$$\left(\frac{\partial f}{\partial\tau}\right)^2 + \frac{\kappa_1^2}{4}\left(f - \frac{\check{\rho}}{f}\right)^2 - \frac{\check{\alpha}_1}{2}(f^2 - \check{\rho})^2 = 0.$$

We have chosen the "constant" of integration according to condition (5.24), i.e., $f^2 \to \check{\rho}$ as $\tau \to \pm\infty$. Now, by the substitution $z = f^2$, the last equation is reduced to its final form:

$$\left(\frac{\partial z}{\partial\tau}\right)^2 = 2\check{\alpha}_1\left(z - \frac{\kappa_1^2}{2\check{\alpha}_1}\right)(z - \check{\rho})^2. \tag{5.43}$$

It is not hard to see that (5.43) has a real bounded solution only under the condition

$$\frac{\kappa_1^2}{2\check{\alpha}_1} < \check{\rho}, \tag{5.44}$$

in which case z runs from $\kappa_1^2/2\breve{\alpha}_1$ to $\breve{\rho}$. Obviously, (5.44) is just another form of condition (5.13).

The elementary integration of (5.43) yields the formula

$$f^2 = \breve{\rho} - \frac{A}{\cosh^2 \beta(\tau + \varphi_1)}, \tag{5.45}$$

where

$$A = \breve{\rho} - \frac{\kappa^2}{2\breve{\alpha}_1}, \quad \beta = \sqrt{A\breve{\alpha}_1/2}, \tag{5.46}$$

and $\varphi_1 = \varphi_1(x)$ denotes, as usual, the "constant" of integration.

Obviously, the function (5.45) satisfies the assumption (5.24). In turn, (5.42) and (5.45) immediately imply that

$$\breve{\sigma} = \arctan(\gamma \tanh \beta(\tau + \varphi_1)) + c, \tag{5.47}$$

where γ is the same as in (5.17) and c is an arbitrary "constant". In subsequent formulas we will not write the additive "constant" φ_1 explicitly, but will just incorporate it into τ.

Let us construct the extension of f and $\breve{\sigma}$ outside Γ_t. First of all, passing to the limit in (5.47) as $\tau \to \pm\infty$, we readily find that

$$c = \frac{1}{2}(\breve{\sigma}^+ + \breve{\sigma}^-), \quad \arctan \gamma = \frac{1}{2}(\breve{\sigma}^+ - \breve{\sigma}^-). \tag{5.48}$$

Hence, introducing the auxiliary function

$$\eta = (\arctan \gamma)^{-1} \arctan(\gamma \tanh \beta\tau),$$

so that $\eta \to \pm 1$ as $\tau \to \pm\infty$, we can rewrite (5.47) as

$$\breve{\sigma} = \frac{\breve{\sigma}^+ + \breve{\sigma}^-}{2} + \frac{\breve{\sigma}^+ - \breve{\sigma}^-}{2}\eta(\tau, x). \tag{5.49}$$

Now the problem of extending f and $\breve{\sigma}$ is solved in an obvious way, because we dealt with extending functions of the form (5.45) repeatedly when constructing soliton-type asymptotics (e.g., for the Korteweg–de Vries equation), while extensions for functions of the form (5.49) were carried through when constructing asymptotics of smoothed shock wave type. Thus we obtain the formula

$$V_0(\tau, x, t) = e^{i\sigma(\tau,x,t)} \sqrt{\rho(x,t) - \frac{A(x)}{\cosh^2(\beta(x)\tau)}},$$

where $\sigma(\tau, x, t)$ is defined by (5.19). This and (5.23) yield formula (5.18) for the leading term of the asymptotic expansion.

Moreover, observe that the second equality in (5.48) is nothing but the condition on the magnitude of the jump of $\sigma^\pm$ on the surface Γ_t. Thus we have obtained the first Hugoniot condition in (5.16).

Nevertheless, both the function φ and the second Hugoniot condition in (5.48) are not determined yet. To find them, we again turn to (5.37) and consider the terms of order ε. It is not hard to check that by (5.7) and (5.14) all these terms decrease exponentially fast as $\tau \to \pm\infty$. Hence, using the Taylor expansion, we obtain from (5.37) the equation

$$\frac{\partial^2 v_1}{\partial\tau^2} + i\kappa_1\frac{\partial v_1}{\partial\tau} - \check{\alpha}_1(2|v_0|^2 - \check{\rho})v_1 - \check{\alpha}v_0^2\overline{v_1} = F_1(\tau, x) \tag{5.50}$$

for the function $v_1(\tau, x) = V_1(\tau, x)\big|_{t=\varphi(x)}$. Here $F_1 = -\check{\mathcal{F}}_1 - G_1 - G_2$,

$$G_1 = \tau\left\{\frac{\partial^3 V_0}{\partial\tau^2\partial t} + i\kappa_1\frac{\partial^2 V_0}{\partial\tau\partial t} - \check{\alpha}_1\left(2|v_0|^2\frac{\partial V_0}{\partial t} + v_0^2\frac{\partial\overline{V}_0}{\partial t} - \check{\rho}\frac{\partial V_0}{\partial t}\right)\right\}\Bigg|_{\tau=\varphi},$$

$$G_2 = \tau\left\{i\frac{\partial v_0}{\partial\tau}\frac{\partial\kappa_1(x,t)}{\partial t} - |v_0|^2 v_0\frac{\partial\alpha_1(x,t)}{\partial t} + v_0\frac{\partial}{\partial t}(\alpha_1(x,t)\rho(x,t))\right\}\Bigg|_{t=\varphi},$$

$\kappa_1(x,t) = (V_\nu(x) - \langle\nu, u(x,t)\rangle)/|\nabla\varphi|$, $\alpha_1(x,t) = \alpha(x,t)/|\nabla\varphi|^2$, and the explicit expression for $\mathcal{F}_1$ was given after (5.37).

It is not hard to establish [**238**] that for solvability of (5.50) in the class of functions stabilizing at $\pm\infty$ it is necessary and sufficient that

$$\operatorname{Re}\int_{-\infty}^{\infty} F_1(\tau, x)\overline{\frac{\partial v_0}{\partial\tau}}\,d\tau = 0. \tag{5.51}$$

If (5.51) is fulfilled, a solution to (5.50) in the class of interest is of the form

$$v_1(\tau, x) = \widetilde{v}_1(\tau, x) + \varphi_2(x)\frac{\partial v_0}{\partial\tau}, \tag{5.52}$$

where $\widetilde{v}$ is a particular solution of (5.50) and φ_2 is an arbitrary "constant" of integration.

Consider the orthogonality condition (5.51). First of all, let us evaluate the integral

$$J_1 = \operatorname{Re}\int_{-\infty}^{\infty} G_1\frac{\partial\overline{v_0}}{\partial\tau}\,d\tau.$$

On integrating by parts we have

$$J_1 = \operatorname{Re}\left\{\int_{-\infty}^{\infty}\frac{\partial V_0}{\partial t}\left(2\frac{\partial^2\overline{v_0}}{\partial\tau^2} - i\kappa_1\frac{\partial\overline{v_0}}{\partial\tau}\right)d\tau - \check{\alpha}_1\int_{-\infty}^{\infty}\tau v_0^2\frac{\partial\overline{v_0}}{\partial\tau}\frac{\partial\overline{V_0}}{\partial t}\,d\tau \right.$$
$$\left. + \int_{-\infty}^{\infty}\tau\frac{\partial V_0}{\partial t}\left(\frac{\partial^3\overline{v_0}}{\partial\tau^3} - i\kappa_1\frac{\partial^2\overline{v_0}}{\partial\tau^2} - \check{\alpha}_1(2|v_0|^2 - \check{\rho})\frac{\partial\overline{v_0}}{\partial\tau}\right)d\tau\right\}\Bigg|_{t=\varphi}.$$

However, on differentiating (5.38) and taking the complex conjugate we obtain

$$\frac{\partial^3\overline{v_0}}{\partial\tau^3} - i\kappa_1\frac{\partial^2\overline{v_0}}{\partial\tau^2} - \check{\alpha}_1(2|v_0|^2 - \check{\rho})\frac{\partial\overline{v_0}}{\partial\tau} = \check{\alpha}_1\overline{v_0}^2\frac{\partial v_0}{\partial\tau}.$$

Therefore

$$J_1 = \operatorname{Re}\int_{-\infty}^{\infty} \frac{\partial V_0}{\partial t}\left(2\frac{\partial^2 \overline{v_0}}{\partial \tau^2} - i\kappa_1 \frac{\partial \overline{v_0}}{\partial \tau}\right) d\tau\bigg|_{t=\varphi}. \tag{5.53}$$

For the subsequent calculations we need the following readily verified indentity:

$$\begin{aligned} G_3 &\overset{\text{def}}{=} \left\{\frac{1}{|\nabla\varphi|^2}\left(\frac{dV_0}{dt} + \Delta\Phi V_0\right) - \kappa_1\frac{\partial V_0}{\partial t}\right\}\bigg|_{t=\varphi} \\ &= \frac{2}{|\nabla\varphi|^2}\left\{\langle\nabla_x\Phi, \nabla_x V_0\rangle + \langle\nabla_x\Phi, \nabla\varphi\rangle\frac{\partial V_0}{\partial t} + \frac{1}{2}\Delta_x\Phi V_0\right\}\bigg|_{t=\varphi}, \end{aligned} \tag{5.54}$$

where the notation ∇_x, Δ_x is used to emphasize that we differentiate functions with independent variables x and t.

Observe that for any smooth function $f(\tau, x, t)$ and any vector g we have

$$\langle g, \nabla\rangle f(\tau, x, \varphi(x)) = \left\{\langle g, \nabla_x\rangle f(\tau, x, t) + \langle g, \nabla\varphi\rangle\frac{\partial f(\tau, x, t)}{\partial t}\right\}\bigg|_{t=\varphi}. \tag{5.55}$$

Hence (5.54) can be readily rewritten as

$$G_3 = \frac{2}{|\nabla\varphi|^2}\left\{\langle\nabla_x\Phi, \nabla\rangle + \frac{1}{2}\Delta_x\Phi\right\}v_0(\tau, x)\bigg|_{t=\varphi}. \tag{5.56}$$

With the aid of (5.53) and (5.56) we find that

$$\begin{aligned} J_2 &\overset{\text{def}}{=} \operatorname{Re}\int_{-\infty}^{\infty} (\check{\mathcal{F}}_1 + G_1)\frac{\partial \overline{v_0}}{\partial \tau}\, d\tau \\ &= \operatorname{Re}\left\{\frac{i}{|\nabla\varphi|^2}\int_{-\infty}^{\infty}\left(\langle u, \nabla v_0\rangle + \frac{1}{2}v_0\Delta_x u\right)\frac{\partial \overline{v_0}}{\partial \tau}\, d\tau \right. \\ &\quad \left. - \int_{-\infty}^{\infty}\frac{\partial \overline{v_0}}{\partial \tau}\left(\frac{1}{|\nabla\varphi|^2}\widehat{\pi}\frac{\partial V_0}{\partial \tau} + 2\frac{\partial^2 V_0}{\partial \tau\partial t}\right) d\tau\right\}\bigg|_{t=\varphi}. \end{aligned}$$

Next, using (5.55) we have

$$\left(\frac{1}{|\nabla\varphi|^2}\widehat{\pi}\frac{\partial V_0}{\partial \tau} + 2\frac{\partial^2 V_0}{\partial \tau\partial t}\right)\bigg|_{t=\varphi} = \frac{2}{|\nabla\varphi|^2}\left(\langle\nabla\varphi, \nabla\rangle + \frac{1}{2}\Delta\varphi\right)\frac{\partial v_0}{\partial \tau}.$$

By means of this equality, (5.57) can be easily transformed into

$$\begin{aligned} J_2 = &- \frac{1}{|\nabla\varphi|^2}\left\langle\nabla, \left\{\int_{-\infty}^{\infty}\left|\frac{\partial v_0}{\partial \tau}\right|^2 d\tau\nabla\varphi\right\}\right\rangle \\ &+ \operatorname{Re}\frac{i}{|\nabla\varphi|^2}\int_{-\infty}^{\infty}\frac{\partial \overline{v_0}}{\partial \tau}\left(\langle u, \nabla\rangle + \frac{1}{2}\Delta_x u\right)v_0\, d\tau\bigg|_{t=\varphi}. \end{aligned} \tag{5.58}$$

For further transformations we use the representation (5.39). We have

$$\begin{aligned} \operatorname{Re}\left(i\frac{\partial \overline{v_0}}{\partial \tau}\frac{d}{dx}v_0\right) &= -\frac{1}{2}\left(\frac{d\check{\sigma}}{dx}\frac{\partial f^2}{\partial \tau} - \frac{\partial\check{\sigma}}{\partial \tau}\frac{df^2}{dx}\right), \\ \operatorname{Re}\left(i\frac{\partial \overline{v_0}}{\partial \tau}v_0\right) &= f^2\frac{\partial\check{\sigma}}{\partial \tau}. \end{aligned}$$

Using this equality and (5.58) and integrating by parts, we obtain

$$
\begin{aligned}
J_2 = &-\frac{1}{|\nabla\varphi|^2}\Big\langle \nabla, \Big\{\int_{-\infty}^{\infty} \Big|\frac{\partial v_0}{\partial\tau}\Big|^2 d\tau \nabla\varphi\Big\}\Big\rangle \\
&+\frac{1}{2|\nabla\varphi|^2}\Big\langle \nabla, \Big\{\int_{-\infty}^{\infty} f^2 \frac{\partial\breve{\sigma}}{\partial\tau}\, d\tau u(x,t)\Big\}\Big\rangle\Big|_{t=\varphi} \\
&-\frac{\breve{\rho}}{2|\nabla\varphi|^2}\langle \breve{u}, \nabla(\breve{\sigma}^+ - \breve{\sigma}^-)\rangle,
\end{aligned}
\tag{5.59}
$$

where $\breve{u} = 2\nabla_x \Phi(x,t)\big|_{t=\varphi}$.

Now we transform the expression

$$
J_3 \stackrel{\text{def}}{=} \operatorname{Re}\int_{-\infty}^{\infty} G_2 \frac{\partial \overline{v_0}}{\partial\tau}\, d\tau.
$$

Integrating by parts, we can easily establish that

$$
\begin{aligned}
J_3 = \Big\{&\frac{1}{4}\frac{\partial\alpha_1(x,t)}{\partial t}\int_{-\infty}^{\infty} \big(|v_0|^4 - \breve{\rho}^2\big)\, d\tau \\
&-\frac{1}{2}\frac{\partial}{\partial t}\big(\alpha_1(x,t)\rho(x,t)\big)\int_{-\infty}^{\infty}\big(|v_0|^2 - \breve{\rho}\big)\, d\tau\Big\}\Big|_{t=\varphi}.
\end{aligned}
\tag{5.60}
$$

Let us evaluate the integrals in (5.59), (5.60). In view of (5.39), (5.42), (5.45), an elementary integration yields

$$
\int_{-\infty}^{\infty} f^2\frac{\partial\breve{\sigma}}{\partial\tau}\, d\tau = \frac{A\kappa_1}{\beta}, \qquad \int_{-\infty}^{\infty} |v_{0\tau}|^2\, d\tau = \frac{2}{3}\frac{\breve{\alpha}_1 A^2}{\beta},
$$

$$
\int_{-\infty}^{\infty} \big(|v_0|^4 - \breve{\rho}^2\big)\, d\tau = 4\frac{A}{\beta}\Big(\frac{1}{3}A - \breve{\rho}\Big), \qquad \int_{-\infty}^{\infty}\big(|v_0|^2 - \breve{\rho}\big)\, d\tau = -2\frac{A}{\beta}.
$$

Moreover, taking the first Hugoniot condition (5.16) into account, we obtain

$$
\breve{\rho}\big\langle \breve{u}, \nabla(\breve{\sigma}^+ - \breve{\sigma}^-)\big\rangle = \frac{\kappa^2}{\breve{\alpha}}\langle \breve{u}, \nabla\gamma\rangle.
\tag{5.61}
$$

Combining the above formulas and bearing in mind that the left-hand side of (5.51) is the sum of J_2 and J_3, we rewrite the orthogonality condition (5.51) as

$$
\begin{aligned}
&-\frac{2}{3}\Big\langle\nabla, \frac{\breve{\alpha}_1 A^2}{\beta}\nabla\varphi\Big\rangle + \frac{1}{2}\Big\langle \nabla, \frac{A\kappa_1}{\beta}u(x,t)\Big\rangle\Big|_{t=\varphi} \\
&\qquad -\frac{1}{2}\frac{\kappa^2}{\alpha}\langle u, \nabla\gamma\rangle + \frac{A}{\beta}\Big(\frac{A}{3}\frac{\partial\alpha}{\partial t} + \alpha\frac{\partial\rho}{\partial t}\Big)\Big|_{t=\varphi} = 0.
\end{aligned}
\tag{5.62}
$$

Using the explicit expressions (5.17) and (5.19) for γ and β and applying (5.55) once more, we can rewrite equation (5.62) as

$$
\begin{aligned}
&-\frac{2}{3}\langle \nabla, A^{3/2}\nu\rangle + \frac{\sqrt{A}}{\kappa}\Big\langle\nabla, \frac{\kappa^2}{2\alpha}\breve{u}\Big\rangle \\
&\qquad + \sqrt{A}\Big\{|\nabla\varphi|\frac{\partial\rho}{\partial t} - \frac{\kappa}{2\alpha}\Big\langle\nabla, \frac{\partial u}{\partial t}\Big\rangle - \frac{1}{3}\frac{A}{\alpha}\langle \nu, \nabla_x\alpha\rangle\Big\}\Big|_{t=\varphi} = 0.
\end{aligned}
\tag{5.63}
$$

It remains to use the equalities $A = \check{\rho} - \kappa^2/2\check{\alpha}$ and $|\nabla\varphi|\kappa = 1 - \langle\nabla\varphi, u\rangle$, and the fact that the functions $\rho(x,t)$ and $u(x,t)$ satisfy the continuity condition (5.8). Then, after some algebra, equation (5.63) transforms into (5.2).

To complete the construction of the leading terms of asymptotics we still have to derive the Hugoniot condition in (5.16). Clearly, for that we have to find the limits of $\mathrm{Re}(\overline{v_0}v_1)$ as $\tau \to \pm\infty$. To avoid solving equation (5.50) explicitly, which is possible, but requires cumbersome calculations, we employ the following observation [**238**]. Multiply the left-hand side of (5.50) by $i\overline{v_0}$ and integrate with respect to τ. Then after some algebra we obtain

$$(5.64)\qquad \begin{aligned} J &\overset{\text{def}}{=} i\int_{-\infty}^{\infty} \overline{v_0}\Big\{\frac{\partial^2 v_1}{\partial\tau^2} + i\kappa_1\frac{\partial v_1}{\partial\tau} - \check{\alpha}_1(2|v_0|^2 - \check{\rho})v_1 - \check{\alpha}_1 v_0^2\overline{v_1}\Big\}\,d\tau \\ &= -\int_{-\infty}^{\infty}\Big\{\kappa_1\frac{\partial}{\partial\tau}(\overline{v_0}v_1) + i\check{\alpha}_1|v_0|^2(\overline{v_0}v_1 + v_0\overline{v_1})\Big\}\,d\tau. \end{aligned}$$

Therefore, using the notation (5.30), we derive from (5.64) that

$$\mathrm{Re}\,J = -\kappa_1(\check{\rho}_1^+ - \check{\rho}_1^-).$$

Hence

$$(5.65)\qquad [\rho_1^\pm] = -\frac{1}{\kappa_1}\,\mathrm{Re}\,i\int_{-\infty}^{\infty}\overline{V_0}F_1(\tau,x)\,d\tau\Big|_{t=\varphi}.$$

To evaluate the right-hand side of (5.65) we use the following easily verifiable equalities:

$$(5.66)\qquad \begin{aligned} \mathrm{Re}\,i\int_{-\infty}^{\infty}\overline{V_0}\widehat{\pi}\frac{\partial V_0}{\partial\tau}\Big|_{t=\varphi} &= -\int_{-\infty}^{\infty}\Big\{\frac{\partial f^2}{\partial\tau}\langle\nabla\varphi, \nabla_x\sigma\rangle + f^2\Big\langle\nabla\varphi, \nabla_x\frac{\partial\sigma}{\partial\tau}\Big\rangle \\ &\qquad + \Big\langle\nabla_x, \Big(f^2\frac{\partial\sigma}{\partial\tau}\Big)\nabla\varphi\Big\rangle\Big\}\,d\tau\Big|_{t=\varphi} \\ &= -\Big\{\Big\langle\nabla_x, \int_{-\infty}^{\infty} f^2\frac{\partial\sigma}{\partial\tau}\,d\tau\nabla\varphi\Big\rangle + \rho\langle\nabla\varphi, \nabla_x[\sigma]\rangle\Big\}\Big|_{t=\varphi}. \end{aligned}$$

Here $f = f(\tau,x,t)$ and $\sigma = \sigma(\tau,x,t)$ are the modulus and the argument of $V_0(\tau,x,t)$, and $[\sigma] = \sigma^+(x,t) - \sigma^-(x,t)$, where, as usual, $\sigma^\pm = \lim_{\tau\to\pm\infty}\sigma(\tau,x,t)$.

Then after some calculations we obtain

$$(5.67)\qquad \begin{aligned} \mathrm{Re}\,i\int_{-\infty}^{\infty}\overline{V_0}\mathcal{F}_1(\tau,x,t)\,d\tau\Big|_{t=\varphi} &= -\frac{1}{|\nabla\varphi|^2}\Big\{\frac{1}{2}\frac{\partial}{\partial t}\int_{-\infty}^{\infty}(|V_0|^2 - \rho)\,d\tau \\ &\quad + \Big\langle\nabla_x, \Big(\int_{-\infty}^{\infty}(|V_0|^2 - \rho)\,d\tau\nabla_x\Phi - \int_{-\infty}^{\infty} f^2\frac{\partial\sigma}{\partial\tau}\,d\tau\nabla\varphi\Big)\Big\rangle \\ &\quad - \rho\langle\nabla\varphi, \nabla_x[\sigma]\rangle\Big\}\Big|_{t=\varphi}. \end{aligned}$$

In the derivation of (5.67) we have used the fact that ρ and $u = 2\nabla\Phi$ satisfy the continuity equation (5.8).

Next we integrate by parts and, after some algebra, obtain the equalities

$$
\begin{aligned}
&\operatorname{Re} i \int_{-\infty}^{\infty} \overline{V_0} G_1 \, d\tau \Big|_{t=\varphi} \\
(5.68) \qquad &= \operatorname{Re} i \int_{-\infty}^{\infty} \left\{ \frac{\partial V_0}{\partial t} \frac{\partial \overline{V_0}}{\partial \tau} - \overline{V_0} \frac{\partial^2 V_0}{\partial \tau \partial t} + i\kappa_1 \tau \frac{\partial}{\partial \tau} \left(\overline{V_0} \frac{\partial V_0}{\partial t} \right) \right\} d\tau \Big|_{t=\varphi} \\
&= \left\{ \frac{\kappa_1}{2} \frac{\partial}{\partial t} \int_{-\infty}^{\infty} (|V_0|^2 - \rho) \, d\tau + \frac{\partial}{\partial t} \int_{-\infty}^{\infty} f^2 \frac{\partial \sigma}{\partial \tau} \, d\tau \right\} \Big|_{t=\varphi},
\end{aligned}
$$

since

$$
\operatorname{Re} i \int_{-\infty}^{\infty} \left(\frac{\partial V_0}{\partial t} \frac{\partial \overline{V_0}}{\partial \tau} - \overline{V_0} \frac{\partial^2 V_0}{\partial \tau \partial t} \right) d\tau = \frac{\partial}{\partial t} \int_{-\infty}^{\infty} f^2 \frac{\partial \sigma}{\partial \tau} \, d\tau.
$$

Finally, elementary calculations lead to the equality

$$
(5.69) \qquad \operatorname{Re} i \int_{-\infty}^{\infty} \overline{V_0} G_2 \, d\tau \Big|_{t=\varphi} = \frac{1}{2} \frac{\partial \kappa_1(x,t)}{\partial t} \int_{-\infty}^{\infty} (|V_0|^2 - \rho) \, d\tau \Big|_{t=\varphi}.
$$

Combining (5.67)–(5.69) and bearing in mind that $F_1 = -\check{\mathcal{F}}_1 - G_1 - G_2$, we transform (5.65) into

$$
\begin{aligned}
(5.70) \quad [\rho_1^{\pm}] = \frac{1}{\kappa |\nabla \varphi|} \Big\{ &-\frac{1}{2} \Big\langle \nabla, \int_{-\infty}^{\infty} (|v_0|^2 - \check{\rho}) \, d\tau \check{u} \Big\rangle \\
&+ \Big\langle \nabla, \Big(\int_{-\infty}^{\infty} f^2 \frac{\partial \sigma}{\partial \tau} \, d\tau \Big|_{t=\varphi} \Big) \nabla \varphi \Big\rangle + \rho \langle \nabla \varphi, \nabla_x [\sigma] \rangle \big|_{t=\varphi} \Big\}.
\end{aligned}
$$

Hence on integrating in (5.70) we obtain

$$
(5.71) \qquad [\rho_1^{\pm}] = \sqrt{2} \frac{V_\nu}{\kappa} \left\langle \nabla, \sqrt{\frac{A}{\check{\alpha}}} |\nabla \varphi| (\check{u} + \kappa \nu) \right\rangle + \frac{\check{\rho}}{\kappa} \left[\frac{\partial \sigma^{\pm}}{\partial \nu} \right].
$$

It remains to decompose the vector $\check{u}$ into the component $\nu \check{u}_\nu$ orthogonal to Γ_t and the tangent components $\check{u}^{\|} = \check{u} - \nu \check{u}_\nu$. Then we immediately obtain from (5.71) the second Hugoniot condition in (5.16).

The construction of the higher order terms of the asymptotic expansion is carried through according to our general scheme of constructing the single-phase asymptotics. The presence of integration "constants" $\varphi_k(x)$, $k = 1, 2, \ldots$, which appear in the previous steps of the asymptotic procedure guarantees the fulfillment of orthogonality conditions similar to (5.51), and evaluating the limiting (as $\tau \to \pm\infty$) values of the modulus and argument of the functions $v_k(\tau, x)$ gives rise to the Hugoniot conditions similar to (5.16). This completes the determination of the solution outside a small neighborhood of Γ_t.

5.2. Multi-phase soliton-type solutions

We continue to consider multi-phase asymptotics. In contrast to the previous section, now we consider the equations for which the process of constructing multi-phase solutions cannot be reduced to several single-phase procedures. We present the method in detail, taking as an example the construction of a two-phase solution to the sine-Gordon equation with variable coefficients

$$\varepsilon^2 \frac{\partial^2 u}{\partial t^2} - \varepsilon^2 \frac{\partial^2 u}{\partial x^2} + \frac{\alpha^2(x,t)}{\gamma(x,t)} \sin(\gamma(x,t)u) = 0, \qquad x \in \mathbb{R}^1,\ t > 0, \tag{5.72}$$

and we conclude this section with results on N-phase solutions to (5.72), as well as to the Korteweg–de Vries equations with variable coefficients and the Kadomtsev–Petviashvili equations.

It is tempting to interpret the formulas to be obtained as describing the interaction of distorted solitons. However, similarly to the exact solutions to the corresponding equations with constant coefficients, distorted solutions exhibit almost no interaction in the leading term. We illustrate this by the formula for the exact two-phase solution of (5.72) with constant coefficients $\alpha = \gamma = 1$ [**270**],

$$u = 4 \arctan \left\{ \frac{e^{\beta_1(S_1(x,t)/\varepsilon)+\lambda_1} + e^{\beta_2(S_2(x,t)/\varepsilon)+\lambda_2}}{1 - \nu e^{\beta_1(S_1(x,t)/\varepsilon)+\beta_2(S_2(x,t)/\varepsilon)+\lambda_1+\lambda_2}} \right\}. \tag{5.73}$$

Here $S_i = x - V_i t - x_i$, the velocities V_i are constant with $|V_i| < 1$, $\beta_i = 1/\sqrt{1 - V_i^2}$, the x_i and λ_i are arbitrary constants, and

$$\nu = \frac{1 - V_1V_2 - \sqrt{(1 - V_1^2)(1 - V_2^2)}}{1 - V_1V_2 + \sqrt{(1 - V_1^2)(1 - V_2^2)}}. \tag{5.74}$$

Since the solution of the sine-Gordon equation has the physical meaning of an angle (the phase increment), in (5.73) and below we neglect additive terms which are multiples of 2π. This corresponds to invariance of equation (5.72) (with $\gamma = 1$) with respect to the change of variable $u' = 2\pi k \pm u$.

The fact of "almost linear" interaction of solitons is well known. Discussing a multi-soliton solution of the Korteweg–de Vries equation, Miura [**174**] describes this phenomenon as follows: "two solitary waves with different amplitudes interact non-linearly, but emerge unaltered from the reaction. The non-linearity of such solutions lies in the fact that the solitons are displaced relative to the position which they would occupy in the absence of any interaction".

Let us analyze the process of interaction of fluxons (kinks) when ε is a small parameter. For definiteness, we will assume that $x_1 < x_2$ and $v_1 > v_2$, and let $x_2 - x_1$ and $v_1 - v_2$ be independent of ε.

Consider formula (5.73) for $t = 0$. Let $x \le x_2 - c$, where $c < x_2 - x_1$. Then $S_2(x,0)/\varepsilon \to -\infty$ as $\varepsilon \to 0$, and we have, with exponential accuracy,

$$u = 4 \arctan \left\{ e^{\beta_1(S_1(x,0)/\varepsilon)+\lambda_1} \right\}.$$

Thus to the left of the point x_2 satisfying $S_2(x_2, 0) = 0$ formula (5.73) describes a solitary fluxon which varies from 0 to 2π in the ε-neighborhood of the point x_1 such that $S_1(x,0) = 0$. The constant λ_1 is related to a small (of magnitude $\varepsilon\lambda_1/\beta_1$) shift of the front of this fluxon. To simplify the formulas, we set $\lambda_1 = 0$.

Consider (5.73) in a neighborhood of the point x_2. Since now $x \ge x_1 + c$, we have $S_1(x,0)/\varepsilon \to \infty$ as $\varepsilon \to 0$. Hence for such x we obtain from (5.73) that, with exponential accuracy,

$$
\begin{aligned}
u &= 4\arctan\Big\{\Big(-\nu e^{\beta_2(S_2(x,0)/\varepsilon)+\lambda_2}\Big)^{-1}\Big\}\\
&= 2\pi + 4\arctan\Big\{\nu e^{\beta_2(S_2(x,0)/\varepsilon)+\lambda_2}\Big\}\\
&\approx 4\arctan\Big\{e^{\beta_2(S_2(x,0)/\varepsilon)+\lambda_2+\log\nu}\Big\},
\end{aligned}
\tag{5.75}
$$

where the sign $\approx$ means that we neglect the term 2π in (5.75). To simplify the formulas, set $\lambda_2 = \log 1/\nu$.

Thus the assumption $|x_1 - x_2| \ge \text{const}$ implies that formula (5.73) for $t = 0$ describes two fluxons whose fast variation is concentrated in small neighborhoods of the points x_1 and x_2 and which do not interact within accuracy $O(\varepsilon^\infty)$.

To analyze the solution for $t > 0$ we rewrite formula (5.73) as

$$
\begin{aligned}
u &= 4\arctan\{e^{\beta_1 S_1/\varepsilon}\} + 4\arctan\{e^{\beta_2 S_2/\varepsilon}\}\\
&\quad + 4\arctan\Big\{(\nu^{-1}-1)e^{\beta_2 S_2/\varepsilon}\big(1 - e^{\beta_1(S_1/\varepsilon)+\beta_2(S_2/\varepsilon)}\big)/P\Big\},
\end{aligned}
\tag{5.76}
$$

where

$$
\begin{aligned}
P &= 1 + e^{2\beta_1 S_1/\varepsilon} + \nu^{-1}e^{2\beta_2 S_2/\varepsilon}\\
&\quad + (\nu^{-1}-1)e^{\beta_1(S_1/\varepsilon)+\beta_2(S_2/\varepsilon)} + e^{2(\beta_1(S_1/\varepsilon)+\beta_2(S_2/\varepsilon))}.
\end{aligned}
$$

Denote by $t^* = (x_2 - x_1)/(V_1 - V_2)$ the time moment when the lines $\Gamma_t^1 = \{(x,t),\ S_1(x,t) = 0\}$ and $\Gamma_t^2 = \{(x,t),\ S_2(x,t) = 0\}$ intersect, and let $\delta \in (0,1)$, $c > 0$ be constants. Then it is not hard to show that for all $t \le t^* - c\varepsilon^{1-\delta}$ the third term in the right-hand side of (5.76) is of magnitude $O(\varepsilon^\infty)$. Thus, up to a small neighborhood of the intersection point of the fronts Γ_t^1 and Γ_t^2, the solution (5.73) describes the dynamics of two non-interacting (up to exponentially small terms) solitary fluxons. Recall that we set $\lambda_1 = 0$ and $\lambda_2 = \log 1/\nu$ in (5.73).

Consider the solution (5.73) for $t \ge t^* + c\varepsilon^{1-\delta}$. For such t we have $S_2(x,t) - S_1(x,t) \ge c(V_1 - V_2)\varepsilon^{1-\delta}$; therefore we obtain from (5.73) that, with exponential accuracy,

$$
u = 4\arctan\Big\{e^{\beta_2(S_2(x,t)/\varepsilon)-\log\nu}\Big\}
\tag{5.77}
$$

in a neighborhood of the line Γ_t^2, and

$$
u = 4\arctan\Big\{-\Big(\nu e^{\beta_1(S_1(x,t)/\varepsilon)}\Big)^{-1}\Big\} \approx 4\arctan\Big\{e^{\beta_1(S_1(x,t)/\varepsilon)+\log\nu}\Big\}
\tag{5.78}
$$

in a neighborhood of the line Γ_t^1. Hence, for the time after the interaction, formula (5.73) should be rewritten as

$$
\begin{aligned}
u &= 4\arctan\{e^{\beta_1(S_1/\varepsilon)+\log\nu}\} + 4\arctan\{e^{\beta_2(S_2/\varepsilon)-\log\nu}\}\\
&\quad - 4\arctan\Big\{(\nu-1)e^{\beta_1 S_1/\varepsilon}\big(1 - e^{\beta_1(S_1/\varepsilon)+\beta_2(S_2/\varepsilon)}\big)/P_1\Big\},
\end{aligned}
\tag{5.79}
$$

where

$$
\begin{aligned}
P_1 &= 1 + \nu e^{2\beta_1 S_1/\varepsilon} + \nu^{-2}e^{2\beta_2 S_2/\varepsilon}\\
&\quad + (\nu^{-1}-1)e^{\beta_1(S_1/\varepsilon)+\beta_2(S_2/\varepsilon)} + e^{2(\beta_1(S_1/\varepsilon)+\beta_2(S_2/\varepsilon))}.
\end{aligned}
$$

Thus for $t \geq t^* + c\varepsilon^{1-\delta}$ the solution (5.73) is also a sum of two non-interacting (up to the exponentially small third term in (5.79)) solitary fluxons (5.77) and (5.78). The interaction results only in the appearance of phase shifts $\pm \log \nu$ of these fluxons. Note that, unlike the case $\varepsilon = 1$, for the solution with a small parameter ε these phase shifts result only in small, of magnitude $O(\varepsilon)$, shifts of the fronts of the fluxons which have interacted. The second observation is that in our case the duration of interaction of the fluxons is also small, of order $O(\varepsilon^{1-\delta})$. These specific features of the solution with a small parameter explain our claim that the soliton (fluxon) waves are almost non-interacting. The same properties enable us to construct easily the leading term of the multi-phase soliton-type solution.

To conclude the analysis of formula (5.73), we remark that for both functions z_i, which are the third terms in the right-hand sides of (5.76) and (5.79), the following bounds, uniform in x, t, hold:

$$\left| \left(\varepsilon \frac{\partial}{\partial t}\right)^k \left(\varepsilon \frac{\partial}{\partial x}\right)^l z_i \right| \leq c_{k,l,i} e^{-\frac{1}{2}(\beta_1 |S_1|/\varepsilon + \beta_2 |S_2|/\varepsilon)}, \tag{5.80}$$

where $i = 1, 2$, $k, l \geq 0$, and $k + l \geq 1$.

Now we turn to the sine-Gordon equation (5.72) with variable coefficients. To simplify the calculations, we make the substitution $v = \gamma u$, which transforms (5.72) into

$$\varepsilon^2 \left(\frac{\partial^2 v}{\partial t^2} - \frac{\partial^2 v}{\partial x^2} \right) + \alpha^2 \sin v = \varepsilon \mathcal{F}(x, t, v, \varepsilon), \tag{5.81}$$

where

$$\mathcal{F}(x, t, v, \varepsilon) = -2\gamma\varepsilon \langle \nabla_r v, \nabla_r \gamma^{-1} \rangle - \varepsilon \gamma v \left(\frac{\partial^2}{\partial t^2} - \frac{\partial^2}{\partial x^2} \right) \gamma^{-1},$$

$$\langle \nabla_r v, \nabla_r \gamma^{-1} \rangle = \frac{\partial v}{\partial t} \frac{\partial \gamma^{-1}}{\partial t} - \frac{\partial v}{\partial x} \frac{\partial \gamma^{-1}}{\partial x}.$$

We restrict ourselves to construction of the leading term of a two-fluxon asymptotic solution to equation (5.81), i.e., we write v as

$$v(x, t, \varepsilon) = v_0 \left(\frac{S_1(x, t, \varepsilon)}{\varepsilon}, \frac{S_2(x, t, \varepsilon)}{\varepsilon}, x, t \right) + \varepsilon v_1(x, t, \varepsilon) \tag{5.82}$$

and we will find the function $v_0(\tau_1, \tau_2, x, t)$, as well as the phases $S_i(x, t, \varepsilon) = x - \varphi_i(t) + \varepsilon c_i(x, t)$ with accuracy $O(\varepsilon)$ only.

Substitute (5.82) into (5.81) and apply the formula

$$\varepsilon^2 \left(\frac{\partial^2}{\partial t^2} - \frac{\partial^2}{\partial x^2} \right) v_0 \left(\frac{S_1}{\varepsilon}, \frac{S_2}{\varepsilon}, x, t \right) = \Big\{ (\widehat{A}^2 - \widehat{B}^2) v_0(\tau_1, \tau_2, x, t)$$
$$+ \varepsilon \widehat{\Pi} v_0(\tau_1, \tau_2, x, t) + \varepsilon^2 \left(\frac{\partial^2}{\partial t^2} - \frac{\partial^2}{\partial x^2} \right) v_0(\tau_1, \tau_2, x, t) \Big\} \Big|_{\tau_i = S_i/e},$$

where

$$\widehat{A} = -\sum_{i=1}^{2} \dot{\varphi}_i \frac{\partial}{\partial \tau_i}, \quad \widehat{B} = \sum_{i=1}^{2} \frac{\partial}{\partial \tau_i}, \quad \dot{\varphi}_i = \frac{\partial \varphi_i}{\partial t},$$

$$\widehat{\Pi} = 2\widehat{A} \left(\widehat{C}_t + \frac{\partial}{\partial t} \right) + \widehat{A}_t - 2\widehat{B} \left(\widehat{C}_x + \frac{\partial}{\partial x} \right),$$

$$\widehat{A}_t = -\sum_{i=1}^{2} \frac{\partial^2 \varphi_i}{\partial t^2} \frac{\partial}{\partial \tau_i}, \quad \widehat{C}_t = -\sum_{i=1}^{2} \frac{\partial c_i}{\partial t} \frac{\partial}{\partial \tau_i}, \quad \widehat{C}_x = -\sum_{i=1}^{2} \frac{\partial c_i}{\partial x} \frac{\partial}{\partial \tau_i}.$$

Then for the determination of v_0 and v_1 we obtain the equations

$$\widehat{\mathcal{L}}\Big(\frac{\partial}{\partial\tau_1},\frac{\partial}{\partial\tau_2},x,t\Big)v_0(\tau_1,\tau_2,x,t)=0, \tag{5.83}$$

$$\widehat{L}\Big(\varepsilon\frac{\partial}{\partial t},\varepsilon\frac{\partial}{\partial x},x,t\Big)v_1(x,t,\varepsilon)=F\Big(\frac{S_1}{\varepsilon},\frac{S_2}{\varepsilon},x,t,\varepsilon\Big). \tag{5.84}$$

Here $\widehat{\mathcal{L}}$ is the nonlinear operator acting by the formula

$$\widehat{\mathcal{L}}v_0=\widehat{A}^2v_0-\widehat{B}^2v_0+\alpha^2\sin v_0,$$

$\widehat{L}$ is the linearization with respect to v_0 of the operator in the left-hand side of (5.81), i.e.,

$$\widehat{L}v_1=\varepsilon^2\frac{\partial^2 v_1}{\partial t^2}-\varepsilon^2\frac{\partial^2 v_1}{\partial x^2}+\alpha^2 v_1\cos v_0,$$

and the function F has the form

$$F=\mathcal{F}(x,t,v_0+\varepsilon v_1,\varepsilon)+\varepsilon G(x,t,v_0,\varepsilon),$$

where εG comprises all the terms arising from substitution of (5.82) into the left-hand side of (5.81) and not included in the left-hand sides of (5.83) and (5.84). We have

$$G=\widehat{\Pi}v_0(\tau_1,\tau_2,x,t)\big|_{\tau_i=S_i/\varepsilon}+O(\varepsilon).$$

Consider equation (5.83). We treat the slow variables x and t as parameters, and assume that

$$\dot\varphi_1\neq\dot\varphi_2. \tag{5.85}$$

Then we pass from the variables (τ_1,τ_2) to new variables (η,ξ),

$$\eta=\frac{\alpha}{\dot\varphi_1-\dot\varphi_2}(\tau_1-\tau_2),\quad \xi=\frac{\alpha}{\dot\varphi_1-\dot\varphi_2}(\dot\varphi_1\tau_2-\dot\varphi_2\tau_1). \tag{5.86}$$

It is not hard to calculate that with this change of variables

$$\widehat{A}=\alpha\frac{\partial}{\partial\eta},\quad \widehat{B}=\alpha\frac{\partial}{\partial\xi}.$$

Hence, denoting $\widetilde{v}(\eta,\xi,x,t)=v_0\big((\dot\varphi_1\eta+\xi)/\alpha,(\dot\varphi_2\eta+\xi)/\alpha,x,t\big)$, we transform equation (5.83) to the form

$$\frac{\partial^2\widetilde{v}}{\partial\eta^2}-\frac{\partial^2\widetilde{v}}{\partial\xi^2}+\sin\widetilde{v}=0. \tag{5.87}$$

Again we have obtained a sine-Gordon equation. However, unlike the initial equation (5.81), the model equation (5.87) does not depend on ε and has constant coefficients. The latter property allows us to use the results obtained by the method of the inverse scattering problem (see, e.g., [**270**]), in particular, to use the exact two-fluxon solution (5.73). Putting $\lambda_1 = 0$, $\lambda_2 = \log 1/\nu$, we obtain the formula

$$\widetilde{v}(\eta, \xi, x, t) = 4\arctan\left\{\frac{e^{\tilde{\beta}_1(\xi - V_1\eta)} + (1/\nu)e^{\tilde{\beta}_2(\xi - V_2\eta)}}{1 - e^{\tilde{\beta}_1(\xi - V_1\eta) + \tilde{\beta}_2(\xi - V_2\eta)}}\right\} \tag{5.88}$$

for the required solution to equation (5.87), where $\tilde{\beta}_i = 1/\sqrt{1 - V_i^2}$, the V_i are arbitrary "constants", $V_1 \neq V_2$, and ν is given by (5.74). We set $V_i = -\dot{\varphi}_i$ and denote

$$\beta_i = \frac{\alpha}{\sqrt{1 - \dot{\varphi}_i^2}}. \tag{5.89}$$

Then, returning to the fast variables τ_1, τ_2, we obtain the solution of (5.83) as

$$v_0(\tau_1, \tau_2, x, t) = 4\arctan\left\{\frac{e^{\beta_1\tau_1} + (1/\nu)e^{\beta_2\tau_2}}{1 - e^{\beta_1\tau_1 + \beta_2\tau_2}}.\right\} \tag{5.90}$$

Here $\nu = \nu(t)$ and $\beta_i = \beta_i(x, t)$ are functions of slow variables given by (5.74) (with V_i replaced by $-\dot{\varphi}_i(t)$) and (5.89). Note that the phases $S_i = \varepsilon\tau_i$, which remain undetermined in this approximation, contain small shifts εc_i playing the same role as the "constants" of integration in the single-phase asymptotics.

Next, the above analysis of (5.73) allows for a refinement of condition (5.85). Namely, this condition is required to hold only for $t = t^*$, where t^* is the time moment when the fronts $\Gamma_t^1 = \{x = \varphi_1(t)\}$ and $\Gamma_t^2 = \{x = \varphi_2(t)\}$ intersect.

Now consider equation (5.84). First of all, we observe that the functions $y_i = \varepsilon^{-1}\partial v_0/\partial c_i$, $i = 1, 2$, satisfy the equalities

$$\begin{aligned}\widehat{L}y_i &= \frac{\partial}{\partial \tau_i}\left\{\frac{\beta_i}{\varepsilon}\widehat{\mathcal{L}}v_0 + \left(\widehat{\Pi} + \varepsilon\left(\frac{\partial^2}{\partial t^2} - \frac{\partial^2}{\partial x^2}\right)\right)\beta_i v_0(\tau_1, \tau_2, x, t)\right\}\Big|_{\tau_i = S_i/\varepsilon} \\ &= \frac{\partial}{\partial \tau_i}\left\{\widehat{\Pi} + \varepsilon\left(\frac{\partial^2}{\partial t^2} - \frac{\partial^2}{\partial x^2}\right)\right\}\beta_i v_0(\tau_1, \tau_2, x, t)\Big|_{\tau_i = S_i/\varepsilon}.\end{aligned} \tag{5.91}$$

Using the explicit expression (5.90) for v_0, it is not hard to establish that the right-hand side of (5.91) is bounded by a constant uniformly in $x, t, \varepsilon \in [0, 1]$ and belongs to the space $C^\infty(0, T; S(\mathbb{R}_x^1))$. Here S is the Schwartz space of functions decreasing faster than any positive power of $1/|x|$ as $|x| \to \infty$. Furthermore, it follows from (5.90) that, uniformly in $t \in [0, T]$,

$$\widehat{L}y_i \sim e^{-|x - \varphi_i|/\varepsilon}.$$

Hence for all finite $T > 0$ we immediately obtain that $\widehat{L}y_i = O(\sqrt{\varepsilon})$ in the norm of $L^2(Q_T)$, and $\widehat{L}y_i = O(\varepsilon)$ in the norm of $L^1(Q_T)$, where $Q_T = [0, T] \times \mathbb{R}_x^1$.

We assume that v_1 for $\varepsilon > 0$ is a smooth function of x and t for which the following bounds hold uniformly in t:

$$\max_x |v_1| \leq \frac{c}{\varepsilon^\mu}, \quad \max_x \left(\left|\frac{\partial v_1}{\partial t}\right| + \left|\frac{\partial v_1}{\partial x}\right|\right) \leq \frac{c}{\varepsilon^{1+\mu}}, \tag{5.92}$$

where $\mu < 1$ is a fixed number and $c > 0$ is a constant independent of ε.

Consider the scalar product of y_i by $\widehat{L}v_1$ in $L^2(Q_T)$. Integrating by parts, we obtain

$$\begin{aligned} \int_0^T \int_{-\infty}^{\infty} y_i \widehat{L} v_1 \, dx\, dt &= \varepsilon^2 \int_{-\infty}^{\infty} \left(y_i \frac{\partial v_1}{\partial t} - v_1 \frac{\partial y_i}{\partial t} \right) dx \bigg|_0^T \\ &\quad + \int_0^T \int_{-\infty}^{\infty} v_1 \widehat{L} y_i \, dx\, dt. \end{aligned} \tag{5.93}$$

Since $y_i = O(1)$ and $\partial y_i/\partial t = O(\varepsilon^{-1})$ in the norm of $L^1(Q_T)$, taking (5.92) into account we obtain that the first term in the right-hand side of (5.93) is of order $O(\varepsilon^{1-\mu})$. By (5.92) and the above analysis of $\widehat{L}y_i$, the same estimate holds for the second term in the right-hand side of (5.93). Thus

$$(y_i, \widehat{L}v_1) = O(\varepsilon^{1-\mu}) \to 0 \qquad \text{as} \quad \varepsilon \to 0. \tag{5.94}$$

This relation yields necessary conditions for solvability of equation (5.84):

$$\lim_{\varepsilon \to 0} (y_i, F) = 0, \quad i = 1, 2. \tag{5.95}$$

For further analysis of the necessary conditions (5.95), observe that the explicit expression for the right-hand side F of (5.84) and assumptions (5.92) allow us to write F as a sum $F = F_0(v_0, x, t, \varepsilon) + F_1(v_0, v_1, x, t, \varepsilon)$, where $\max_{x,t} |F_1| \to 0$ as $\varepsilon \to 0$.

Consider the scalar product (y_1, F_0). Denote by t^* the time moment when the curves $x = \varphi_1(t)$ and $x = \varphi_2(t)$ intersect. If $t^* \notin [0, T]$, then our asymptotic solution on the time interval at hand is a sum of two non-interacting single-phase asymptotics, which case was studied in detail in Chapter 1. Hence we will assume that $t^* \in [0, T]$. Consider $t \leq t^*$ and use the representation

$$v_0(\tau_1, \tau_2, x, t) = v_0^{(1)}(\tau_1, x, t) + v_0^{(2)}(\tau_2, x, t) + v_0^{(12)}(\tau_1, \tau_2, x, t), \tag{5.96}$$

where $v_0^{(1)}, v_0^{(2)}$, and $v_0^{(12)}$ are given by the same formulas as the corresponding terms in the right-hand side of (5.76). Then

$$\begin{aligned} F_0 &= F_0^{(1)}(\tau_1, x, t) + F_0^{(2)}(\tau_2, x, t) + F_0^{(12)}(\tau_1, \tau_2, x, t)\big|_{\tau_i = S_i/\varepsilon}, \\ y_1 &= y_1^{(1)}(\tau_1, x, t) + y_1^{(12)}(\tau_1, \tau_2, x, t, \varepsilon)\big|_{\tau_i = S_i/\varepsilon}. \end{aligned} \tag{5.97}$$

Thus

$$\int_0^{t^*} \int_{-\infty}^{\infty} F_0 y_1 \, dx\, dt = \int_0^{t^*} \int_{-\infty}^{\infty} F_0^{(1)} y_1^{(1)} \, dx\, dt + \int_0^{t^*} \int_{-\infty}^{\infty} f_1 \, dx\, dt, \tag{5.98}$$

where $f_1 = (F_0^{(2)} + F_0^{(12)}) y_1^{(1)} + F_0 y_1^{(12)}$.

Consider the first term in the right-hand side of (5.98). Making the change of variables $x = \varphi_1 - \varepsilon c_1 + \varepsilon \tau_1$, we obtain

$$\begin{aligned} I &\stackrel{\text{def}}{=} \int_{-\infty}^{\infty} \left\{ F_0^{(1)} \left(\frac{x - \varphi_1 + \varepsilon c_1}{\varepsilon}, x, t \right) \frac{\beta_1(x,t)}{\varepsilon} \frac{\partial v_0^{(1)}}{\partial \tau_1} \bigg|_{\tau_1 = (x - \varphi_1 + \varepsilon c_1)/\varepsilon} \right\} dx \\ &= \int_{-\infty}^{\infty} \left\{ F_0^{(1)}(\tau_1, z, t) \beta_1(z, t) \frac{\partial v_0^{(1)}}{\partial \tau_1}(\tau, z, t) \bigg|_{z = \varphi_1 - \varepsilon c_1 + \varepsilon \tau_1} \right\} d\tau_1. \end{aligned} \tag{5.99}$$

We use the fact that $|\tau_1^k \partial v_0^{(1)}/\partial\tau_1| \le c_k$ uniformly in $\tau_1 \in \mathbb{R}^1$ for all $k \ge 0$. Hence we expand the functions involved in (5.99) by the Taylor formula about $z = \varphi_1(t)$ and pass to the limit as $\varepsilon \to 0$ to obtain

$$\lim_{\varepsilon\to 0} I = \beta_1(\varphi_1, t) \int_{-\infty}^{\infty} F_0^{(1)}(\tau_1, \varphi_1, t) \frac{\partial v_0^{(1)}}{\partial \tau_1}(\tau_1, \varphi_1, t)\, d\tau_1. \tag{5.100}$$

Consider the second term in the right-hand side of (5.98). It follows from the above analysis that the integrand is of order $O(\varepsilon^\infty)$ outside a small neighborhood of the point (x^*, t^*), where $x^* = \varphi_1(t^*) = \varphi_2(t^*)$, and in this neighborhood f_1 admits a bound similar to (5.80). Hence, again making the substitution $x = \varphi_1 - \varepsilon c_1 + \varepsilon\tau_1$, we obtain

$$\begin{aligned} \Big| \int_0^{t^*} \int_{-\infty}^{\infty} f_1\, dx\, dt \Big| &\le \frac{C_1}{\varepsilon} \int_0^{t^*} \int_{-\infty}^{\infty} e^{-(\beta_1/2)|x-\varphi_1|/\varepsilon - (\beta_2/2)|x-\varphi_2|/\varepsilon}\, dx\, dt \\ &\le C_2 \int_0^{t^*} e^{-(\beta_2/2)|\varphi_1-\varphi_2|/\varepsilon} \int_{-\infty}^{\infty} e^{-(\beta_2/2)|\tau_1|}\, d\tau_1\, dt \\ &\le C_3\varepsilon. \end{aligned} \tag{5.101}$$

For $t \ge t^*$ the function v_0 still has a representation of the form (5.96), where $v_0^{(1)}, v_0^{(2)}$, and $v_0^{(12)}$ are the corresponding terms in the right-hand side of the expansion (5.79). Hence, using the notation (5.97), we have

$$\int_{t^*}^{T} \int_{-\infty}^{\infty} F_0 y_1\, dx\, dt = \int_{t^*}^{T} \int_{-\infty}^{\infty} F_0^{(1)} y_1^{(1)}\, dx\, dt + \int_{t^*}^{T} \int_{-\infty}^{\infty} f_2\, dx\, dt, \tag{5.102}$$

where all notation has the same meaning as in (5.98).

In a similar way to the above, we obtain

$$\begin{aligned} II &\stackrel{\text{def}}{=} \int_{-\infty}^{\infty} \Big\{ F_0^{(1)}(\tau_1(x,\varepsilon), x, t) \frac{\beta_1}{\varepsilon} \frac{\partial v_0}{\partial \tau_1}\Big|_{\tau_1 = \tau_1(x,\varepsilon)} \Big\}\, dx \\ &= \int_{-\infty}^{\infty} \Big\{ F_0^{(1)}(\tau_1 + c', z, t) \beta_1(z,t) \frac{\partial v_0}{\partial \tau_1}(\tau_1 + c', z, t)\Big|_{z = \varphi_1 - \varepsilon c_1 + \varepsilon\tau_1} \Big\}\, d\tau_1, \end{aligned} \tag{5.103}$$

where we denote

$$\tau_1(x,\varepsilon) = \frac{x - \varphi_1 + \varepsilon c_1}{\varepsilon} + \frac{\log\nu}{\beta_1}, \qquad c' = \frac{\log\nu}{\beta_1(z,t)}.$$

We expand the functions in (5.103) by the Taylor formula about $z = \varphi_1(t)$ and make the change of variables $\tau_1' = \tau_1 + \beta_1^{-1}(\varphi_1, t)$. Then (5.103) implies

$$\lim_{\varepsilon\to 0} II = \beta_1(\varphi_1, t) \int_{-\infty}^{\infty} F_0^{(1)}(\tau_1', \varphi_1, t) \frac{\partial v_0^{(1)}}{\partial \tau_1'}(\tau_1', \varphi_1, t)\, d\tau_1'. \tag{5.104}$$

In order to estimate the second term in the right-hand side of (5.102), it is sufficient to use (5.80) and to observe that the phase shifts $\pm\log\nu$, which appear after the interaction of the fluxons, affect only the constants c_{kl2}. Hence, similarly to (5.101), we have

$$\left|\int_{t^*}^{T}\int_{-\infty}^{\infty} f_2\,dx\,dt\right| \le C_4\varepsilon \tag{5.105}$$

with constant C_4 independent of ε.

Now, combining (5.98)–(5.105), we see that

$$\lim_{\varepsilon\to 0}(y_1, F) = \int_0^T \beta_1(\varphi_1, t)\int_{-\infty}^{\infty} F_0^{(1)}(\tau_1, \varphi_1, t)\frac{\partial v_0^{(1)}}{\partial \tau_1}(\tau_1, \varphi_1, t)\,d\tau_1\,dt. \tag{5.106}$$

Obviously, for (y_2, F) we will obtain an equality similar to (5.106).

Since $\beta_1 > 0$, these relations and (5.95) imply necessary conditions for solvability of equation (5.84):

$$\int_{-\infty}^{\infty} F_0^{(i)}(\tau_i, \varphi_i, t)\frac{\partial v_0^{(i)}}{\partial \tau_i}(\tau_i, \varphi_i, t)\,d\tau_i = 0, \quad i = 1, 2. \tag{5.107}$$

These equalities mean that the fluxon trajectories in the leading terms remain unchanged after their interaction. Indeed, it is not hard to calculate that

$$\begin{aligned}
\frac{\partial v_0^{(i)}}{\partial \tau_i}(\tau_i, \varphi_i, t) &= \frac{2\beta_i(\varphi_i, t)}{\cosh(\beta_i(\varphi_i, t)\tau_i)},\\
F_0^{(i)}(\tau_i, \varphi_i, t) &= \left\{2\dot\varphi_i\frac{\partial}{\partial t} + \ddot\varphi_i - \frac{2}{\gamma}(\dot\varphi_i\gamma_t - \gamma_x)\right\}\frac{\partial v_0^{(i)}}{\partial \tau_i} + 2\frac{\alpha\alpha_x}{\beta_i^2}\tau_i\frac{\partial^2 v_0^{(i)}}{\partial \tau_i^2}\bigg|_{x=\varphi_i}.
\end{aligned}$$

Hence, after some algebra, we can rewrite (5.107) as the differential equation

$$\frac{d}{dt}\left(\frac{\beta_i\dot\varphi_i}{\gamma^2}\bigg|_{x=\varphi_i}\right) + \frac{\alpha}{\beta_i}\frac{d}{dx}\left(\frac{\alpha}{\gamma^2}\right)\bigg|_{x=\varphi_i} = 0, \quad i = 1, 2. \tag{5.108}$$

Finally, rewriting (5.89) as

$$\left(\frac{d\varphi_i}{dt}\right)^2 = 1 - \left(\frac{\alpha}{\beta_i}\right)^2\bigg|_{x=\varphi_i}, \quad i = 1, 2, \tag{5.109}$$

we obtain a closed system of equations for the fluxon fronts $\Gamma_t^i = \{x = \varphi_i(t)\}$, which precisely coincide with the corresponding equations (1.145) in §1.5 for solitary fluxons. At the same time, the present problem differs in principle from the single-phase setting studied in Chapter 1. Indeed, in the single-phase case the model equation (1.152) is an ordinary differential equation, so that there are only two independent functions z_1 and z_2 that solve the equation $\widehat{L}z = 0$. One of them, $z_1 = \partial v_0/\partial\tau$, gives rise to the orthogonality condition (1.153), similar to (5.107). The second function grows exponentially, and hence must be discarded because it does not belong to the space dual to $S \ni \mathcal{F}$. Hence (1.153) is the necessary and sufficient condition for solvability of the linearized equations (1.152) of Chapter 1 in this class of functions. In the multi-phase setup we deal with a model partial differential equation, and we only have the relation $\widehat{L}z_i \to 0$ as $\varepsilon \to 0$, rather than the equality $\widehat{L}z_i = 0$. Hence, instead of (1.153) of Chapter 1 we have the condition (5.95), where for y_i we can take any element of the subspace $\mathcal{N} = \{z \in C^\infty(Q) \cap C^\infty(0, T; S'(R'_x)),\ \widehat{L}z = o(1),\ \|z\|_{L^1} = 1\}$, with S' being the dual space to S. Passing to the variables τ_1, τ_2, we see that

$$\widehat{L} = \widehat{L}_0 + O(\varepsilon),$$

where

$$\widehat{L}_0 = \widehat{A}^2 - \widehat{B}^2 + \alpha^2 \cos \widetilde{v}_0,$$

or, making the substitution (5.86),

$$\widehat{L}_0 = \alpha^2 \left(\frac{\partial^2}{\partial \eta^2} - \frac{\partial^2}{\partial \xi^2} + \cos v_0 \right). \tag{5.110}$$

This implies that the subspace $\mathcal{N}$ is infinite-dimensional. It is well known (see, e.g., [**112, 169**]) that the formal kernel $\widetilde{\mathcal{N}}$ of the operator $\widehat{L}_0$ as in (5.110) can be represented as a linear span of a finite-dimensional discrete subspace $\widetilde{\mathcal{N}}_d$ and an infinite-dimensional continuous subspace $\widetilde{\mathcal{N}}_c$, and in the N-fluxon case the dimension of $\widetilde{\mathcal{N}}_d$ equals $2N$. Accordingly, discarding exponentially growing functions, we obtain that $\mathcal{N}$ is the linear span of subspaces $\mathcal{N}_d$ and $\mathcal{N}_c$. This discrimination is related to the fact that a solution to the equation

$$\widehat{L}_0 v_1 = F_0 \tag{5.111}$$

in general contains secular terms growing linearly in η, and they disappear only if F_0 is orthogonal to $\mathcal{N}_d$. It is not hard to see that, in addition to $y_i = \varepsilon^{-1} \partial v_0 / \partial c_i$, $\mathcal{N}_d$ contains also the functions $\partial v_0 / \partial \tau_i$, $\partial \widetilde{v} / \partial V_i |_{V_i = \dot{\varphi}_i}$. However, in the leading terms all of them are linear combinations of y_1 and y_2. In particular, using the representation (5.88) one can easily find that

$$\left. \frac{\partial \widetilde{v}}{\partial V_i} \right|_{V_i = \dot{\varphi}_i} = \frac{1}{\varepsilon} \left\{ \alpha \frac{\varphi_1 - \varphi_2}{\dot{\varphi}_1 - \dot{\varphi}_2} \frac{\partial v_0}{\partial c_i} + O(\varepsilon) \right\}.$$

Hence taking these functions into consideration does not give any new conditions compared to (5.95).

Thus the necessary conditions (5.95) ensure orthogonality (in the leading term) of F to $\mathcal{N}_d$. However, F is not orthogonal to $\mathcal{N}_c$. What is more, the function v_0 does not contain any free parameters which would allow us to make F orthogonal to $\mathcal{N}_c$ (arbitrariness in the choice of c_i means only that the φ_i can be changed by $O(\varepsilon)$). This results in the fact that the solution of (5.111) is not a purely multi-fluxon function, but will contain an additional oscillating term, the so-called "radiation".

The correction v_1 can be constructed by methods related to the inverse scattering problem or to the Bäcklund transformation. Namely, passing to the variables η, ξ, equation (5.84), up to higher order terms, can be rewritten in the form (5.111), where $\widehat{L}_0$ has the form (5.110) and the right-hand side F_0 is a function of the derivatives of $\widetilde{v}$ which depends on x and t as parameters. In its turn, the solution of (5.111) can be found in an explicit form using the Green function for the operator $\widehat{L}_0$. It is important here that the leading term of the asymptotic solution $\widetilde{v}$ given by (5.88) as a function of ξ and η contains constant unequal velocities V_i. The corresponding constructions are carried through in detail, e.g., in [**112, 169**].

To avoid a detailed description of very specific methods related to the inverse scattering problem and the Bäcklund transformation, we refer the reader to those papers and restrict our consideration here to construction of the leading term of the asymptotic expansion. We only mention that construction and analysis of the correction v_1 do not lead to any change of the Hugoniot type conditions (5.108), (5.109), and that the multi-phase solution to the nonlinear model equation (5.83) can be found by elementary methods, in particular, by the method of Hirota [**2, 82, 83**].

Summarizing the above constructions, we see that the leading term (up to $o(1)$) of the two-fluxon solution to equation (5.72) has the form

$$u = \gamma^{-1}(x,t)v_0\Big(\frac{S_1}{\varepsilon}, \frac{S_2}{\varepsilon}, x, t\Big),$$

where $v_0(\tau_1, \tau_2, x, t)$ is given by (5.90), $S_i = x - \varphi_i(t) + O(\varepsilon)$, and the functions φ_i describing the fluxon fronts within accuracy $O(\varepsilon)$ are found from equations (5.108) and (5.109), the same as for the solitary fluxons.

The leading term of an N-phase solution describing N_1 fluxons and $N_2 = N - N_1$ antifluxons is constructed similarly to the above scheme. The only new point here is that for $N > 2$ instead of (5.86) we pass from the variables $\tau = (\tau_1, \dots, \tau_N)$ to the new variables $\tau' = (\xi, \eta, \zeta_1, \dots, \zeta_{N-2})$ related by $\tau = (1/\alpha)K\tau'$, where the $N \times N$ matrix K is composed of the vectors $k_1 = (1, \dots, 1)$, $k_2 = (\dot{\varphi}_1, \dots, \dot{\varphi}_N)$, and some linearly independent vectors k_i, $i = 3, \dots, N$. Then we seek a particular solution which does not depend on ζ_i, $i = 1, \dots, N-2$. Clearly, we again obtain equation (5.87) for the function $\widetilde{v}_0 = \widetilde{v}_0(\xi, \eta, x, t)$. Now we observe that the exact N-soliton solution $\widetilde{v}_0(\xi, \eta)$ of the sine-Gordon equation depends only on the linear combinations of ξ and η of the form $\xi - V_i\eta$ with arbitrary "constants" V_i, $|V_i| < 1$ ([**270**]),

$$\widetilde{v}_0 = 2i \log \Phi(\xi - V_1\eta, \dots, \xi - V_N\eta).$$

Here i is the imaginary unit,

$$\Phi = \frac{\det \|I + B(\xi,\eta)\|}{\det \|I - B(\xi,\eta)\|},$$

I is the identity $N \times N$ matrix, and B is the $N \times N$ matrix with entries

$$B_{nm} = i2\frac{a_n}{a_n + a_m} \exp\left((b_n + b_m)\xi + \Big(a_n - \frac{1}{16a_n}\Big)\eta + c_n + \lambda_n\right),$$

where

$$b_n = a_n + \frac{1}{16a_n}, \quad a_n = \frac{1}{4}\sqrt{\frac{1+V_n}{1-V_n}},$$

and c_n, λ_n are arbitrary constants.

We choose $V_i = \dot{\varphi}_i$, $i = 1, \dots, N$, and let $\zeta_i = 0$, to obtain

$$v_0 = 2i \log \Phi(\tau_1, \dots, \tau_N).$$

As we pointed out, such a solution is actually an N-phase solution only in a small neighborhood of the point (x^*, t^*), where $x^* = \varphi_1(t^*) = \cdots = \varphi_N(t^*)$. We impose the condition

$$\dot{\varphi}_l(t^*) \neq \dot{\varphi}_j(t^*), \qquad l, j = 1, \ldots, N, \quad l \neq j, \tag{5.112}$$

which is a natural analog of (5.85), and assume that $\dot{\varphi}_1(t^*) > \dot{\varphi}_2(t^*) > \cdots > \dot{\varphi}_N(t^*)$ and that before interaction v_0 was a superposition of N_1 fluxons with indices $j = 1, \ldots, N_1$ and N_2 antifluxons with indices $j = N_1 + 1, \ldots, N$. Then for determination of the "constants" λ_j we obtain a natural normalization condition:

$$v_0 \to 4 \arctan\{\exp(\pm\beta_j \tau_j)\} \qquad \text{as } \tau_l \to \infty,\ l = 1, \ldots, j-1,$$
$$\text{and } \tau_k \to -\infty,\ k = j+1, \ldots, N,$$

where the sign $+$ is taken for $j \leq N_1$ and the sign $-$ for $j > N$.

It remains to remark that the above scheme can be easily extended to other equations similar to nonlinear equations integrated by the method of the inverse scattering problem. In particular, for the Korteweg–de Vries equation with variable coefficients (1.86), (1.87) the leading term of the N-phase soliton-type solution is given by the formula

$$u = u_0(x, t) + 2\rho_3(x)\rho_2^{-1}(x) V_0\Big(\frac{S_1}{\varepsilon}, \ldots, \frac{S_N}{\varepsilon}, t\Big),$$

where u_0 is the solution to the degenerate Korteweg–de Vries equation (1.92), $S_i = x - \varphi_i(t) + O(\varepsilon)$, and

$$V_0(\tau_1, \ldots, \tau_N, t) = 2\frac{\partial^2}{\partial \xi^2} \log \Phi \Big|_{\xi - \beta_k^2 \eta = \tau_k}.$$

Here

$$\Phi = \det \Big\| \delta_{n,m} + \frac{b_n}{\beta_n + \beta_m} \exp\big(-(\beta_n + \beta_m)\xi + 2\beta_n^3 \eta\big) \Big\|,$$

$$b_n = 2\beta_n \exp(\lambda_n - 2c_n), \quad \beta_n = \sqrt{\frac{3A_n}{4H^3(\varphi_n)}},$$

$\delta_{n,m}$ is the Kronecker delta, the functions φ_n and $A_n = A_n(t)$ are determined with the aid of the Hugoniot type condition (1.118) subject to condition (5.112), and the parameters λ_n are found from the normalization condition

$$2\rho_3(\varphi_n)\rho_2^{-1}(\varphi_n) V_0(\tau_1, \ldots, \tau_n, t) \to v_0(\tau_n, t)$$

as $\tau_j \to \infty$, $j < n$, and $\tau_k \to -\infty$, $k > n$, where $v_0(\tau, t)$ is a distorted soliton of the form (1.98).

Regarding the Kadomtsev–Petviashvili equation (2.1) of Chapter 2, it is to be noted that the front of each distorted soliton forms a two-dimensional surface $\mathcal{T}$

in the space $\mathbb{R}^3_{x,y,t}$. Hence the intersection of the surfaces T_1 and T_2 is now a one-dimensional curve γ, rather than a point as it was for the Korteweg–de Vries and sine-Gordon equations. Nevertheless, for the distorted soliton $v(\tau, y, t)$ we have

$$\int_0^T \int_{-\infty}^{\infty} \int_{-\infty}^{\infty} v\Big(\frac{x-\varphi(y,t)}{\varepsilon}, y, t\Big)\eta(y)\,dx\,dy\,dt = O(\varepsilon),$$

where η is the truncating function, whereas

$$\int_0^T \int_{-\infty}^{\infty} \int_{-\infty}^{\infty} v^{(12)}\Big(\frac{x-\varphi_1(y,t)}{\varepsilon}, \frac{x-\varphi_2(y,t)}{\varepsilon}, y, t\Big)\eta(y)\,dx\,dy\,dt = O(\varepsilon^2),$$

where $v^{(12)}$ is the additional term arising in expansions of the form (5.96), localized in a neighborhood of the curve γ: $\varphi_1(y,t) = \varphi_2(y,t)$. Therefore the interaction of the solitons also does not affect their dynamics (in the leading term). Using the formula for the exact N-soliton solution to the Kadomtsev–Petviashvili equation and the results of Chapter 2, we obtain the formula for the leading term of the N-phase soliton-type asymptotics:

$$u = u_0(x,y,t) + V_0\Big(\frac{S_1}{\varepsilon}, \dots, \frac{S_N}{\varepsilon}, y, t\Big),$$

where u_0 is the solution to the degenerate Kadomtsev–Petviashvili equation (2.3),

$$V_0(\tau_1, \dots, \tau_N, y, t) = 2\frac{d^2}{d\xi^2}\log\Phi\Big|_{\xi+\psi_k\zeta+\mu_k\nu=\tau_k},$$

$$\Phi = \det\Big\|\delta_{n,m} + \frac{b_n}{\kappa_n+\eta_m}\exp(-(\kappa_n+\eta_m)\xi)\Big\|,$$

$$b_n = \exp\big\{(\eta_n^2 - \kappa_n^2)\zeta + (\kappa_n^3 + \eta_n^3)\nu + c_n + \lambda_n\big\},$$

$$\psi_k = \kappa_k - \eta_k, \quad \mu_k = -\eta_k^2 + \kappa_k\eta_k - \kappa_k^2,$$

$$\kappa_k = \sqrt{\frac{A_k}{2}} + \frac{\varphi_{ky}}{2}, \quad \eta_k = \sqrt{\frac{A_k}{2}} - \frac{\varphi_{ky}}{2},$$

$S_k = x - \varphi_k(t,y) + O(\varepsilon)$, the functions φ_k and $A_k = A_k(t,y)$ are found from the Hugoniot conditions (2.4), (2.5) and satisfy conditions similar to (5.112), and the parameters λ_k are found by means of normalization conditions in the same way as for the Korteweg–de Vries equations.

5.3. Dynamics of the breather in a medium with variable parameters

One more nontrivial example of localized solutions is given by the so-called bion, or breather, which is a particular solution to the sine-Gordon equation (5.72) with constant coefficients ($\alpha = \gamma = 1$) of the following form:

$$u = 4\arctan\Big\{\beta_0\frac{\cos\big(\beta_2(t - Vx - t_0)/\varepsilon\big)}{\cosh\big(\beta_1(x - Vt - x_0)/\varepsilon\big)}\Big\}. \tag{5.113}$$

Here

$$\beta_0 = \tan\nu, \quad \beta_1 = \frac{\sin\nu}{\sqrt{1-V^2}}, \quad \beta_2 = \frac{\cos\nu}{\sqrt{1-V^2}},$$

where $\nu \in (0, \pi/2)$ and $V \in (-1, 1)$, and the shifts x_0, t_0 are constant. The breather is usually regarded as a connected couple of interacting fluxon and antifluxon (see, e.g., [**169, 205**]). Formula (5.113) shows that the breather is concentrated in an ε-neighborhood of the trajectory $x = Vt + x_0$, and on this line its amplitude

$$4 \arctan \left\{ \beta_0 \cos \left(\frac{\sqrt{1 - V^2} \cos \nu}{\varepsilon} t + t_0' \right) \right\}$$

oscillates with frequency $\omega_\varepsilon = (\sqrt{1 - V^2}/\varepsilon) \cos \nu$. It is clear that the main parameters characterizing the breather for a given ε are the velocity V of the motion of the envelope function and the number ν, which determines the frequency ω_ε and period $T_p = 2\pi/\omega_\varepsilon$ of oscillations. Of course, these quantities can be constant only when there are no perturbations.

Our goal in this section is to derive Hugoniot type conditions describing the dynamics of these parameters in the presence of perturbations related to slow variations of characteristics of the medium. For simplicity, we restrict our consideration to the sine-Gordon equation with variable coefficients in the form (5.72). Nevertheless, it is easily seen from the subsequent constructions that taking into consideration the perturbations related to the presence of a small exciting force would bring about only minor technical changes in the calculations. Moreover, let us point out that the theory of perturbations related to a distorted breather was developed in [**112, 114, 139, 169**], based on the method of the inverse scattering problem. We will show that the ideas of the previous section, with minor modifications, enable us to construct the leading term of the asymptotics by elementary methods.

Before we proceed to specific constructions, we will discuss the nature of the solution to be found. It is clear that the function with properties similar to the breather (5.113) must be a two-phase one, because the behavior of the breather is essentially different in different "directions" (along the trajectory of its motion and "across" it). Hence, instead of (5.113), we have to consider

$$u = U\left(\frac{S_1(x,t)}{\varepsilon}, \frac{S_2(x,t)}{\varepsilon}, x, t \right), \tag{5.114}$$

where $U(\tau_1, \tau_2, x, t)$ is a smooth function T_p-periodic in τ_2 and exponentially decreasing together with all its derivatives as $\tau_1 \to \pm\infty$. Let us discuss the choice of the phases S_1 and S_2. As in the single-phase soliton-type setting, localization of the solution in a neighborhood of the curve $\Gamma_t = \{x,\ S_1(x,t) = 0\}$ immediately entails that without loss of generality we can take for S_1 the function

$$S_1 = x - \varphi(t).$$

Moreover, the same argument implies that neglecting the dependence of the leading term $U(\tau_1, \tau_2, x, t)$ on the slow variable x results in a change of U of magnitude $O(\varepsilon)$. Therefore (5.114) can be rewritten as

$$u = U_1\left(\frac{x - \varphi(t)}{\varepsilon}, \frac{S_2(x,t)}{\varepsilon}, t \right) + O(\varepsilon), \tag{5.115}$$

where $U_1(\tau_1, \tau_2, t) = U(\tau_1, \tau_2, \varphi(t), t)$.

Next, we expand the phase S_2 by the Taylor formula about the point $x = \varphi(t)$:

$$S_2(x,t) = \left\{ S_2(z,t) + (x-\varphi)S'_{2z}(z,t) + \frac{1}{2}(x-\varphi)^2 S''_{2zz}(z,t) \right\}\Big|_{z=\varphi(t)} + O((x-\varphi)^3).$$

Denote $x - \varphi = \varepsilon\tau_1$ and expand the function U_1:

$$\begin{aligned} &U_1\left(\tau_1, \frac{S_2(\varphi,t)}{\varepsilon} + \tau_1 S'_2(\varphi,t) + \frac{\varepsilon}{2}\tau_1^2 S''_2(\varphi,t) + O(\varepsilon^2\tau_1^3), t\right) \\ &\quad = U_1(\tau_1,\zeta,t) + \varepsilon\left\{\frac{1}{2}\tau_1^2 S''_2(\varphi,t) + O(\varepsilon\tau_1^3)\right\} \frac{\partial U_1}{\partial \tau_2}\Big|_{\tau_2=\zeta} \\ &\quad + \frac{1}{2}\varepsilon^2 \left\{\frac{1}{2}\tau_1^2 S''_2(\varphi,t) + O(\varepsilon\tau_1^3)\right\}^2 \frac{\partial^2 U_1}{\partial \tau_2^2}\Big|_{\tau_2=\theta}, \end{aligned} \tag{5.116}$$

where $\zeta = \varepsilon^{-1}S_2(\varphi,t) + \tau_1 S'_2(\varphi,t)$ and θ is an intermediate point. By the assumption on exponential decay of U_1 and its derivatives as $\tau_1 \to \pm\infty$, the second and third terms in the right-hand side of (5.116) are of order $O(\varepsilon)$ and $O(\varepsilon^2)$, respectively. Such corrections can be incorporated into higher order terms of the asymptotic expansion (5.115). Hence, without loss of generality (5.115) can be rewritten as

$$u = U_2\left(\frac{x-\varphi(t)}{\varepsilon}, \frac{\psi(t)}{\varepsilon}, t\right) + O(\varepsilon), \tag{5.117}$$

where $\psi(t) = S_2(\varphi(t), t)$ is a new unknown function,

$$U_2(\tau_1, \tau_2, t) = U_1(\tau_1, \tau_2 + g(t)\tau_1, t),$$

and the coefficient $g(t) = S'_{2x}(\varphi(t),t)$ is included in the dependence of $U_2(\tau_1,\tau_2,t)$ on the slow time variable t.

To conclude our preliminary arguments, we recall that the period T_p of oscillations in the fast variable must be constant. This is a well-known property of the solution, which immediately results from assuming that $T_p = T_p(t)$, differentiating the equality $U_2(\tau_1, \tau_2 + kT_p(t), t) = U_2(\tau_1,\tau_2,t)$ with respect to t, and then letting $k \to \infty$. Further, the function ψ in (5.117) is so far arbitrary. Hence without loss of generality we can take any constant for T_p. In this problem, as in many others dealing with periodic solutions (see, e.g., [**262**]), the answer takes the simplest form for $T_p = 2\pi$.

Now we proceed to the derivation of Hugoniot type conditions for the distorted breather. We make the change of variables $v = \gamma u$ and rewrite (5.72) in the form (5.81). We write the required two-phase asymptotic solution in a form similar to (5.82):

$$v(x,t,\varepsilon) = v_0\left(\frac{x-\varphi(t)}{\varepsilon}, \frac{\psi(t)}{\varepsilon}, t\right) + \varepsilon v_1(x,t,\varepsilon), \tag{5.118}$$

where φ and ψ are smooth functions to be determined, such that $\dot\varphi$ and $\dot\psi$ do not vanish (the dot denotes differentiation with respect to time), and the function

$v_0(\tau_1, \tau_2, t)$ is 2π-periodic in τ_2 and decreases exponentially fast together with all its derivatives as $\tau_1 \to \pm\infty$.

To find the leading term of the asymptotics, substitute (5.118) into (5.81) and expand the coefficient $\alpha(x,t)$ by the Taylor formula about $x = \varphi(t)$. Then we obtain equations for v_0 and v_1 similar to (5.83), (5.84):

$$\widehat{\mathcal{L}} v_0(\tau_1, \tau_2, t) = 0, \tag{5.119}$$

$$\widehat{L} v_1(x, t, \varepsilon) = F. \tag{5.120}$$

In our case

$$\widehat{\mathcal{L}} v_0 = \widehat{A}^2 v_0 - \widehat{B}^2 v_0 + \check{\alpha}^2 \sin v_0,$$

$$\widehat{A} = -\dot{\varphi}\frac{\partial}{\partial \tau_1} + \dot{\psi}\frac{\partial}{\partial \tau_2}, \quad \widehat{B} = \frac{\partial}{\partial \tau_1}, \quad \check{\alpha} = \alpha(\varphi(t), t),$$

the linearized operator $\widehat{L}$ is of the same form as in (5.84), i.e.,

$$\widehat{L} = \varepsilon^2 \frac{\partial^2}{\partial t^2} - \varepsilon^2 \frac{\partial^2}{\partial x^2} + \alpha^2(x,t)\cos v_0,$$

and the leading term F_0 of the right-hand side F is given by

$$F_0 = -\widehat{\Pi} v_0 + 2\frac{\check{\alpha}_x}{\check{\alpha}}\tau_1(\widehat{A}^2 - \widehat{B}^2)v_0 + 2\frac{\check{\gamma}_t}{\check{\gamma}}\widehat{A} v_0 - 2\frac{\check{\gamma}_x}{\check{\gamma}}\frac{\partial v_0}{\partial \tau_1}. \tag{5.121}$$

Here

$$\widehat{\Pi} = 2\widehat{A}\frac{\partial}{\partial t} + \widehat{A}_t, \quad \widehat{A}_t = -\ddot{\varphi}\frac{\partial}{\partial \tau_1} + \ddot{\psi}\frac{\partial}{\partial \tau_2},$$

and for a function $f(x,t)$ we denote by $\check{f}_t$ or $\check{f}_x$ the corresponding partial derivative taken for $x = \varphi(t)$.

Now let us consider equation (5.119). After the change of variables

$$\eta = \frac{\check{\alpha}}{\dot{\psi}}\tau_2, \quad \xi = \frac{\check{\alpha}}{\dot{\psi}}(\dot{\psi}\tau_1 + \dot{\varphi}\tau_2),$$

similar to (5.86), equation (5.119) becomes the standard form (5.87) of the sine-Gordon equation with constant coefficients for

$$\widetilde{v}(\eta, \xi, t) = v_0\left(\frac{\xi - \dot{\varphi}\eta}{\check{\alpha}}, \frac{\dot{\psi}\eta}{\check{\alpha}}, t\right).$$

Hence, invoking the formula for the exact solution (5.113) with $\varepsilon = 1$, we find $\widetilde{v}$ as

$$\widetilde{v} = 4\arctan\left\{\widetilde{\beta}_0 \frac{\cos\widetilde{\beta}_2(\eta - V\xi + \widetilde{c}_2)}{\cosh\widetilde{\beta}_1(\xi - V\eta + \widetilde{c}_1)}\right\}, \tag{5.122}$$

where the "constants" $\widetilde{c}_1$, $\widetilde{c}_2$ are related to the invariance of (5.87) with respect to translations in η and ξ, the coefficients $\widetilde{\beta}_i$ are found by the same formulas as the β_i in (5.113), and $\nu \in (0, \pi/2)$, $V \in (-1, 1)$ are arbitrary "constants".

We put $V = \dot{\varphi}$ and return to the variables τ_1 and τ_2. Then we immediately obtain from (5.122) the formula

$$v_0 = 4\arctan\left\{\beta_0 \frac{\cos\beta_2((1-\dot{\varphi}^2)\tau_2 - \dot{\varphi}\dot{\psi}\tau_1 + c_2')}{\cosh\beta_1(\tau_1 + c_1)}\right\}, \tag{5.123}$$

where

$$\beta_0 = \tan\nu, \quad \beta_1 = \breve{\alpha}\frac{\sin\nu}{\sqrt{1-\dot{\varphi}^2}}, \quad \beta_2 = \breve{\alpha}\frac{\cos\nu}{\dot{\psi}\sqrt{1-\dot{\varphi}^2}},$$

$c_1 = \widetilde{c}_1/\breve{\alpha}$, $c_2' = \widetilde{c}_2\dot{\psi}/\breve{\alpha}$, and we assume the natural condition

$$|\dot{\varphi}| < 1. \tag{5.124}$$

Now we have to take into account the condition of 2π-periodicity of $v_0(\tau_1, \tau_2, t)$ in τ_2. Obviously, this yields the equality

$$\dot{\psi} = \breve{\alpha}\sqrt{1-\dot{\varphi}^2}\cos\nu, \tag{5.125}$$

which relates the frequency $\dot{\psi}$, the velocity $\dot{\varphi}$, and the auxiliary function $\nu = \nu(t)$. Note that the right-hand side of (5.125) could contain a factor of $1/n$, where $n \neq 0$ is an arbitrary integer. However, using the arbitrariness in the choice of ν, which is still available, we let $n = 1$.

Now from (5.123) and (5.125) we obtain the final formula

$$v_0 = 4\arctan\left\{\beta_0 \frac{\cos(\tau_2 - g\tau_1 + c_2)}{\cosh\beta_1(\tau_1 + c_1)}\right\} \tag{5.126}$$

for the leading term of the asymptotics (5.118). Here the coefficient

$$g = \breve{\alpha}\frac{\dot{\varphi}\dot{\psi}}{1-\dot{\varphi}^2}$$

has the same meaning as $S_{2x}'(\varphi, t)$, which appeared in (5.116) when we analyzed the localized periodic solution of general form, and $c_2 = \beta_2 c_2'$ is a new unknown function.

Let us proceed to the derivation of necessary conditions for the solvability of equation (5.120). It is not hard to check directly that the functions

$$y_1 = \varepsilon^{-1}\left(\frac{\partial v_0}{\partial c_1} - g\frac{\partial v_0}{\partial c_2}\right) \quad \text{and} \quad y_2 = \varepsilon^{-1}\frac{\partial v_0}{\partial c_2}$$

satisfy the equation $Ly = 0$ (mod $O(\varepsilon)$ in the sense of $L^1(Q_T)$). Furthermore, the derivatives of v_0 with respect to all other parameters and variables differ from linear combinations of y_1 and y_2 by $O(\varepsilon)$ in the sense of $L^1(Q_T)$. Finally, it is not hard to establish similarly to the above that, under the conditions (5.92),

$$\int_0^T \int_{-\infty}^{+\infty} y_j \widehat{L} v_1 \, dx dt \to 0 \qquad \text{as} \quad \varepsilon \to 0, \quad j = 1, 2.$$

Therefore we again arrive at necessary conditions of the form (5.95), where, as before, the right-hand side F can be replaced by its leading term $F_0(v_0, x, t)$. Thus

$$\lim_{\varepsilon\to 0}\int_0^T \int_{-\infty}^{\infty} y_j F_0 \, dx\, dt = 0. \tag{5.127}$$

Next we observe that the products $R_j = y_j F_0(v_0, x, t)$ are periodic in τ_2. Hence, expanding the R_j into Fourier series, we obtain

$$R_j(x,t,\varepsilon) = \frac{1}{\varepsilon}\sum_{k=-\infty}^{\infty} r_{j,k}(\tau_1, t) e^{ik\psi/\varepsilon}\Big|_{\tau_1=(x-\varphi)/\varepsilon}. \tag{5.128}$$

However, according to our assumption, $\dot\psi \neq 0$ for all t. Hence, integrating by parts, we can easily see that for all nonzero k

$$\frac{1}{\varepsilon}\int_0^T\int_{-\infty}^{\infty} r_{j,k}\left(\frac{x-\varphi}{\varepsilon}, t\right) e^{ik\psi/\varepsilon}\, dx dt = \int_{-\infty}^{\infty}\int_0^T r_{j,k}(\tau_1, t) e^{ik\psi/\varepsilon}\, dt d\tau_1 = O(\varepsilon).$$

Thus the only nontrivial term in (5.128) is the zero term of the Fourier series,

$$r_{j,0} = \frac{1}{2\pi}\int_0^{2\pi} (y_j F_0)\left(\frac{x-\varphi}{\varepsilon}, \tau_2, t\right) d\tau_2,$$

i.e., the product $y_j F_0$ averaged over oscillations.

The above analysis, the arbitrariness of T, and the substitution $x = \varphi(t) + \varepsilon\tau_1$ allow us to rewrite the necessary conditions (5.127) as

$$\int_0^{2\pi}\int_{-\infty}^{\infty} \frac{\partial v_0}{\partial \tau_j} F_0 \, d\tau_1 d\tau_2 = 0, \quad j = 1, 2. \tag{5.129}$$

Consider (5.129) for $j = 1$. With the aid of (5.121) this equality can easily be transformed into

$$\begin{aligned}
&-\frac{d}{dt}\left\{\dot\varphi \int_0^{2\pi}\int_{-\infty}^{\infty}\left(\frac{\partial v_0}{\partial \tau_1}\right)^2 d\tau_1 d\tau_2\right\} \\
&\quad + \left\{\frac{\breve\alpha_x}{\breve\alpha}(\dot\varphi^2 - 1) + 2\frac{\breve\gamma_t\dot\varphi + \breve\gamma_x}{\breve\gamma}\right\}\int_0^{2\pi}\int_{-\infty}^{\infty}\left(\frac{\partial v_0}{\partial \tau_1}\right)^2 d\tau_1 d\tau_2 \\
&\quad + 2\dot\psi\int_0^{2\pi}\int_{-\infty}^{\infty}\frac{\partial v_0}{\partial \tau_1}\frac{\partial^2 v_0}{\partial \tau_2 \partial t} d\tau_1 d\tau_2 \\
&\quad - \left(\ddot\psi - 2\frac{\breve\gamma_t}{\breve\gamma}\dot\psi\right)\int_0^{2\pi}\int_{-\infty}^{\infty}\frac{\partial v_0}{\partial \tau_1}\frac{\partial v_0}{\partial \tau_2} d\tau_1 d\tau_2 \\
&\quad - \dot\psi^2\frac{\breve\alpha_x}{\breve\alpha}\int_0^{2\pi}\int_{-\infty}^{\infty}\left(\frac{\partial v_0}{\partial \tau_2}\right)^2 d\tau_1 d\tau_2 = 0.
\end{aligned} \tag{5.130}$$

For further transformation of (5.130) we have to take into account the equalities

$$\int_0^{2\pi}\int_{-\infty}^{\infty}\frac{\partial v_0}{\partial\tau_1}\frac{\partial v_0}{\partial\tau_2}\,d\tau_1 d\tau_2=-g\int_0^{2\pi}\int_{-\infty}^{\infty}\left(\frac{\partial v_0}{\partial\tau_2}\right)^2 d\tau_1 d\tau_2,$$

$$\begin{aligned}\frac{\partial^2 v_0}{\partial\tau_2\partial t}&=\frac{\dot\beta_0}{\beta_0}\frac{\partial v_0}{\partial\tau_2}+\left(\frac{\dot\beta_1}{\beta_1}g-\dot g\right)\tau_1\frac{\partial^2 v_0}{\partial\tau_2^2}\\&\quad+\frac{\dot\beta_1}{\beta_1}\tau_1\frac{\partial^2 v_0}{\partial\tau_1\partial\tau_2}+8\beta_0^2\dot\beta_0\frac{\sin z\cos^2 z\cosh\beta_1\tau_1}{(\cosh^2\beta_1\tau_1+\beta_0^2\cos^2 z)^2}\bigg|_{z=\tau_2-g\tau_1},\end{aligned}\tag{5.131}$$

$$\begin{aligned}\int_0^{2\pi}\int_{-\infty}^{\infty}\left(\frac{\partial v_0}{\partial\tau_1}\right)^2 d\tau_1 d\tau_2&=(g^2-\beta_1^2)\int_0^{2\pi}\int_{-\infty}^{\infty}\left(\frac{\partial v_0}{\partial\tau_2}\right)^2 d\tau_1 d\tau_2\\&\quad+16\beta_0^2\beta_1^2\int_0^{2\pi}\int_{-\infty}^{\infty}\frac{\cosh^2\beta_1\tau_1-\cos^2 z}{(\cosh^2\beta_1\tau_1+\beta_0^2\cos^2 z)^2}\bigg|_{z=\tau_2-g\tau_1} d\tau_1 d\tau_2,\end{aligned}$$

which can be easily verified by direct calculations.

Denote

$$J_1=\int_0^{2\pi}\int_{-\infty}^{\infty}\frac{\sin^2 z\cosh^2\tau}{(\cosh^2\tau-\beta_0^2\cos^2 z)^2}\,d\tau dz,$$

$$J_2=\int_0^{2\pi}\int_{-\infty}^{\infty}\frac{\cosh^2\tau-\cos^2 z}{(\cosh^2\tau-\beta_0^2\cos^2 z)^2}\,d\tau dz,$$

and observe that

$$4\beta_0\dot\beta_0\int_0^{2\pi}\int_{-\infty}^{\infty}\frac{\sin^2 z\cos^2 z\cosh^2\tau}{(\cosh^2\tau+\beta_0^2\cos^2 z)^3}\,d\tau dz=-\frac{dJ_1}{dt}.$$

Hence, after quite elementary transformations, the equality (5.130) can be rewritten as

$$\begin{aligned}&-\frac{d}{dt}\left\{\dot\varphi(g^2-\beta_1^2)\frac{\beta_0^2}{\beta_1}J_1+\dot\varphi\beta_0^2\beta_1 J_2\right\}\\&\quad+\left\{\left(\frac{\breve\alpha_x}{\breve\alpha}(\dot\varphi^2-1)+2\frac{\breve\gamma_t\dot\varphi+\breve\gamma_x}{\breve\gamma}\right)(g^2-\beta_1^2)\right.\\&\quad\left.-\frac{d}{dt}(g\dot\psi)+g\dot\psi\left(2\frac{\dot\beta_0}{\beta_0}-\frac{\dot\beta_1}{\beta_1}\right)+\dot\psi^2\frac{\breve\alpha_x}{\breve\alpha}-2\dot\psi g\frac{\breve\gamma_t}{\breve\gamma}\right\}\frac{\beta_0^2}{\beta_1}J_1\\&\quad+\left\{\frac{\breve\alpha_x}{\breve\alpha}(\dot\varphi^2-1)+2\frac{\breve\gamma_t\dot\varphi+\breve\gamma_x}{\breve\gamma}\right\}\beta_0^2\beta_1 J_2-\dot\psi g\frac{\beta_0^2}{\beta_1}\frac{dJ_1}{dt}=0.\end{aligned}\tag{5.132}$$

Further transformations, taking into account the explicit form of the coefficients β_0, β_1, g, the relation (5.125), and the equality

$$\frac{d\gamma(\varphi,t)}{dt}=\left\{\frac{\partial\gamma(x,t)}{\partial t}+\dot\varphi\frac{\partial\gamma(x,t)}{\partial x}\right\}\bigg|_{x=\varphi},\tag{5.133}$$

allow us to write down (5.132) in a more compact form:

$$\begin{aligned}&\frac{d}{dt}\left\{\frac{\breve\alpha\dot\varphi\sin\nu}{\breve\gamma^2\sqrt{1-\dot\varphi^2}}[(1-\beta_0^2)J_1+\beta_0^2J_2]\right\}\\&\quad+\frac{\breve\alpha}{\breve\gamma^2}\sin\nu\sqrt{1-\dot\varphi^2}\left\{\frac{\breve\alpha_x}{\breve\alpha}[(1-\beta_0^2)J_1+\beta_0^2J_2]\right.\\&\qquad\left.+2\frac{\breve\gamma_x}{\breve\gamma}\beta_0^2(J_1-J_2)\right\}=0.\end{aligned}\tag{5.134}$$

Now we consider condition (5.129) for $j = 2$. First of all, rewrite it as follows:

$$\begin{aligned}
&\frac{d}{dt}\left\{\dot{\psi}\int_0^{2\pi}\int_{-\infty}^{\infty}\left(\frac{\partial v_0}{\partial\tau_2}\right)^2 d\tau_1 d\tau_2\right\} \\
&\quad - 2\dot{\varphi}\frac{d}{dt}\int_0^{2\pi}\int_{-\infty}^{\infty}\frac{\partial v_0}{\partial\tau_1}\frac{\partial v_0}{\partial\tau_2}\,d\tau_1 d\tau_2 + 2\dot{\varphi}\int_0^{2\pi}\int_{-\infty}^{\infty}\frac{\partial v_0}{\partial\tau_1}\frac{\partial^2 v_0}{\partial\tau_2\partial t}\,d\tau_1 d\tau_2 \\
&\quad - g\left\{2\dot{\psi}\left(\frac{\breve{\alpha}_x}{\breve{\alpha}} + \frac{\breve{\gamma}_t}{\breve{\gamma}}\right) - \ddot{\varphi} + 2\frac{\breve{\alpha}_x}{\breve{\alpha}}(\dot{\varphi}^2 - 1) + 2\frac{\breve{\gamma}_t\dot{\varphi} + \breve{\gamma}_x}{\breve{\gamma}}\right\} \\
&\quad \times\int_0^{2\pi}\int_{-\infty}^{\infty}\left(\frac{\partial v_0}{\partial\tau_2}\right)^2 d\tau_1 d\tau_2 = 0.
\end{aligned} \tag{5.135}$$

Then, using (5.131) and (5.133) and making some transformations, we reduce (5.135) to the form

$$\frac{d}{dt}\left\{\frac{\beta_0}{\breve{\gamma}^2}J_1\right\} = 0.$$

Thus

$$\tan\nu(t)J_1(t) = \mu\gamma^2(\varphi, t), \tag{5.136}$$

where $\mu > 0$ is a constant.

In its turn, the last equality allows us to transform (5.134) to the final form

$$\begin{aligned}
&\frac{d}{dt}\left\{\breve{\alpha}\frac{\dot{\varphi}\cos\nu}{\sqrt{1-\dot{\varphi}^2}}\left(\mu + \frac{I}{\breve{\gamma}^2}\tan^3\nu\right)\right\} \\
&\qquad + \cos\nu\sqrt{1-\dot{\varphi}^2}\left\{\mu\breve{\alpha}_x + I\tan^3\nu\left(\frac{\breve{\alpha}}{\gamma^2}\right)_x\right\} = 0,
\end{aligned} \tag{5.137}$$

where

$$I = \int_0^{2\pi}\int_{-\infty}^{\infty}\frac{\cos^2 z\sinh^2\tau}{(\cosh^2\tau + \tan^2\nu\cos^2 z)^2}\,d\tau dz.$$

It is easily seen that equations (5.125), (5.136), and (5.137) form a closed system of Hugoniot type conditions for finding the functions $\varphi(t)$, $\psi(t)$, and $\nu(t)$, which determine the leading terms of the trajectory of the breather and the frequency of its oscillations. The assumption (5.124) is implicitly incorporated into (5.137) as the smoothness condition on the coefficients of the equation, and the restrictions $\dot{\psi} \neq 0$ and $\nu \in (0, \pi/2)$ follow from (5.125), (5.137), (5.124), and the boundedness of the coefficients α and γ.

In conclusion, we point out three particular cases where the Hugoniot type conditions for the distorted breather take a simpler form.

1. Let $\gamma = \text{const}$. Then (5.136) implies that $\nu = \text{const}$. In this case (5.137) becomes

$$\frac{d}{dt}\left\{\frac{\breve{\alpha}\dot{\varphi}}{\sqrt{1-\dot{\varphi}^2}}\right\} + \breve{\alpha}_x\sqrt{1-\dot{\varphi}^2} = 0.$$

By comparing this equation with (1.45) in §1.5, we see that the trajectory of the distorted breather in this case is the same as that of the distorted fluxon. In other words, for γ = const the frequency of oscillation, despite its not being a linear function of t (by (5.125)), does not affect the trajectory $x = \varphi(t)$.

2. Suppose the functions α and γ do not depend on x. Then the equation (5.137) can be integrated, and we obtain

$$\dot{\varphi} = \pm \frac{E}{\sqrt{E^2 + \alpha^2 \rho^2}}, \tag{5.138}$$

where $\rho = \mu + I\gamma^{-2} \tan^3 \nu$ and $E > 0$ are constants.

3. For α = const, γ = const we immediately obtain from (5.125), (5.136), and (5.138) that ν, $\dot{\varphi}$, and $\dot{\psi}$ are constants. In this case the leading term of the asymptotics (5.126) coincides with the exact solution (5.113).

5.4. The structure of a two-scale distorted solitary wave. Isothermic jump

In this section we return to the problem of constructing asymptotic solutions of a smoothed shock wave type for gas dynamics equations in the case of large Reynolds numbers Re and small Prandtl numbers Pr. It was shown in §2.6 that for such a relation between these parameters the asymptotic solution admits a single-phase representation only for sufficiently small Mach numbers (computed ahead of the front of the shock wave):

$$M^2 < \frac{3\gamma - 1}{\gamma(3-\gamma)}. \tag{5.139}$$

If this condition is violated, the overturning effects arising due to large velocities can be compensated only by taking the viscous terms in the equation of motion into account. As we pointed out in §2.6, this necessitates constructing a three-scale asymptotic solution.

The setting arising when (5.139) fails, which in the physical literature [**125, 251**] is called an the isothermic jump, is the subject of this section.

Taking into account the indicated relation between the Reynolds and Prandtl numbers, we rewrite the gas dynamics equations (2.107) as

$$\begin{aligned}
&\frac{\partial \rho}{\partial t} + \langle \nabla, \rho u \rangle = 0, \\
&\frac{\partial}{\partial t}(\rho u_{(i)}) + \langle \nabla, \rho u u_{(i)} \rangle + \frac{\partial p}{\partial x_i} = \varepsilon^2 \frac{\partial}{\partial x_j} \sigma_{ij}, \\
&\frac{\partial}{\partial t}\left(\frac{1}{2}\rho u^2 + \frac{p}{\gamma - 1}\right) \\
&\qquad + \left\langle \nabla, \left(\frac{1}{2}\rho u^2 + \frac{\gamma}{\gamma-1} p\right) u - \varepsilon^2 \sigma \cdot u - \varepsilon \kappa_0 \nabla T \right\rangle = 0.
\end{aligned} \tag{5.140}$$

Here ε is a small parameter inversely proportional to the square root of the Reynolds number, and we assume that

$$\mathrm{Pr} \sim (\mathrm{Re})^{-1/2} \ll 1.$$

To simplify the notation, we set $\Pr = \varepsilon$. All other notations have the same meaning as in §2.6. For simplicity we will assume that the coefficients of heat conductivity κ_0 and viscosity η_0, ζ_0 are constant.

Let us discuss the structure of the asymptotic solution. The presence of three scales (of magnitude $1/\varepsilon^2$, $1/\varepsilon$, and 1) entails the two-phase structure of the asymptotic expansion. At the same time, it is clear that the problem on the motion of a smoothed shock wave involves a single geometric object, namely, the front of the limiting (as $\varepsilon \to 0$) shock wave. Hence, unlike, say, the problem on interaction of solitons, the phases appearing here may differ from one another only by higher order terms in ε. As before, we will assume that the front Γ_t of the limiting shock wave can be described as the zero level surface of some smooth function $S(x,t)$. Moreover, let the velocity of the front motion have a nonzero normal component. Then we have the standard representation for Γ_t:

$$\Gamma_t = \{x,\ t = \varphi(x)\}, \tag{5.141}$$

with a nondegenerate smooth function $\varphi(x)$. It is also clear that in order to describe the limiting solution as $\varepsilon \to 0$ we have to obtain the equations of ideal gas dynamics (2.161) in the domains ahead of and after the shock wave front, as well as the standard Rankine–Hugoniot conditions (2.127), (2.128) on the wave front.

Moreover, note that in the treatment of the problem with a small but nonvanishing ε in the three-scale setting we have to take into account small corrections to the function φ in the description of the smoothed wave front. Hence it is natural to use the following representation instead of (5.141):

$$\Gamma_{t,\varepsilon} = \{x,\ t = \varphi(x) + \varepsilon\varphi_1(x) + \varepsilon^2\varphi_2(x) + \cdots\}, \tag{5.142}$$

where the φ_i are smooth functions. It remains to observe that all fast variations of the solution in a neighborhood of the front are directed essentially along the normal to Γ_t. These preliminary arguments suggest that the required asymptotic solution can be represented as a function

$$F\left(\frac{S_1(x,t)}{\varepsilon^2}, \frac{S_2(x,t)}{\varepsilon}, x, t, \varepsilon\right)$$

with phases of special form

$$S_1 = t - \varphi(x) - \varepsilon\varphi_1(x), \quad S_2 = t - \varphi(x).$$

We proceed to our construction of the asymptotic solution. We write it in the form of an expansion

$$\begin{aligned} u &= \mathcal{U}^{(0)}(\eta,\tau,x,t) + \varepsilon\mathcal{U}^{(1)}(\eta,\tau,x,t) + \dots, \\ \rho &= \mathcal{R}^{(0)}(\eta,\tau,x,t) + \varepsilon\mathcal{R}^{(1)}(\eta,\tau,x,t) + \dots, \\ p &= \mathcal{P}^{(0)}(\eta,\tau,x,t) + \varepsilon\mathcal{P}^{(1)}(\eta,\tau,x,t) + \dots, \\ T &= \mathcal{T}^{(0)}(\eta,\tau,x,t) + \varepsilon\mathcal{T}^{(1)}(\eta,\tau,x,t) + \dots, \end{aligned} \tag{5.143}$$

where $\eta = S_1/\varepsilon^2$, $\tau = S_2/\varepsilon$,

$$\begin{aligned} \mathcal{U}^{(i)}(\eta,\tau,x,t) &= \mathcal{U}_i(\tau,x,t) + W_i(\eta,\tau,x,t), \\ \mathcal{R}^{(i)}(\eta,\tau,x,t) &= \mathcal{R}_i(\tau,x,t) + r_i(\eta,\tau,x,t), \\ \mathcal{P}^{(i)}(\eta,\tau,x,t) &= \mathcal{P}_i(\tau,x,t) + \pi_i(\eta,\tau,x,t), \\ \mathcal{T}^{(i)}(\eta,\tau,x,t) &= \mathcal{T}_i(\tau,x,t) + \Theta_i(\eta,\tau,x,t). \end{aligned} \tag{5.144}$$

The most rapidly varying functions $W_i, r_i, \pi_i, \Theta_i$ are assumed to stabilize exponentially fast at infinity in η, i.e.,

$$\begin{aligned} W_i(\eta,\tau,x,t) &\to W_i^+(\tau,x,t), & r_i(\eta,\tau,x,t) &\to r_i^+(\tau,x,t), \\ \pi_i(\eta,\tau,x,t) &\to \pi_i^+(\tau,x,t), & \Theta_i(\eta,\tau,x,t) &\to \Theta_i^+(\tau,x,t), \end{aligned} \tag{5.145}$$

as $\eta \to +\infty$, and they satisfy the normalization condition

$$W_i, r_i, \pi_i, \Theta_i \to 0 \qquad \text{as} \quad \eta \to -\infty. \tag{5.146}$$

The energy equation in (5.140), in contrast to the first two, is not a regularly degenerate one with respect to the rapidly varying (in the scale $\sim \varepsilon^2$) solution, i.e., it cannot be represented as

$$A\left(\varepsilon^2 \frac{\partial}{\partial t} f_1(\rho,u,p), \varepsilon^2 \frac{\partial}{\partial x} f_2\left(\rho,u,p,\varepsilon^2 \frac{\partial u}{\partial x}, \varepsilon^2 \frac{\partial T}{\partial x}\right)\right) = 0.$$

Hence it is clear that the leading term of the asymptotic expansion for temperature Θ_0 cannot be a function (in the scale $\sim \varepsilon^2$) of smoothed shock wave type. Nevertheless we allow for appearance of a regularization of the singularity of type η_+ in the expansion for T. Hence we will assume that

$$\Theta_0(\eta,\tau,x,t) = f(\tau,x,t)\Psi(\eta,x), \tag{5.147}$$

where f and Ψ are smooth functions satisfying the conditions

$$\begin{gathered} f(\varphi_1,x,t) = 0, \\ \Psi \to 1 \quad \text{as} \quad \eta \to \infty, \qquad \Psi \to 0 \quad \text{as} \quad \eta \to -\infty. \end{gathered} \tag{5.148}$$

Next, the functions $\mathcal{U}_i, \mathcal{R}_i, \mathcal{P}_i, \mathcal{T}_i$, which may be viewed as a background with respect to $W_i, \dots, \Theta_i$, are also rapidly varying, but in the scale $\sim \varepsilon$. Denote

$$\begin{aligned} \mathcal{U}_i(\tau,x,t) &= u_i(x,t) + U_i(\tau,x,t), \\ \mathcal{R}_i(\tau,x,t) &= \rho_i(x,t) + R_i(\tau,x,t), \\ \mathcal{P}_i(\tau,x,t) &= p_i(x,t) + P_i(\tau,x,t), \\ \mathcal{T}_i(\tau,x,t) &= T_i(x,t) + \theta_i(\tau,x,t), \end{aligned} \tag{5.149}$$

where the $U_i, \dots, \theta_i$ are functions stabilizing as $\tau \to \pm\infty$,

$$\begin{aligned} &U_i(\tau,x,t) \to U_i^+(x,t), \quad R_i(\tau,x,t) \to R_i^+(x,t), \\ &P_i(\tau,x,t) \to P_i^+(x,t), \\ &\theta_i(\tau,x,t) \to \theta_i^+(x,t) \qquad \text{as} \quad \tau \to +\infty, \\ &U_i, R_i, P_i, \theta_i \to 0 \qquad \text{as} \quad \tau \to -\infty. \end{aligned} \tag{5.150}$$

Moreover, the equation of state directly implies relationships between $\mathcal{P}^{(i)}, \mathcal{R}^{(j)}$, and $\mathcal{T}^{(j)}$. In particular,

$$\mathcal{P}^{(0)} = \mathcal{R}^{(0)}\mathcal{T}^{(0)}, \quad \pi_0 = \mathcal{R}_0\theta_0 + r_0(\mathcal{T}_0 + \theta_0). \tag{5.151}$$

Now that we have introduced the basic notation, we can execute our standard scheme of constructing the asymptotics.

Substitute the expansion (5.143) into equation (5.140) and consider the terms of order ε^{-2}. Since

$$\tau = \varepsilon\eta + \varphi_1, \quad t = \varphi + \varepsilon\varphi_1 + \varepsilon^2\eta,$$

and the derivatives of $W_0, \dots, \Theta_0$ with respect to η tend to zero exponentially fast, we can use the Taylor expansions with respect to τ about the point $\tau = \varphi_1$ and with respect to t about the point $t = \varphi$. Next, we have

$$\frac{1}{\varepsilon^3}\frac{\partial^2 \Theta_0}{\partial\eta^2} = \frac{1}{\varepsilon^2}\frac{\partial^2\Psi}{\partial\eta^2}\left\{\eta\frac{\partial f}{\partial\tau} + \varphi_1\frac{\partial f}{\partial t}\right\}\bigg|_{\tau=\varphi_1,\, t=\varphi} + O(\varepsilon^{-1}).$$

Therefore this expression is of order $O(\varepsilon^{-2})$. To simplify the notation, we will use, as before, the sign $\check{}$ for functions "stopped" at the point $\tau = \varphi_1$ and $t = \varphi$,

$$\check{F} \overset{\text{def}}{=} F(\eta, \tau, x, t)\big|_{\tau=\varphi_1,\, t=\varphi}. \tag{5.152}$$

Now, equating to zero the terms of order ε^{-2}, integrating with respect to η, and choosing the constants of integration according to (5.146), we obtain the equalities

$$(V_\nu - \langle\nu, \check{\mathcal{U}}^{(0)}\rangle)\check{\mathcal{R}}^{(0)} = (V_\nu - \langle\nu, \check{\mathcal{U}}_0\rangle)\check{\mathcal{R}}_0, \tag{5.153}$$

$$\begin{aligned}(V_\nu - \langle\nu, \check{\mathcal{U}}^{(0)}\rangle)&\check{\mathcal{R}}^{(0)}\check{\mathcal{U}}^{(0)} - \nu\check{\mathcal{P}}^{(0)} \\ &= |\nabla\varphi|\frac{\partial}{\partial\eta}\left\{\eta_0\check{W}_0 - \left(\zeta_0 + \frac{\eta_0}{3}\right)\nu\langle\nu, \check{W}_0\rangle\right\} \\ &\quad + (V_\nu - \langle\nu, \check{\mathcal{U}}_0\rangle)\check{\mathcal{R}}_0\check{\mathcal{U}}_0 - \nu\check{\mathcal{P}}_0,\end{aligned} \tag{5.154}$$

$$\begin{aligned}(V_\nu &- \langle\nu, \check{\mathcal{U}}^{(0)}\rangle)\left(\frac{1}{2}\check{\mathcal{R}}^{(0)}\check{\mathcal{U}}^{(0)2} + \frac{\check{\mathcal{P}}^{(0)}}{\gamma-1}\right) - \langle\nu, \check{\mathcal{U}}^{(0)}\rangle\check{\mathcal{P}}^{(0)} \\ &= |\nabla\varphi|\frac{\partial}{\partial\eta}\left\{\kappa_0(\check{\Theta}_1 + \varphi_1\check{f}_t\check{\Psi}) + \frac{1}{2}\left(\eta_0\check{\mathcal{U}}^{(0)2} + \left(\zeta_0 + \frac{\eta_0}{3}\right)\langle\nu, \check{\mathcal{U}}^{(0)}\rangle^2\right)\right\} \\ &\quad + |\nabla\varphi|\kappa_0\check{f}_\tau\left(\eta\frac{\partial\check{\Psi}}{\partial\eta} + \check{\Psi}\right) - \langle\nu, \check{\mathcal{U}}_0\rangle\check{\mathcal{P}}_0 \\ &\quad + (V_\nu - \langle\nu, \check{\mathcal{U}}_0\rangle)\left(\frac{1}{2}\check{\mathcal{R}}_0\check{\mathcal{U}}_0^2 + \frac{\check{\mathcal{P}}_0}{\gamma-1}\right).\end{aligned} \tag{5.155}$$

Here, as before, we use the notation

$$V_\nu = \frac{1}{|\nabla\varphi|}, \quad \nu = \frac{\nabla\varphi}{|\nabla\varphi|}$$

for the velocity of the front and the normal vector to Γ_t.

Obviously, letting $\eta \to \infty$, we obtain conditions analogous to the Rankine–Hugoniot conditions (2.127), (2.128). We will formulate them later as relations between the background $\check{\mathcal{U}}_0, \dots, \check{\mathcal{P}}_0$ and the limiting values $\check{W}_0^+, \dots, \check{\pi}_0^+$.

Let us simplify the equations (5.153)–(5.155). First of all, it is seen from (5.154) that the vector $\check{W}_0$ is directed along the normal ν. Denote

$$a = V_\nu - \langle \nu, \check{\mathcal{U}}_0 \rangle, \quad \check{W}_0 = a(1 - w(\eta, x))\nu, \tag{5.156}$$

where w is an unknown scalar function, and a may be viewed as the relative velocity of the gas motion before the jump, which propagates over the "background" $\mathcal{U}_0(\tau, x, t)$. Then we readily find from (5.153) that

$$\check{r}_0 = \check{\mathcal{R}}_0 \frac{1 - w}{w}. \tag{5.157}$$

In its turn, equation (5.154) can be easily transformed into

$$a^2 \check{\mathcal{R}}_0 (1 - w) - \check{\pi}_0 + a|\nabla \varphi| \Big(\zeta_0 + \frac{4}{3}\eta_0 \Big) \frac{\partial w}{\partial \eta} = 0. \tag{5.158}$$

Since

$$\check{\pi}_0 = \check{\mathcal{T}}_0 \check{r}_0, \tag{5.159}$$

(5.158) implies the equation

$$\frac{(1 - w)(w - w^+)}{w} = -\delta \frac{\partial w}{\partial \eta}. \tag{5.160}$$

Similarly to §2.5, we denote here

$$\delta = \frac{|\nabla \varphi|}{a \check{\mathcal{R}}_0} \Big(\zeta_0 + \frac{4}{3}\eta_0 \Big), \quad w^+ = \frac{1}{\gamma m^2}, \quad m^2 = \frac{a^2}{\gamma \check{\mathcal{T}}_0}. \tag{5.161}$$

The meaning of m is the intermediate Mach number, taken ahead of the front of the wave and moving over the "background" $\mathcal{U}_0(\tau, x, t)$. Assuming that the flow $\check{\mathcal{U}}_0$ is supersonic, we obtain $1 > w^+ > 0$. Hence integration of (5.160) leads us to the formula similar to (2.140):

$$\frac{1 - w}{(w - w^+)^{w^+}} = e^{\beta(\eta + \varphi_2)}, \tag{5.162}$$

where $\beta = (1 - w^+)/\delta$ and $\varphi_2 = \varphi_2(x)$ is the constant of integration.

Now we consider the equation (5.155). After some algebra we rewrite it as

$$\begin{aligned} \check{\mathcal{R}}_0(1 - w)\Big\{ a^2 \Big(\mathcal{U}_{0\nu} + \frac{a}{2}(1 - w) \Big) - \frac{V_\nu \check{\mathcal{T}}_0}{w} \Big\} \\ = |\nabla \varphi| \frac{\partial}{\partial \eta} \Big\{ \kappa_0 (\check{\Theta}_1 + (\varphi_1 \check{f}_t + \eta \check{f}_\tau) \Psi) \\ + \frac{a^2}{2} \Big(\zeta_0 + \frac{4}{3}\eta_0 \Big) \Big(\frac{2}{a} \check{\mathcal{U}}_{0\nu} + 1 - w \Big)(1 - w) \Big\}. \end{aligned} \tag{5.163}$$

As before, the subscript ν denotes the projection of a vector on the normal ν.

Recall that when constructing a regularization of a singularity of type η_+, we impose only conditions (5.148) on the auxiliary function Ψ. Therefore this function may be chosen in a fairly arbitrary way, so that (5.163) is an equation for Θ_1. In its turn, (5.163) has a bounded solution only if both left- and right-hand sides tend to zero as $\eta \to \pm\infty$. This and (5.162) imply that

$$\check{f}_\tau = -\frac{a^3}{2}\frac{\check{\mathcal{R}}_0}{\kappa_0|\nabla\varphi|}(1-(w^+)^2). \tag{5.164}$$

Hence $\check{f}_\tau \neq 0$, so that the function Θ_0, which is the most rapidly varying term in the expansion for temperature, is indeed a regularization of a singularity of type η_+.

Now, assuming that (5.164) is fulfilled and Ψ is given, we uniquely determine $\check{\Theta}_1$ from (5.163) and the normalization condition (5.146). It is not hard to calculate that the limit of $\check{\Theta}_1$ as $\varepsilon \to 0$ is a singularity of Heaviside function form.

Thus the analysis of the terms of order ε^{-2} resulted in formulas (5.156), (5.157), (5.162), which determine $\check{r}_0, \check{W}_0$, and $\check{\pi}_0$, as well as in the relations

$$w^+ = \frac{\check{\mathcal{T}}_0}{a^2}, \quad \check{r}_0^+ = \check{\mathcal{R}}_0\frac{1-w^+}{w^+}, \quad \check{\pi}_0^+ = \check{\mathcal{T}}_0\check{r}_0^+, \tag{5.165}$$

and (5.164), which are an analog of the Rankine–Hugoniot conditions.

Let us introduce notation similar to (2.144):

$$\chi = \frac{1-w(\eta,x)}{1-w^+}, \quad \chi_1 = \frac{w^+}{w(\eta,x)}\chi(\eta,x). \tag{5.166}$$

Then we obtain formulas for $\check{W}_0$ and $\check{r}_0$:

$$\check{W}_0(\eta,x) = \check{W}_0^+(x)\chi(\eta,x), \quad \check{r}_0(\eta,x) = \check{r}_0^+(x)\chi_1(\eta,x), \tag{5.167}$$

where

$$\check{W}_0^+ = a(1-w^+)\nu.$$

We put $\Psi = \chi(\eta,x)$ and seek the extension of w_0, r_0 and Θ_0 in forms similar to (2.145), i.e.,

$$\begin{aligned} W_0(\eta,\tau,x,t) &= W_0^+(\tau,x,t)\chi(\eta,x), \\ r_0(\eta,\tau,x,t) &= r_0^+(\tau,x,t)\chi_1(\eta,x), \\ \Theta_0(\eta,\tau,x,t) &= f(\tau,x,t)\chi(\eta,x). \end{aligned} \tag{5.168}$$

The equation of state and (5.168) imply a formula for the pressure π_0:

$$\begin{aligned} \pi_0(\eta,\tau,x,t) = {}& \mathcal{T}_0(\tau,x,t)r_0(\eta,\tau,x,t) \\ & + (\mathcal{R}_0(\tau,x,t) + r_0(\eta,\tau,x,t))f(\tau,x,t)\chi(\eta,x). \end{aligned} \tag{5.169}$$

Consider the terms of order ε^{-1} which result from substituting the expansion (5.143) into (5.140):

$$\begin{aligned}
&\frac{\partial}{\partial\eta}A(\mathcal{R}^{(1)},\mathcal{U}^{(1)},\eta,\tau,x,t)+\frac{\partial}{\partial\tau}\{(1-\langle\nabla\varphi,\mathcal{U}^{(0)}\rangle)\mathcal{R}^{(0)}\}=\mathcal{F}_\rho^{(1)},\\
&\frac{\partial}{\partial\eta}B(\mathcal{R}^{(1)},\mathcal{U}^{(1)},\mathcal{P}^{(1)},\eta,\tau,x,t)\\
&\quad+\frac{\partial}{\partial\tau}\{(1-\langle\nabla\varphi,\mathcal{U}^{(0)}\rangle)\mathcal{R}^{(0)}\mathcal{U}^{(0)}-\nabla\varphi\mathcal{P}^{(0)}\}=\mathcal{F}_u^{(1)},\\
&\frac{\partial}{\partial\eta}C(\mathcal{R}^{(1)},\mathcal{U}^{(1)},\mathcal{P}^{(1)},\Theta_2,\eta,\tau,x,t)\\
&\quad+\frac{\partial}{\partial\tau}\Big\{\frac{1}{2}(1-\langle\nabla\varphi,\mathcal{U}^{(0)}\rangle)\mathcal{R}^{(0)}{\mathcal{U}^{(0)}}^2\\
&\quad+(1-\gamma\langle\nabla\varphi,\mathcal{U}^{(0)}\rangle)\frac{\mathcal{P}^{(0)}}{\gamma-1}-\kappa_0|\nabla\varphi|^2\frac{\partial}{\partial\tau}(\theta_0+\Theta_0)\Big\}=\mathcal{F}_T^{(1)}.
\end{aligned}\tag{5.170}$$

Here A, B, C are linearizations of the corresponding expressions in (5.153)–(5.155); in particular,

$$A(\mathcal{R}^{(1)},\mathcal{U}^{(1)},\eta,\tau,x,t)=(1-\langle\nabla\varphi,\mathcal{U}^{(0)}\rangle)\mathcal{R}^{(1)}-\langle\nabla\varphi,\mathcal{U}^{(1)}\rangle\mathcal{R}^{(0)},$$

and $\mathcal{F}_\rho^{(1)},\mathcal{F}_u^{(1)},\mathcal{F}_T^{(1)}$ are functions of $\mathcal{U}^{(0)},\dots,\mathcal{T}^{(0)}$, exponentially decreasing as $\eta\to\pm\infty$. We do not state the explicit expressions of B, C, and the right-hand sides $\mathcal{F}^{(1)}$ because they are very cumbersome.

It is not hard to see that for solvability of equation (5.170) in the class of functions we are interested in, it is necessary that the following relations hold as $\eta\to\pm\infty$:

$$\begin{aligned}
&\frac{\partial}{\partial\tau}\{(1-\langle\nabla\varphi,\mathcal{U}^{(0)}\rangle)\mathcal{R}^{(0)}\}\to 0,\\
&\frac{\partial}{\partial\tau}\{(1-\langle\nabla\varphi,\mathcal{U}^{(0)}\rangle)\mathcal{R}^{(0)}\mathcal{U}^{(0)}-\nabla\varphi\mathcal{P}^{(0)}\}\to 0,\\
&\frac{\partial}{\partial\tau}\Big\{\frac{1}{2}(1-\langle\nabla\varphi,\mathcal{U}^{(0)}\rangle)\mathcal{R}^{(0)}{\mathcal{U}^{(0)}}^2+(1-\gamma\langle\nabla\varphi,\mathcal{U}^{(0)}\rangle)\frac{\mathcal{P}^{(0)}}{\gamma-1}\\
&\qquad-\kappa_0|\nabla\varphi|^2\frac{\partial}{\partial\tau}(\theta_0+\Theta_0)\Big\}\to 0.
\end{aligned}\tag{5.171}$$

Consider equations (5.171) as $\eta\to-\infty$. We take into account the relationship between τ and η, and write down the Taylor expansion with respect to t about the point $t=\varphi(x)$. Neglecting the terms of order $O(\varepsilon)$ (i.e., incorporating them into higher order terms), we obtain equations valid for $\tau\le\varphi_1$ and $t=\varphi(x)$:

$$\frac{\partial}{\partial\tau}\{(1-\langle\nabla\varphi,\check{\mathcal{U}}_0\rangle)\check{\mathcal{R}}_0\}=0,\tag{5.172}$$

$$\frac{\partial}{\partial\tau}\{(1-\langle\nabla\varphi,\check{\mathcal{U}}_0\rangle)\check{\mathcal{R}}_0\check{\mathcal{U}}_0-\nabla\varphi\check{\mathcal{P}}_0\}=0,\tag{5.173}$$

$$\begin{aligned}
&\frac{\partial}{\partial\tau}\Big\{\frac{1}{2}(1-\langle\nabla\varphi,\check{\mathcal{U}}_0\rangle)\check{\mathcal{R}}_0\check{\mathcal{U}}_0^2+(1-\gamma\langle\nabla\varphi,\check{\mathcal{U}}_0\rangle)\frac{\check{\mathcal{P}}_0}{\gamma-1}\\
&\qquad-\kappa_0|\nabla\varphi|^2\frac{\partial\check{\theta}_0}{\partial\tau}\Big\}=0.
\end{aligned}\tag{5.174}$$

Here we again use the sign $\check{}$ to denote functions considered on the front. However, unlike (5.153)–(5.155), in equations (5.172)–(5.174) only t is fixed, whereas τ varies from $-\infty$ to φ_1.

Denote by

$$\alpha = V_\nu - \langle \nu, \check{u}_0 \rangle \tag{5.175}$$

the relative velocity of the front Γ_t, and invoke the normalization condition (5.150) for $\tau \to -\infty$. Then (5.173) immediately implies that the vector $\check{U}_0$ is directed along the normal ν. Put

$$\check{U}_0 = \alpha(1 - \omega(\tau, x))\nu. \tag{5.176}$$

Then, integrating (5.172) and (5.173), we get

$$\check{R}_0 = \check{\rho}_0 \frac{1-\omega}{\omega}, \quad \check{P}_0 = \alpha^2 \check{\rho}_0 (1 - \omega). \tag{5.177}$$

Now we integrate (5.174) once to obtain the equation

$$(1 - \omega)(\omega - \omega^+) = \frac{\delta_1}{2\alpha^2} \frac{\partial \check{\theta}_0}{\partial \tau}, \tag{5.178}$$

where

$$\omega^+ = \frac{2 + (\gamma - 1)M^2}{(\gamma + 1)M^2}, \quad \delta_1 = 4\frac{\gamma - 1}{\gamma + 1} \frac{\kappa_0 |\nabla \varphi|}{\alpha \check{\rho}_0}, \quad M^2 = \frac{\alpha^2}{\gamma \check{T}_0}. \tag{5.179}$$

Here M (unlike m) is the Mach number ahead of the wave front. Recall that the intermediate Mach number m as in (5.161) is computed for the "interior" wave whose fast variation is concentrated within a distance $\sim \varepsilon^2$ and which moves over the background $\mathcal{U}_0 = u_0(x, t) + U_0((x - \varphi)/\varepsilon, x, t)$. The quantity m has no physical meaning for the limiting solution as $\varepsilon \to 0$, whereas M as in (5.179) coincides with the Mach number for the limiting shock wave.

Now we use the relation

$$\check{P}_0 = \check{R}_0 \check{T}_0 + \check{\theta}_0 (\check{\rho}_0 + \check{R}_0), \tag{5.180}$$

which follows from the equation of state. Let

$$\check{\theta}_0 = \check{T}_0 (\psi - 1). \tag{5.181}$$

Then the second equality in (5.177) and (5.180) imply the relationship

$$\frac{\psi}{\omega} = 1 + \gamma M^2 (1 - \omega) \tag{5.182}$$

between ω and the new unknown function ψ. Taking this equality into account, we can easily transform (5.178) into

$$(1 - \omega)(\omega - \omega^+) + \delta_1 (\omega - \omega^*) \frac{\partial \omega}{\partial \tau} = 0, \tag{5.183}$$

where

$$\omega^* = \frac{\gamma M^2 + 1}{2\gamma M^2}.$$

It is not hard to see that under the condition (5.139) we have $\omega^+ > \omega^*$, and hence this equation has a bounded solution for all $\tau \in \mathbb{R}^1$. We consider the opposite case where

$$\omega^* > \omega^+. \tag{5.184}$$

In this case integration of (5.183) yields

$$(1-\omega)(\omega - \omega^+)^{\sigma^-} = e^{\beta_1^-(\tau + \varphi_2^-)}, \tag{5.185}$$

where $\varphi_2^- = \varphi_2^-(x)$ is the constant of integration, and

$$\sigma^- = \frac{\omega^* - \omega^+}{1 - \omega^*}, \quad \beta_1^- = \frac{1}{\delta_1}\frac{1-\omega^+}{1-\omega^*}. \tag{5.186}$$

We will assume that

$$\gamma M^2 > 2, \tag{5.187}$$

since this inequality certainly follows from (5.184) for $M > 1$, for both monatomic ($\gamma = 5/3$) and diatomic ($\gamma = 1,4$) gases. Then $\sigma^- > 0$, $\beta_1^- > 0$, and we see that the function ω tends to 1 at an exponential rate as $\tau \to -\infty$. Recall that we consider the system (5.172)–(5.174) only for $\tau \le \varphi_1$. Thus formula (5.185) yields the required solution, which remains bounded on the entire semi-axis $\tau \le \varphi_1$.

Now we consider the equations (5.171) for $\eta \to +\infty$. Since W_0, r_0, Θ_0 stabilize as $\eta \to \infty$, all the "middle-scale" terms $U_0, \dots, P_0$ of the expansion (5.149) get additions $W_0^+, \dots, \pi_0^+$. Hence it is natural to introduce the notation

$$\widetilde{U}_0 = U_0 + W_0^+, \quad \widetilde{R}_0 = R_0 + r_0^+, \quad \widetilde{\theta}_0 = \theta_0 + f, \tag{5.188}$$

as well as to denote, taking the equation of state into account,

$$\widetilde{P}_0 = P_0 + \pi_0^+ = T_0\widetilde{R}_0 + (\rho_0 + \widetilde{R}_0)\widetilde{\theta}_0. \tag{5.189}$$

As before, we denote the limiting values of $\widetilde{U}_0, \dots \widetilde{P}_0$ as $\tau \to \infty$ by the superscript $+$. Next we observe that the sum $(u_0 + \widetilde{U}_0^+)|_{t=\varphi}$ differs only by $O(\varepsilon)$ from the total velocity $u_0 + U_0(S_2/\varepsilon, x, t) + W_0(S_1/\varepsilon^2, S_2/\varepsilon, x, t)$ computed at the points within distance $O(\varepsilon)$ from Γ_t. The same correspondence holds for density, pressure, and temperature as well. Hence we denote

$$\begin{aligned} \check{u}^+(x) &= (u_0 + \widetilde{U}_0^+)\big|_{t=\varphi}, \quad \check{\rho}^+(x) = (\rho_0 + \widetilde{R}_0^+)\big|_{t=\varphi}, \\ \check{T}^+(x) &= (T_0 + \widetilde{\theta}_0^+)\big|_{t=\varphi}, \quad \check{p}^+(x) = (p_0 + \widetilde{P}_0^+)\big|_{t=\varphi}. \end{aligned} \tag{5.190}$$

Moreover, let

$$\alpha_1 = V_\nu - \check{u}_\nu^+(x). \tag{5.191}$$

Take the limit of (5.171) as $\eta \to \infty$. Then we integrate the resulting system with respect to τ and choose the constants of integration by the first conditions (5.150). Using the notation we introduced and the Taylor expansion with respect to t, we obtain the system

$$\{\alpha_1 - (\widetilde{U}_0 - \widetilde{U}_0^+)\}\{\breve{\rho}^+ + (\widetilde{R}_0 - \widetilde{R}_0^+)\}\big|_{t=\varphi} = \alpha_1 \rho^+, \tag{5.192}$$

$$\alpha_1 \rho^+ (\widetilde{U}_0 - \widetilde{U}_0^+)\big|_{t=\varphi} = \nu(\widetilde{P}_0 - \widetilde{P}_0^+)\big|_{t=\varphi}, \tag{5.193}$$

$$\begin{aligned}
&\frac{1}{2}\alpha_1\rho^+\{2\breve{u}_\nu^+ + (\widetilde{U}_0 - \widetilde{U}_0^+)_\nu\}(\widetilde{U}_0 - \widetilde{U}_0^+)_\nu\big|_{t=\varphi} \\
&\quad - \frac{\gamma\breve{p}^+}{\gamma - 1}(\widetilde{U}_0 - \widetilde{U}_0^+)\Big|_{t=\varphi} \\
&\quad + \{V_\nu - \gamma(\breve{u}_\nu^+ + (\widetilde{U}_0 - \widetilde{U}_0^+)_\nu)\}\frac{\widetilde{P}_0 - \widetilde{P}_0^+}{\gamma - 1}\bigg|_{t=\varphi} \\
&= \kappa_0 |\nabla\varphi| \frac{\partial\widetilde{\theta}_0}{\partial\tau}\bigg|_{t=\varphi}.
\end{aligned} \tag{5.194}$$

It follows from (5.193) that the vector $(\widetilde{U}_0 - \widetilde{U}_0^+)|_{t=\varphi}$ is parallel to ν. Obviously, we can simplify the system (5.192)–(5.194) by introducing the new unknown functions

$$\begin{aligned}
(\widetilde{U}_0 - \widetilde{U}_0^+)\big|_{t=\varphi} &= \alpha_1(1 - v(\tau, x))\nu, \\
(\widetilde{R}_0 - \widetilde{R}_0^+)\big|_{t=\varphi} &= Z(\tau, x), \\
(\widetilde{\theta}_0 - \widetilde{\theta}_0^+)\big|_{t=\varphi} &= \breve{T}^+(X(\tau, x) - 1), \\
(\widetilde{P}_0 - \widetilde{P}_0^+)\big|_{t=\varphi} &= Y(\tau, x).
\end{aligned} \tag{5.195}$$

Using this notation we obtain from the equation of state that

$$Y = \breve{T}^+(X(\breve{\rho}^+ + Z) - \breve{\rho}^+), \tag{5.196}$$

and derive from (5.192) and (5.193) the equalities

$$Z = \breve{\rho}^+ \frac{1 - v}{v}, \quad Y = \alpha_1^2 \breve{\rho}^+ (1 - v). \tag{5.197}$$

Now we can rewrite (5.194) as

$$(1 - v)(v - v^+) = \frac{\delta_2}{2\alpha_1^2}\frac{\partial\widetilde{\theta}_0}{\partial\tau}, \tag{5.198}$$

where

$$v^+ = \frac{2 - (\gamma - 1)M_1^2}{(\gamma + 1)M_1^2}, \quad \delta_2 = 4\frac{\gamma - 1}{\gamma + 1}\frac{\kappa_0|\nabla\varphi|}{\alpha_1\breve{\rho}^+}, \quad M_1^2 = \frac{\alpha_1^2}{\gamma\breve{T}^+}. \tag{5.199}$$

Obviously, M_1 is the Mach number computed behind the wave front.

It remains to express X as a function of v using (5.195) and (5.196):

$$\frac{X}{v} = 1 + \gamma M_1^2 (1 - v), \tag{5.200}$$

to see that the equations (5.178) and (5.198) differ only in notation. Hence the solution of (5.198) has a form similar to (5.185), but with a negative coefficient β_1^+ in the exponent,

$$\beta_1^+ = -\frac{1}{\delta_2}\frac{v^+ - 1}{1 - v^*} < 0, \tag{5.201}$$

where

$$v^* = \frac{1 + \gamma M_1^2}{2\gamma M_1^2}.$$

The inequality (5.201) is derived using the fact that M_1 is the Mach number computed behind the wave front. Therefore, as is well known (see, e.g., [**251**]), M_1 satisfies the inequalities $1 > M_1 > (\gamma - 1)/2\gamma$. Hence we easily derive that $v^+ > 1$ and $v^* < 1$.

Thus we obtain that the function v, being bounded on the half-line $\tau \geq \varphi_1$, tends to 1 as $\tau \to \infty$ exponentially fast.

Now we turn from the model functions $\omega(\tau + \varphi_2^-, x)$ and $v(\tau + \varphi_2^+, x)$ to the leading term of the asymptotic solution. For $\tau \leq \varphi_1$ we have the functions $\check{U}_0, \check{R}_0, \check{P}_0$, and $\check{\theta}_0$ defined by (5.176), (5.177), (5.180)–(5.182), while for $\tau \geq \varphi_1$ we have the functions $\widetilde{U}_0|_{t=\varphi}$, $\widetilde{R}_0|_{t=\varphi}$, $\widetilde{P}_0|_{t=\varphi}$, and $\widetilde{\theta}_0|_{t=\varphi}$ defined by (5.195)–(5.197), (5.200). It is clear that these functions remain bounded also for all finite τ greater than φ_1 and smaller than φ_1 respectively. Now we have to employ condition (5.165), which is an analog of the Hugoniot conditions, and condition (5.164). As is easily seen, conditions (5.165) specify the magnitude of jumps of functions $U_0 + W_0|_{t=\varphi}, \dots, P_0 + \pi_0|_{t=\varphi}$ when passing the point $\eta = 0$. These conditions imply the equality

$$\left(\widetilde{U}_0\big|_{t=\varphi} - \check{U}_0\right)\Big|_{\tau=\varphi_1} = W_0^+, \tag{5.202}$$

which may be regarded as a continuity condition at the point $\tau = \varphi_1$ of the globally defined function U_0 involved in the asymptotic expansion (5.143)–(5.149). Note that the functions $\check{U}_{0\nu}$ and $\check{R}_0$, $\check{P}_0$, $\check{\theta}_0$ are related by algebraic identities. Similar identities hold for $\widetilde{U}_{0\nu}, \dots \widetilde{\theta}_0$, as well as for $\check{W}_0^+, \dots, \check{\pi}_0^+$. These arguments and the way these functions are defined imply that, provided the equality (5.202) holds, similar equalities will hold also for all other functions.

The equality (5.164) means a condition on the magnitude of the jump of the derivative of the temperature that occurs when passing the point $\eta = 0$. Obviously, this condition can be written as

$$\left(\frac{\partial \widetilde{\theta}_0}{\partial \tau}\bigg|_{t=\varphi} - \frac{\partial \check{\theta}_0}{\partial \tau}\right)\bigg|_{\tau=\varphi_1} = \frac{\partial f}{\partial \tau}\bigg|_{\tau=\varphi_1,\, t=\varphi}. \tag{5.203}$$

We will analyze conditions (5.202), (5.203) later on, but now we match the functions $\check{U}_0, \dots$ and $\widetilde{U}_0, \dots$. For that, let $\check{U}_0^c, \dots, \widetilde{U}_0^c, \dots$ be smooth extensions of these functions to the entire τ-axis, where, say, $\check{U}_0^c$ is defined so that $\check{U}_0^c = \check{U}_0$ for $\tau \leq \varphi_1$, $\check{U}_0^c$ is given by (5.176) for $\varphi_1 < \tau \leq \varphi_1 + \varepsilon^{3/2} < \infty$, and $\check{U}_0^c$ is an arbitrary bounded smooth function for $\tau \geq \varphi_1 + \varepsilon^{3/2}$. We define $\check{R}_0^c, \dots$ for $\tau \geq \varphi_1$ and $\widetilde{U}_0^c, \dots$ for $\tau \leq \varphi_1$ in a similar way. Now we define the function $\check{\Phi}_u$ as

$$\check{\Phi}_u(\eta, \tau, x) = \check{U}_0^c(\tau, x) + (\widetilde{U}_0^c(\tau, x) - \check{U}_0^c(\tau, x))\chi(\eta, x), \tag{5.204}$$

where χ is given by (5.166). It is easily seen that

$$\begin{array}{llllll}\check{\Phi}_u \to \check{U}_0^c(\tau,x) & \text{as} \quad \eta \to -\infty & \text{and} & \check{\Phi}_u \to 0 & \text{as} & \eta \to -\infty,\ \tau \to -\infty,\\ \check{\Phi}_u \to \widetilde{U}_0^c(\tau,x) & \text{as} \quad \eta \to \infty & \text{and} & \check{\Phi}_u \to \widetilde{U}_0^+ & \text{as} & \eta \to \infty,\ \tau \to \infty.\end{array}$$

The functions $\check{\Phi}_\rho(\eta,\tau,x), \check{\Phi}_T(\eta,\tau,x)$, and $\check{\Phi}_p(\eta,\tau,x)$ are defined in a similar way. We normalize these functions by letting

$$\begin{aligned}\widetilde{\chi}(\eta,\tau,x) &= \frac{\check{\Phi}_u(\eta,\tau,x)}{\widetilde{U}_0^+(x)}, \quad \widetilde{\chi}_1(\eta,\tau,x) = \frac{\check{\Phi}_\rho(\eta,\tau,x)}{\widetilde{R}_0^+(x)},\\ \widetilde{\chi}_2(\eta,\tau,x) &= \frac{\check{\Phi}_T(\eta,\tau,x)}{\widetilde{\theta}_0^+(x)}, \quad \widetilde{\chi}_3(\eta,\tau,x) = \frac{\check{\Phi}_p(\eta,\tau,x)}{\widetilde{P}_0^+(x)}.\end{aligned} \tag{5.205}$$

Since $\widetilde{\chi},\ldots,\widetilde{\chi}_3$ tend to zero as $\tau,\eta \to -\infty$ and to one as $\tau,\eta \to \infty$, we see that the functions

$$\check{\Phi}_u\left(\frac{S_1}{\varepsilon^2},\frac{S_2}{\varepsilon},x\right) = \widetilde{U}_0^+(x)\widetilde{\chi}\left(\frac{S_1}{\varepsilon^2},\frac{S_2}{\varepsilon},x\right),\ldots$$

are regularized Heaviside functions with amplitudes $\widetilde{U}_0^+(x),\ldots$. Define the extensions of $\check{\Phi}_u,\ldots,\check{\Phi}_p$ as follows:

$$\Phi_u(\eta,\tau,x,t) = \widetilde{u}(x,t)\widetilde{\chi}(\eta,\tau,x),\ldots,\Phi_p(\eta,\tau,x,t) = \widetilde{p}(x,t)\widetilde{\chi}_3(\eta,\tau,x), \tag{5.206}$$

where $\widetilde{u},\ldots,\widetilde{p}$ are smooth functions such that

$$\widetilde{u}\big|_{t=\varphi} = \widetilde{U}_0^+(x),\ldots,\widetilde{p}\big|_{t=\varphi} = \widetilde{P}_0^+(x). \tag{5.207}$$

Now, putting

$$u = u_0(x,t) + \Phi_u(\eta,\tau,x,t),\ldots,p = p_0(x,t) + \Phi_p(\eta,\tau,x,t), \tag{5.208}$$

we obtain the leading term of the asymptotic expansion (5.143)–(5.149). In particular, we see that the components U_0, W_0, Θ_0 of the leading term of the asymptotics are given by

$$\begin{aligned}U_0(\tau,x,t) &= \frac{\widetilde{u}(x,t)}{\widetilde{U}_0^+(x)}U_0^c(\tau,x),\\ W_0(\eta,\tau,x,t) &= \frac{\widetilde{u}(x,t)}{\widetilde{U}_0^+(x)}(\widetilde{U}_0^c(\tau,x) - U_0^c(\tau,x))\chi(\eta,x),\\ \Theta_0(\eta,\tau,x,t) &= \frac{\widetilde{T}(x,t)}{\widetilde{\theta}_0^+(x)}(\widetilde{\theta}_0^c(\tau,x) - \theta_0^c(\tau,x))\chi_2(\eta,x)\end{aligned}$$

and satisfy all our conditions. Let us point out once again the distinction between the functions W_0, r_0, π_0 on the one hand and Θ_0 on the other hand. According to conditions (5.165) the functions χ involved in W_0, r_0, π_0 have factors that do not vanish at the point $\tau = \varphi_1$. Hence these functions get a jump as $\varepsilon \to 0$. The coefficient of χ_2 equals zero for $\tau = \varphi_1$, but its derivative does not vanish by (5.203). Hence Θ_0 is a regularization of a weak discontinuity.

Now we define the functions $u_0, \widetilde{u}, \dots, T_0, \widetilde{T}$. To this end, we consider the terms of order ε^0 which result when substituting (5.143)–(5.149) into (5.140). Passing to the limit as $\tau, \eta \to \infty$ and $\tau, \eta \to -\infty$ in these terms, we obtain the equations (2.161) of ideal gas dynamics. The equations treated in the domain ahead of the front involve the functions $u_0, \dots, T_0$, and those behind the front involve combinations of the functions $u_0 + \widetilde{u}, \dots, T_0 + \widetilde{T}$. As in §2.5, denote these functions by the superscripts $\pm$, i.e., $u_0 = u^-$, $u_0 + \widetilde{u} = u^+$, and so on. Then the leading term of the asymptotics (5.208) can be rewritten in terms of $u^\pm, \dots$. For example,

$$u = u_c^-(x,t) + (u_c^+(x,t) - u_c^-(x,t))\widetilde{\chi}(\eta, \tau, x),$$

where the subscript c denotes the standard extension of functions $u^\pm$ in the domains ahead of and behind the front respectively. Recall that outside an ε-neighborhood of the front we can take for this extension an arbitrary smooth function.

Since the equations (2.161) are completed by the Rankine–Hugoniot conditions (2.126), (2.127), to complete the construction of the leading term it remains to consider conditions (5.202) and (5.203).

Taking into account formulas (5.156), (5.165) and (5.188), (5.190), (5.195), we can rewrite (5.202) as

$$(\alpha\omega - \alpha_1 v)\big|_{\tau=\varphi_1} = a(1 - w^+)\big|_{\tau=\varphi_1}. \tag{5.209}$$

Next, it is not hard to see that

$$\alpha_1 = \alpha - [u_\nu], \quad a\big|_{\tau=\varphi_1} = \alpha\omega\big|_{\tau=\varphi_1}.$$

Hence (5.209) can be rewritten as

$$\alpha_1 v\big|_{\tau=\varphi_1} = \alpha\omega w^+\big|_{\tau=\varphi_1}.$$

By (5.165), (5.181), (5.182) this equality transforms itself into

$$\left(\frac{\alpha_1}{\alpha} v + \omega\right)\Big|_{\tau=\varphi_1} = c, \tag{5.210}$$

where $c = 1 + 1/\gamma M^2$.

Let us transform the condition (5.203). To evaluate the derivatives of $\widetilde{\theta}_0, \theta_0$, and f we use (5.165), (5.181), (5.198) and the fact that $\alpha_1 \breve{\rho}^+ = \alpha \breve{\rho}^-$ in view of the Rankine–Hugoniot conditions. Then (5.203) implies

$$\begin{aligned} &\left\{\left(\frac{\alpha_1}{\alpha}\right)^2 (1-v)(v-v^+) - (1-\omega)(\omega - \omega^+)\right\}\Big|_{\tau=\varphi_1} \\ &\qquad = -\frac{\gamma-1}{\gamma+1}\frac{\omega^3}{\breve{\rho}^-}\mathcal{R}_0\big(1-(w^+)^2\big)\big|_{\tau=\varphi_1}. \end{aligned} \tag{5.211}$$

Next we find by (5.165), (5.177), (5.181), (5.182) that

$$\omega\mathcal{R}_0 = \rho^-, \quad \omega w^+ = c - \omega.$$

Hence from (5.211) we obtain

$$\left\{\left(\frac{\alpha_1}{\alpha}\right)^2(1-v)(v-v^+)-(1-\omega)(\omega-\omega^+)\right\}\Big|_{\tau=\varphi_1} = -\frac{\gamma-1}{\gamma+1}c(2\omega-c)\big|_{\tau=\varphi_1}. \tag{5.212}$$

Finally, (5.179) and (5.199) imply

$$\omega^+=2\frac{\gamma c}{\gamma+1}-1,\quad v^+=2\frac{\gamma c_1}{\gamma+1}-1,\quad c_1=1+\frac{1}{\gamma M_1^2}.$$

Formally, the functions $\omega|_{\tau=\varphi_1}$ and $v|_{\tau=\varphi_1}$ contain three free parameters: φ_1, φ_2^-, and φ_2^+. However, conditions (5.210) and (5.212) actually involve only two parameters, $\varphi_1' = \varphi_1+\varphi_2^-$ and $\Delta\varphi = \varphi_2^+-\varphi_2^-$. Moreover, as follows from the equations for ω and v, the derivatives of these functions with respect to τ at the point $\tau=\varphi_1$ do not vanish. Hence conditions (5.210) and (5.212) make the problem well posed. The physical meaning of these conditions is quite transparent: they determine the position of the front of a "fast" smoothed shock wave (of scale ε^2) propagating over a background of scale ε.

Construction of higher order terms of the asymptotic expansion is carried out along the same lines as above. However, the actual realization of this scheme is very complicated, and we do not present it.

We conclude this section with a comment on the term "isothermic jump", which is used in the physical literature for the phenomenon we discussed. As we see, this term is not quite precise because the temperature is not constant in the scale ε^2. Actually it varies, though more slowly than velocity: in the scale ε^2 the velocity behaves like a regularization of a strong discontinuity, whereas the temperature behaves like a regularization of a weak discontinuity.

CHAPTER 6

Asymptotics of Stationary Solutions to the Navier–Stokes Equations Describing Stretched Vortices

This chapter, written jointly with A. I. Shafarevich, is devoted to stationary asymptotic solutions of the Navier–Stokes equations for large Reynolds numbers, which are concentrated in a small neighborhood of a curve $\Gamma \subset \mathbb{R}^3$. From the viewpoint of hydrodynamics, they describe the so-called localized vortices, or high-vorticity regions which are thin stretched formations arising in a flow of fluid. Such structures apparently play an essential role in various processes in hydrodynamics. In particular, they are used in some scenarios of evolving turbulence; see [**177**] and references therein. Near such a formation the field of velocities may substantially differ from the "outer" field of flow velocities. Then these jets, despite their small width, may contribute substantially to the redistribution of energy, and hence into the evolution of the entire volume of fluid as a whole. The role of stretched vortices is discussed, e.g., by Moffatt, Kida, and Ohkitani [**177**]; they call such formations "the sinews of turbulence" because they, like sinews of a human body, can accumulate energy and forces and translate them from one part of the flow to another.

From the viewpoint of mathematics, it is natural to treat the formations of stretched vortex type as soliton-type solutions concentrated in a neighborhood of a curve, i.e., related to the case $\operatorname{codim} \Gamma = 2$.

The first question in this context is whether the soliton-type waves in the Navier–Stokes model can appear in a natural way—in other words, whether they can evolve from small perturbations.

An answer to this question was obtained in [**54, 55**], where linearized Navier–Stokes equations were studied. Having constructed a parametrix (an expansion of the fundamental solution with respect to smoothness) and then performed additionally its expansion with respect to a small parameter of viscosity, Dobrokhotov and Shafarevich [**54, 55**], as well as Friedlander and Vishik [**71**], found a simple and efficient formula for the leading term of the resolvent for the Cauchy problem. This formula, due to its constructive nature, allowed for a detailed study of the solution to the Cauchy problem with soliton-type initial data (in all the three cases $\operatorname{codim} \Gamma = 1, 2$, and 3), as well as for its investigation for large times [**223**]. It was established that, unless the outer flow has a trivial geometry (is plane-parallel), the amplitude of soliton-type perturbations grows exponentially with time. Moreover, the "width" of perturbations for which this instability occurs was found to be inversely proportional to the square root of the Reynolds number.

These studies were continued in [**164, 165**], where the case of small nonlinearity was treated. It is an intermediate setting between the linear and essentially nonlinear ones, when the phase of the solution remains the same as in the linear case, but the amplitude is computed taking nonlinearity into account [**148, 153**]. The most important conclusion which we draw from the results of these papers is that soliton-type solutions to the Navier–Stokes equations do exist even for small nonlinearities, and that for a nontrivial outer flow their amplitude grows with time.

Hence there are good reasons to ask about the existence of soliton-type solutions of finite amplitude, because such solutions can indeed evolve from infinitesimal perturbations as a result of instability.

The essentially nonlinear rapidly varying solutions to the Navier–Stokes equations, as well as to related equations of gas dynamics, have been studied traditionally in the case $\operatorname{codim}\Gamma = 1$. As a classical example we mention the solution of boundary layer type concentrated near a surface with the fluid flowing over the surface. Another example is given by smoothed shock waves and tangential discontinuities, if we bear in mind that these are their first derivatives that have soliton form (in this sense the kinks of the sine-Gordon equation are often referred to as solitons). As was shown above (see §§2.5–2.7 of this book), these solutions correspond to single-phase asymptotics, whose construction is comparatively simple. A more interesting setting arises when dealing with the so-called coherent structures, i.e., single-phase asymptotic solutions which are periodic with respect to the fast variable $\tau = S(x,t)/\varepsilon$. The corresponding asymptotic procedure for the Navier–Stokes equations was developed in [**146, 147**] and extended to equations of a viscous gas and to magnetohydrodynamics in [**154–159**]. These studies revealed the following important property, which is characteristic of the asymptotic theory of hydrodymamics equations: the model nonlinear equation (the equation for the leading term of the asymptotics) possesses an infinite-dimensional family of exact solutions, so that the condition for solvability of the first approximation linear equations gives rise to infinitely many equations, or, equivalently, to equations involving a "redundant" independent variable. In the case of coherent structures these equations turn out to be an extension of the Prandtl equations in the boundary layer theory containing notrivial additional terms, while the "redundant" variable is simply the only "fast" variable τ (see [**146–147, 154-159**] and §2.7).

As for the solutions of stretched vortex type, their mathematical representation is given by a two-phase asymptotic expansion. Let us emphasize the qualitative distinction between such solutions and those considered in Chapter 5: here we deal with a soliton solution whose front is of codimension two, whereas the multi-phase asymptotics in Chapter 5, as well as the exact multi-soliton solutions of integrable equations, describe the interaction (of various forms) of solutions with $\operatorname{codim}\Gamma = 1$.

As we pointed out in the Introduction, the planar or radially symmetric localized vortices with rectilinear axis have been considered by a number of authors [**3, 177, 179**]. The assumption that the longitudinal component of velocity in the vortex is much larger than the transversal component actually reduces the problem to a single-phase one. However, as pointed out in [**177**], these are transversal circulations in the vortex that are important for evolving the turbulence. Such a stretched vortex in the radially-symmetric case was described for the simplest (linear) outer flow in [**177**].

The goal of this chapter is to construct the asymptotic solution of stretched vortex type for an arbitrary geometry of the outer flow and the front Γ, which are assumed to be smooth. Here we explain the problems that arise, and outline the main stages of the procedure.

We write the required solution in the form that is standard for soliton-type solutions:

$$\mathcal{U} = \mathcal{U}_0(x) + \mathcal{V}_0\Big(\frac{S(x)}{\varepsilon}, x\Big) + \varepsilon\Big(\mathcal{U}_1(x) + \mathcal{V}_1\Big(\frac{S(x)}{\varepsilon}, x\Big)\Big) + \dots,$$

where, for simplicity, $\mathcal{U}$ denotes the vector of velocity and pressure, $S = (S_1, S_2)$, and $\mathcal{U}_i(x)$, $\mathcal{V}_i(\tau, x)$ are smooth functions, $\tau = (\tau_1, \tau_2)$ with $\mathcal{V}_i \to 0$ as $|\tau| \to \infty$. It is natural to select the phases S_1, S_2 as coordinates in the normal plane to the curve Γ around which the solution is concentrated.

We put this expansion into the stationary Navier–Stokes equations and carry out calculations similar to those in Chapter 5, taking into account the fact that $\varepsilon \sim 1/\sqrt{\mathrm{Re}}$. Then for the outer flow (smooth background) $\mathcal{U}_0$ we obtain the three-dimensional Euler equations with respect to the slow variables x, while for $\mathcal{V}_0$ and $\mathcal{V}_1$ we obtain the standard equation with respect to the fast variables τ_1, τ_2. A natural change of variables (or, equivalently, the special choice of phases S_1, S_2 indicated above) enables us to rewrite the standard equations as two-dimensional Euler equations (for $\mathcal{V}_0$) and nonhomogeneous equations in variations (for $\mathcal{V}_1$).

This elucidates the central problem: we have to study the properties of the solution (more precisely, of an infinite-dimensional family of solutions) to the nonlinear standard partial differential equation, and also derive conditions for solvability of linearized equations for the correction.

It is clear that, due to the essential distinction between the cases of $\operatorname{codim} \Gamma = 2$ and $\operatorname{codim} \Gamma = 1$, we cannot exploit in this problem the fact that the soliton-type solution is "thin" in a single direction and circumvent the difficulty mentioned above in the same way as it was done in Chapter 5.

The first step towards resolving this problem was made in [**170**], where in the study of a related problem of constructing a multi-phase coherent structure five conditions for solvability of equations for the first correction were found. Nevertheless, due to infinite dimensionality of the cokernel, this problem remained open until appearance of the work of A. I. Shafarevich [**224–228**]. This breakthrough was made possible due to a novel, "topological" approach to the problem. The fact is that there are strong reasons to consider the topological invariants of a two-dimensional vector field of zero divergence as a kind of parametrization of the set of solutions to a stationary Euler equation (see [**11, 12, 175, 176, 232**] and §6.1). Hence we can associate with a solution of the standard equation a pair consisting of a Reeb graph and a Bernoulli function (integral) defined on this graph. This enables us to "select" a solution to the Euler equations by specifying fairly arbitrarily the topological structure of the two-dimensional vector field, i.e., by specifying the Reeb graph with an as yet arbitrary Bernoulli function on it. This construction imitates integration of the ordinary (with respect to the fast variable) standard equation, which arises in the single-phase case.

In its turn, such specification of a solution to the Euler equation allows for a description of the cokernel of the operator in variations [**224–228**]. Furthermore, utilizing the connection between solutions to the Euler equations and their topological invariants, the orthogonality conditions for the first correction can be rewritten as differential equations on the graph and the boundary conditions that complete them. As in the single-phase case, these equations are designed for final determination of the leading term of the asymptotic expansion. However, in the problem at hand, not only does the dependence of the solution on the slow variables remain indefinite in the first stage, but so does one of the topological characteristics of the solution, namely, the rapidly varying Bernoulli function. Thus, as an analog of the Hugoniot type condition, we obtain equations containing derivatives with respect to the slow variables as well as to a certain fast variable, namely, the so-called "action" variable (a certain combination of τ_1 and τ_2) parametrizing the graph.

Of course, the resulting equations are very complicated. Nevertheless, their structure has been analyzed to reveal that, surprisingly, these are again equations of Prandtl type. More precisely, they are an extension of the Prandtl equations, deeper than those in [**146, 147, 154**–**159**]. The equations obtained turned out to possess some interesting properties. In particular, two integrals of these equations were found and, moreover, an infinite series of integral relations of energy dissipation equations type was constructed. In the small viscosity limit all these relations turn into integrals, so that they form an *infinite series of conservation laws.*

It was established also that the curve Γ next to which the solution is concentrated is completely determined by the outer flow; namely, this curve must be one of its trajectories. This assertion can be reversed. Namely, we can take for Γ any smooth curve subject to the condition that it is a trajectory of a vector field satisfying three-dimensional stationary Euler equations.

It is important that, for an outer flow of simple geometry, the equations can be substantially simplified. In particular, in the radially symmetric inviscid case they reduce to a single ordinary equation, while in the radially symmetric case with a rectilinear outer flow we arrive precisely at the solution found in [**177**].

The scheme outlined above is set forth in detail in the following sections, which are based on the results of [**224**–**228**].

6.1. Asymptotic method of describing the stretched vortices

6.1.1. Equations of zero approximation: Two-dimensional Euler equations and related Reeb graphs. The stationary field of velocities $u(x)$ of an incompressible fluid (which is a vector field in $\mathbb{R}^3$) satisfies the Navier–Stokes equations

$$\begin{cases} (u, \nabla)u + \nabla \mathcal{P} = \nu \Delta u, \\ (\nabla, u) = 0. \end{cases} \tag{6.1}$$

Here u is the three-dimensional field of the fluid velocities, $\mathcal{P}$ is the pressure, and ν is the coefficient of viscosity ($\nu = \mathrm{Re}^{-1}$, where Re is the Reynolds number). Suppose a smooth outer stationary flow is given, i.e., a solution $\mathbf{V}(x,\nu), \mathbf{P}(x,\nu)$ to these equations, which is a smooth function of its arguments $x \in \mathbb{R}^3$, $\nu \in \mathbb{R}_+$ bounded together with all its derivatives. We will consider a perturbation of this flow localized in a small neighborhood of some curve Γ (unknown *a priori*) in the three-dimensional space. The characteristic width ε of this neighborhood is assumed to be small as compared to the characteristic scale of the outer field $\mathbf{V}$, so that it will play the role of a small parameter in our problem. Such a perturbation can be naturally viewed as a thin jet (or a stretched vortex) flowing in the fluid. We do not assume that the amplitude of the perturbation is small as compared to the field $\mathbf{V}$, so that we consider high-vorticity jets that can be very clearly distinguished from the outer flow. It is well known that such rapidly varying solutions may exist only if the viscosity is sufficiently small (see, e.g., §2.7 and [**146**], where this fact is established for rapidly oscillating "coherent" structures), more precisely, if

$\nu = O(\varepsilon^2)$; otherwise the viscosity destroys the jet. Hence in the sequel we let $\nu = \varepsilon^2\nu_0$, where ε is the small parameter and ν_0 is a constant of magnitude $O(1)$.

Now we construct the asymptotic solution. Let $x = \mathcal{R}(s)$ be the parametric equations of the curve Γ (we will assume for definiteness that the parameter s is the curve length); denote by $n_1(s), n_2(s)$ an orthonormal basis in the plane normal to Γ satisfying the condition $(dn_1(s)/ds, n_2(s)) = 0$. We take $n_1(0)$ and $n_2(0)$ to be the normal and the binormal to this curve at the point $s = 0$. Consider a neighborhood Ω' of Γ which does not depend on ε (but is small enough), and introduce the coordinates $\{s, y_1, y_2\}$ in this neighborhood by

$$x = \mathcal{R}(s) + y_1 n_1(s) + y_2 n_2(s).$$

We will seek the solution $u, \mathcal{P}$ to the three-dimensional Navier–Stokes equation defined in a smaller neighborhood $\Omega \subset \Omega'$ in the form

$$u = \mathbf{V}(x, \varepsilon^2) + \mathbf{U}\Big(\frac{y}{\varepsilon}, x, \varepsilon\Big), \quad \mathcal{P} = \mathbf{P}(x, \varepsilon^2) + \pi\Big(\frac{y}{\varepsilon}, x, \varepsilon\Big), \tag{6.2}$$

where $\mathbf{V}(x,\varepsilon^2)$, $\mathbf{P}(x,\varepsilon^2)$, $\mathbf{U}(\tau, x, \varepsilon)$, $\pi(\tau, x, \varepsilon)$ are smooth functions of all their arguments for $\varepsilon \in [0, \varepsilon_0)$, $\tau \in \mathbb{R}^2$, $x \in \Omega$ with $\mathbf{U}(\tau, x, \varepsilon), \pi(\tau, x, \varepsilon) \to 0$ as $|\tau| \to \infty$ and $\varepsilon \to 0$. Here and subsequently we denote by τ the two-dimensional vector of "stretched" coordinates in the normal plane to Γ, $\tau = y/\varepsilon$. We multiply the functions $\mathbf{U}, \pi$ outside Ω by a smooth truncating function which vanishes outside Ω'.

We expand the functions in (6.2) by the Taylor formula about the point $\varepsilon = 0$:

$$\begin{aligned} \mathbf{V}(x,\varepsilon) &= V(x) + \varepsilon^2 V_1(x) + \dots, & \mathbf{P}(x,\varepsilon) &= P(x) + \varepsilon^2 P_1(x) + \dots, \\ \mathbf{U}(\tau, x, \varepsilon) &= U(\tau, s) + \varepsilon U_1(\tau, x) + \dots, & \pi(\tau, x, \varepsilon) &= \pi_0(\tau, s) + \varepsilon\pi_1(\tau, x) + \dots. \end{aligned}$$

We can assume that the leading terms U, π_0, similarly to the leading terms in the soliton asymptotics, see, e.g., §1.1 of this book, do not involve the "slow" dependence on the variables y (otherwise, replacing y with $\varepsilon\tau$ and performing additional Taylor expansions, we carry it over to the higher order terms). Now we substitute (6.2) into the Navier–Stokes equations (6.1). Consider first the resulting equalities outside the neighborhood Ω of the curve Γ. Equating to zero the coefficient of ε^0, we obtain the three-dimensional Euler equations for the vertor field $V(x)$:

$$\begin{cases} (V, \nabla)V + \nabla P = 0, \\ (\nabla, V) = 0. \end{cases} \tag{6.3}$$

Henceforth we assume that $V(x)$ is a smooth solution to these equations.

We rewrite the Navier–Stokes equations inside Ω in coordinates (y, s):

$$\begin{cases} u_0 \dfrac{\partial}{\partial s} u_0 - k(s) u_0 (u_n, e(s)) + \big(1 - k(s)\big)(y, e(s)) \Big(u_n, \dfrac{\partial}{\partial y}\Big) u_0 \\ \qquad + \dfrac{\partial}{\partial s}\mathcal{P} = \varepsilon^2 \nu_0 \Big(\dfrac{d}{ds}\mathcal{R}, \Delta u\Big), \\ u_0 \dfrac{\partial}{\partial s} u_n - k(s) u_0^2 e(s) \\ \qquad + \big(1 - k(s)\big)(y, e(s)) \left(\Big(u_n, \dfrac{\partial}{\partial y}\Big) u_n + \dfrac{\partial}{\partial y}\mathcal{P}\right) = \varepsilon^2 \nu_0 \Delta u, \\ \dfrac{\partial}{\partial s} u_0 - k(s)(e(s), u_n) + \big(1 - k(s)\big)(e(s), y)\Big(\dfrac{\partial}{\partial y}, u_n\Big) = 0. \end{cases} \tag{6.4}$$

Here u_0 and u_n are the tangent and normal (to Γ) components of the vector u:

$$u = u_0 \frac{d\mathcal{R}}{ds} + \sum_{j=1}^{2} u_n^j n_j, \quad u_n = \begin{pmatrix} u_n^1 \\ u_n^2 \end{pmatrix},$$

$k(s)$ and $\varkappa(s)$ are the curvature and the torsion of the curve Γ, $e(s)$ is the result of turning the vector $(1;0)$ in the plane (n_1, n_2) at the angle $\int_0^s \varkappa(s_1)ds_1$, and n is the 2×3 matrix composed of the column-vectors n_1, n_2.

Now we substitute (6.2) into (6.4), set $y_j = \varepsilon\tau_j$, expand the resulting relations by the Taylor formula about the point $\varepsilon = 0$, and equate to zero the coefficients of the various powers of ε. As it was done when constructing soliton-type solutions, we can "freeze" the slow variables in terms of order ε^{-1} on the curve Γ. Equating the sum of these terms to zero, we obtain

$$\text{(6.5)} \qquad \begin{cases} (v, \nabla_\tau)v + \nabla_\tau \pi_0 = 0, \\ (\nabla_\tau, v) = 0, \end{cases}$$

$$\text{(6.6)} \qquad (v, \nabla_\tau)w = 0,$$

where $\nabla_\tau = (\partial/\partial\tau_1, \partial/\partial\tau_2)$, $v = (v_1, v_2)$, $v_j = (V|_\Gamma + U, n_j)$, $w = (U, d\mathcal{R}/ds)$, i.e., v is the projection of the principal part of the field of velocities on the normal plane to Γ and w is its projection on the tangent. Thus the two-dimensional vector field v in the plane of "fast" variables τ satisfies the two-dimensional Euler equations, while the function w is constant on the trajectories of this field. Henceforth we consider a solution v to the Euler equations in the plane satisfying the following conditions:

1) The vector field v has at most finitely many singular points, all of them are nondegenerate (since the field is divergence-free, a singular point in this case may be either a center or a saddle-point), and on each separatrix there is precisely one singular point.

2) Almost all trajectories of the field v are closed.

The second condition and the fact that $|U| \to 0$ as $|\tau| \to \infty$ imply that the vector field v decreases at infinity, and hence

$$(V|_\Gamma, n_j) = 0,$$

i.e., $V(x)$ is tangent to Γ. Thus we arrive at the following conclusion.

THEOREM 6.1.1. *Let the functions* (6.2) *satisfy* mod $O(1)$ *the Navier–Stokes equations* (6.1). *Then:*

a) *The vector field $v(\tau)$ satisfies the two-dimensional Euler equations in the plane.*

b) *The function w is constant on the trajectories of the field v.*

c) *If, in addition, conditions* 1) *and* 2) *are fulfilled, the curve Γ is a trajectory (a streamline) of the vector field $V(x)$.*

REMARK. We will consider the problem on the stretched vortex locally, without taking into account boundary conditions or conditions at infinity. For this problem one can consider the "problem on extension of the vortex" (similar to the problem on extension of the boundary layer, see, e.g., [**193**]) assuming that the field of the form (6.2) is defined on a surface $\mathcal{M}$ which crosses the curve Γ transversally. However, in contrast to the boundary layer problem, this "initial" field cannot be specified arbitrarily; rather, it must satisfy the equations (6.5), (6.6) in fast variables τ.

Henceforth we consider the vector fields v satisfying 1) and 2) and, moreover, decreasing at a rate of $o(|\tau|^{-3})$ as $|\tau| \to \infty$. By $o(\tau^{-n})$ as $\tau \to \infty$ we mean the validity of an upper bound $\operatorname{const} \tau^{-n-\delta}$ for some $\delta > 0$.

According to the general ideas of the asymptotic method employed here, construction of the solution describing a stretched vortex is based on a family of exact solutions to equations (6.5), (6.6) depending on some parameters (which may be functions of the "slow" variable s). Apparently, it is natural to parametrize such families by topological characteristics of the vector field v. This conjecture is based on the following reasons.

1. *Moffatt's theory of magnet relaxation.* Moffatt [**175, 176**] has developed a formal scheme for obtaining solutions to the Euler equations (6.5) with a preassigned topology of streamlines. These scheme is based on the properties of magnetohydrodynamics equations for an ideally conducting fluid, which are of the form

$$\begin{cases} \dfrac{\partial u}{\partial t} + (u, \nabla_\tau)u + \nabla_\tau P = \mu\Delta_\tau u - H \times \operatorname{curl}_\tau H, \\ \dfrac{\partial H}{\partial t} + (u, \nabla_\tau)H - (H, \nabla_\tau)u = 0, \\ (\nabla_\tau, u) = (\nabla_\tau, H) = 0, \end{cases}$$

where $H(\tau, t)$ is the magnetic field, $u(\tau, t)$ is the field of velocities of the conducting fluid, and μ is its coefficient of viscosity. By the vector product $r \times q$ of two-dimensional vectors r and q (in particular, $r = \nabla_\tau$) is meant throughout the product of the three-dimensional vectors $(r, 0)$ and $(q, 0)$. Note that in this case the vector $H \times \operatorname{curl}_\tau H \equiv H \times (\nabla_\tau \times H)$ turns out to lie in the τ-plane. The idea of Moffatt's scheme is as follows. Consider the Cauchy problem for the above system of equations:

$$u|_{t=0} = 0, \quad H|_{t=0} = H_0(\tau), \quad (\nabla_\tau, H_0) = 0.$$

If the viscosity μ is sufficiently large, it is natural to assume for physical reasons that the velocity of the fluid decays with time, i.e., $u(\tau, t) \to 0$ as $t \to \infty$. Then the second equation in the system of magnetohydrodynamics implies that the vector field H (the magnetic field) converges as $t \to \infty$ to a limiting field $H_\infty(\tau)$. On the other hand, we see from the first equation that H_∞ satisfies the equation

$$H_\infty \times \operatorname{curl}_\tau H_\infty = -\nabla_\tau P,$$

which is just another form of the stationary Euler equation (6.5) (in order to pass from one form to another, make the substitution $v = H_\infty$, $\pi_0 = -P - H_\infty^2/2$ in (6.5)). Thus, for large times the magnetic field H, evolving in accordance with the equations of magnetohydrodynamics, converges to a solution of the Euler equations. On the other hand, the second equation of magnetohydrodynamics means that the vector field $H(\tau, t)$ is "frozen" into the fluid, i.e., it is carried over by the flow of the field u; therefore (since this latter field is divergence-free) the field $H(\tau, t)$ is obtained from $H(\tau, 0)$ by an area-preserving transformation of the plane. Thus the topological characteristics of the divergence-free field H cannot vary with time, and so the limiting field H_∞ (satisfying the Euler equations) must have the same topological characteristics as the initial field H_0 (which is chosen arbitrarily). Of course, these are only formal arguments (none of the theorems on the behavior of solutions to magnetohydrodynamics equations for large times needed for their justification have yet been proved).

2. *Variational principle for the Euler equations* [**11**]. Denote by $\mathfrak{K}$ the space of divergence-free vector fields v which are obtained from one another (or from a fixed field v_0) by area-preserving transformations (diffeomorphisms) of the plane. Consider on this space the energy integral

$$\mathfrak{J} = \frac{1}{2}\int_{\mathbb{R}^2} v^2(\tau)\,d\tau.$$

In [**11**] the following assertion has been proved.

The equations (6.5) *are the equations of extremals of the energy funtional* $\mathfrak{J}$ *on the space* $\mathfrak{K}$.

The set of spaces $\mathfrak{K}$ of vector fields which are obtained from one another by area-preserving transformations of the plane can be parametrized by topological invariants of divergence-free vector fields (since any such invariant takes one and the same value for all fields in the same space $\mathfrak{K}$). Hence these "parameters" may be expected to retain in the solutions of the corresponding variational problems. In other words, if the functional $\mathfrak{J}$ attains its minimum on a space $\mathfrak{K}$ related to some values of invariants, we can naturally expect that a minimum will also exist on the spaces specified by close sets of these invariants.

Now we introduce by an explicit construction the topological characteristics of the vector field v which do not change under the area-preserving transformations of the plane. The first such invariant is the type of the phase portrait of the field v; in other words, the information on how many equilibrium states this field has and how they are placed with respect to one another. By knowledge of the mutual placement of the equilibrium states we mean that for each pair of stationary points we know whether or not they are separated by separatrices of the vector field v (i.e., whether they can be joined by a continuous curve on the plane without intersecting any separatrix). It is clear that the phase portrait remains unchanged under smooth transformations of the plane. The type of the phase portrait can be conveniently coded by a set whose points are the trajectories (streamlines) of the field v. This set is a graph (to be denoted by $\mathcal{G}$), i.e., a set of intervals, some of which have joint end-points (the end-points form the vertices of the graph, and the interior points of the intervals constitute its edges; the order of a vertex is the number of edges incident with that vertex). In our setting this graph can have vertices of only two types, viz., or order 1 and 3; the vertices of order 1 correspond to singular points of the field v of type center, and the vertices of order 3 to saddle points. The edges of the graph $\mathcal{G}$ are associated with rings in the plane (τ_1, τ_2) fibered into closed trajectories of the field v; two vertices are joined by an edge if and only if the corresponding equilibrium states are not separated by separatrices (see Figure 6.1). Define on each edge a natural parameter I, which is the action variable of the divergence-free field v in the corresponding ring. Namely, to each point of an edge of the graph (i.e., to each closed trajectory l of the field v) there corresponds the number

$$I = \frac{1}{2\pi}\oint_l \tau_1\,d\tau_2 = \frac{1}{2\pi}S(l),$$

where $S(l)$ is the area of the domain encircled by the curve l. In particular, we can assign to each edge the direction along which I grows. It is clear that at each vertex of order 1 the parameter I is zero, while at the vertices of order 3 the Kirchhoff equality

$$I_3 = I_1 + I_2$$

holds, where I_1, I_2 are the limits of I taken when approaching the vertex along the incoming edges (i.e., such that I grows when moving towards the vertex), and I_3 is the limit of I along the outgoing edge (such that I decreases when moving towards the vertex). The parametrization of the graph with parameter I remains unchanged under area-preserving transformations (since I is just the area). An example of the phase portrait of a vector field v and the corresponding parametrized graph $\mathcal{G}$ is presented in Figure 6.1.

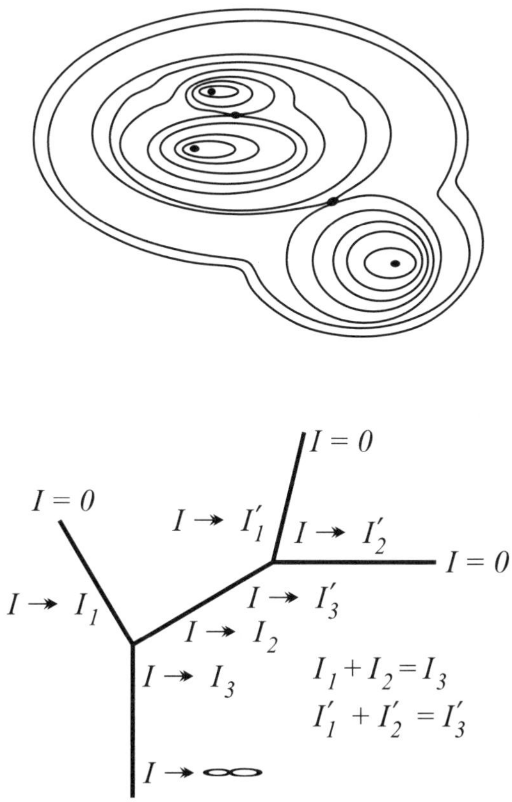

FIGURE 6.1. The graph $\mathcal{G}$ can be briefly defined as the quotient of the plane with respect to the trajectories of the field v.

REMARK. Each smooth divergence-free vector field on a plane is a Hamiltonian field, $v = \operatorname{sgrad}\psi$, where ψ is a smooth function (Hamiltonian), which in hydrodynamics is usually referred to as a *stream function*. We denote here by sgrad the skew gradient of a function,

$$\operatorname{sgrad}\psi = \left\{ -\frac{\partial \psi}{\partial \tau_2}, \frac{\partial \psi}{\partial \tau_1} \right\}.$$

The trajectories of the field v are the level lines of the stream function, while the singular points of v are its critical points. If the field satisfies condition 1), then ψ is the Morse function. The graph $\mathcal{G}$ (the set of level lines of the Morse function) is called its *Reeb graph.* This is a well-known object in Morse theory (and its generalizations), see, e.g., [**23**]. The parameter I is the action variable of the Hamilton system with Hamiltonian ψ; recall that in any ring fibered into closed trajectories there exist canonical action-angle coordinates I, φ.

Besides the graph just constructed, the field v has one more obvious invariant, which remains unchanged under transformations of the plane. This is the frequency (or period) of this field on each closed trajectory. It is clear that the frequency is a function of the parameter I, which indexes the trajectories. This function is smooth on each edge of the graph $\mathcal{G}$, which represents a domain in the plane free of singular points. Moreover, the frequency has the same limits along edges incident with the same vertex. In other words, the frequency is a continuous function on the graph $\mathcal{G}$.

Thus the topological characteristics of the divergence-free field v which remain unchanged under area-preserving diffeomorphisms comprise a parametrized graph $\mathcal{G}$ and a continuous function on this graph which is smooth on its edges. For the field v satisfying equations (6.5), instead of frequency, it will be convenient to use the Bernoulli function

$$B_0(\tau) = \frac{v^2}{2} + \pi_0,$$

which is well known in hydrodynamics. This function is constant on the trajectories of the field v (hence it is often called the Bernoulli integral), so that it is defined and continuous on the graph $\mathcal{G}$. If the solution v, π_0 to the Euler equations is smooth (which is the only kind of solutions we consider), then B_0 is a smooth function of I on each edge of $\mathcal{G}$. Thus with each smooth solution to the Euler equations (6.5) we associate a pair which consists of a parametrized graph $\mathcal{G}$ and a continuous function B_0 on its edges.

CONJECTURE. *There exists an open (in an appropriate sense) subset in the set of all pairs $\mathcal{G}, B_0$, where $\mathcal{G}$ is a parametrized graph with the properties stated above and B_0 is a continuous function on this graph, smooth on its edges, such that for each pair from this open subset there is a smooth solution v, π_0 to the Euler equations* (6.5) *for which the graph $\mathcal{G}$ is the quotient of the plane τ with respect to the trajectories of the field v, and the function B_0 is the Bernoulli integral.*

Now we consider the equation (6.6). This equation implies that the function w is constant on the trajectories of the field v; hence it is also defined and continuous on the graph $\mathcal{G}$ and is smooth on its edges.

Finally, we note that the equations (6.5), (6.6) are invariant under parallel translations and rotations of the plane (see, e.g., [**7**]). Therefore, along with a pair $v(\tau), w(\tau)$, this system is satisfied by all pairs of the form $T(\alpha)v(T^*(\alpha)\tau + c)$, $w(T^*(\alpha)\tau + c)$, where $T(\alpha)$ is the matrix of rotation at the angle α and $c \in \mathbb{R}^2$.

Thus we have parametrized the solutions of the system (6.5), (6.6) by two functions (B_0 and w) on the graph $\mathcal{G}$ and three additional real-valued parameters, $\alpha \in S^1, c \in \mathbb{R}^2$. In the next subsections we will obtain equations specifying the dependence of w, B_0, α on the "slow" variable s.

6.1.2. Solvability conditions for the first approximation equations. Now we write down the equations which are obtained by equating to zero the coefficient of ε^0 resulting upon substitution of the functions (6.2) into the Navier–Stokes equations (6.1). We have

$$-(v, \nabla_\tau) w_1 - (v^1, \nabla_\tau) w = H(\tau, s); \tag{6.7}$$

$$\begin{cases} -(v, \nabla_\tau) v^1 - (v^1, \nabla_\tau) v - \nabla_\tau \pi_1 = F(\tau, s), \\ -(\nabla_\tau, v^1) = G(\tau, s). \end{cases} \tag{6.8}$$

Here

$$\begin{aligned} H(\tau, s) &= (W + w)\frac{\partial}{\partial s}(W + w) - \frac{1}{2}\frac{\partial}{\partial s}W^2 - k(s)(e(s), v)(w + W) \\ &\quad + (b, v) + (\mathcal{A}\tau, \nabla_\tau)w + \frac{\partial \pi_0}{\partial s} - \nu_0 \Delta_\tau w, \\ F(\tau, s) &= (W + w)\frac{\partial}{\partial s}v + k(s)e(s)(W^2 + 2Ww) \\ &\quad + \mathcal{A}v + (\mathcal{A}\tau, \nabla_\tau)v - \nu_0 \Delta_\tau v, \\ G(\tau, s) &= \frac{\partial w}{\partial s} - k(s)(e(s), v), \end{aligned}$$

and we have used the notation

$$w_1 = \Big(U_1, \frac{d\mathcal{R}}{ds}\Big), \quad W(s) = \Big(V, \frac{d\mathcal{R}}{ds}\Big)\Big|_\Gamma, \quad \Delta_\tau = \sum_{j=1}^{2} \frac{\partial^2}{\partial \tau_j^2}.$$

Moreover, v^1, b, and $\mathcal{A}$ are the two-dimensional vectors and the 2×2 matrix with elements

$$v_j^1 = (U_1, n_j), \quad b_j(s) = \Big(\frac{\partial V}{\partial y_j}, \frac{d\mathcal{R}}{ds}\Big)\Big|_\Gamma, \quad \mathcal{A}_{ij}(s) = \Big(\frac{\partial V}{\partial y_j}, n_i\Big)\Big|_\Gamma.$$

This system consists of the linearized equations (6.5), (6.6) with right-hand sides H, F, G. The dependence of B_0, w, and α on the coordinate s is determined from the conditions for solvability of this system. In accordance with the parametrization of the solutions to equations (6.5), (6.6) introduced in the previous subsection, these conditions split into two groups. The first group of conditions is "infinite-dimensional"; it corresponds to the variation of arbitrary functions B_0 and w. The second group is three-dimensional; it is related to symmetries of the Euler equations (note that the group of symmetries has been found in [**7**]) and corresponds to the variation of parameters α, c. Before we write down these conditions, we state a proposition on solutions of adjoint equations to the linearized Euler equations (i.e., on the structure of the cokernel of the linearized Euler operator). These equations for a two-dimensional vector-function ξ and a scalar function χ have the form

$$\begin{cases} (v, \nabla_\tau)\xi - \frac{\partial v^*}{\partial \tau}\xi = \nabla_\tau \chi, \\ (\nabla_\tau, \xi) = 0. \end{cases} \tag{6.9}$$

Here $\partial v^*/\partial \tau$ is the adjoint matrix to the matrix $\partial v/\partial \tau$ of derivatives of the field v.

PROPOSITION 6.1.1. *Let $v(\tau)$ be a smooth solution to equations* (6.5) *with properties* 1), 2). *Then the following smooth vector fields ξ satisfy the equations* (6.9):

1. *Any divergence-free field which commutes with $v(\tau)$.*
2. *Any constant vector field on the plane.*
3. *The vector field $\xi = \operatorname{sgrad}(\tau_1^2 + \tau_2^2)/2 = J\tau$, where $J = \begin{pmatrix} 0 & -1 \\ 1 & 0 \end{pmatrix}$.*

REMARK. Obviously, divergence-free vector fields of the first type constitute an infinite-dimensional linear space which contains, in particular, all fields of the form $J\nabla_\tau f(\psi_0(\tau))$, where $f(t)$ is an arbitrary smooth function and $\psi_0(\tau)$ is the stream function (the Hamiltonian) of the field v, i.e., $v = J\nabla_\tau\psi_0(\tau)$. In the cokernel of the linearized Euler operator this space is generated by the variation of an arbitrary function $B_0(\tau)$ and may be interpreted as a space of functions on the graph $\mathcal{G}$. Indeed, if we introduce the action-angle variables I, φ in an arbitrary ring partitioned into smooth trajectories of the field v, then the field itself will have the form $\omega(I)\partial/\partial\varphi$, where $\omega(I)$ is the frequency on the trajectory corresponding to the parameter I (here we identify in the usual way the vector field with the differentiation operator along this field). On the other hand, any divergence-free field ξ which commutes with v has in the same coordinates the form $b(I)\partial/\partial\varphi$ (where $b(I)$ is an arbitrary smooth function). Thus the set of all divergence-free fields commuting with v can be identified with the set of functions b of the single variable I running over the graph $\mathcal{G}$. This is an argument in favor of the conjecture stated above on parametrization of the set of solutions to the Euler equation.

The other fields listed in Proposition 6.1.1 are generated by symmetries of the Euler equations, namelly, the presence of the two-dimensional space of constant fields in the cokernel is related to invariance of these equations with respect to parallel translations, and the presence of the field in assertion 3 is a consequence of rotational invariance.

PROOF. Consider an arbitrary smooth field ξ of zero divergence commuting with v, and substitute it into the left-hand side of equations (6.9). We are to prove that in this way we obtain a gradient field, i.e., that the operator curl nullifies it. Replace $(v, \nabla_\tau)\xi$ by $(\xi, \nabla_\tau)v$ and apply curl_τ to the left-hand side of (6.9). Then after obvious transformations we obtain the vector $A\, d\mathcal{R}/ds$ with coefficient

$$A \stackrel{\text{def}}{=} (\nabla_\tau, \xi(\text{curl}_\tau v)_3) = (\xi, \nabla_\tau)(\text{curl}_\tau v)_3 + (\text{curl}_\tau v)_3\langle\nabla_\tau, \xi\rangle,$$

where

$$(\text{curl}_\tau v)_3 = \frac{\partial v_2}{\partial \tau_1} - \frac{\partial v_1}{\partial \tau_2}.$$

The second term is equal to zero because ξ is divergence-free. The first term also equals zero. Indeed, as is well known, $(\text{curl}_\tau v)_3$ for two-dimensional fields v is constant on the trajectories of v. Hence in a neighborhood of any point which is not an equilibrium state of v and does not belong to a separatrix of this field the function $(\text{curl}_\tau v)_3$ depends only on the action variable I, while the field ξ commuting with v has the form $\sigma(I)\partial/\partial\varphi$. Therefore the function $(\xi, \nabla_\tau)(\text{curl}_\tau v)_3$ vanishes at any such point, and hence everywhere (because of smoothness). The assertions 2 and 3 can be directly verified by an elementary calculation: the function χ in the first case equals $-(Cv)$ (where C is a constant two-dimensional field), and in the second case (τ, Jv). $\square$

Now we state the solvability conditions for the system of equations (6.7), (6.8).

PROPOSITION 6.1.2. *Let the system* (6.7), (6.8) *have a sufficiently smooth solution* (v^1, w_1, π_1) *bounded together with all its derivatives. Then the right-hand* *sides $H(\tau, s), F(\tau, s), G(\tau, s)$ satisfy the equalities*

$$
\left\{
\begin{aligned}
&\frac{1}{2\pi}\oint_l (F, v)\, d\varphi + a_0(I, s)\frac{\partial B_0}{\partial I} = 0,\\
&\frac{1}{2\pi}\oint_l G\, d\varphi + \frac{\partial a_0}{\partial I} = 0,\\
&\frac{1}{2\pi}\oint_l H\, d\varphi + a_0(I, s)\frac{\partial w}{\partial I} = 0;
\end{aligned}
\right. \tag{6.10}
$$

$$\int_{\mathbb{R}^2} (F + Gv)\, d\tau = 0; \tag{6.11}$$

$$\int_{\mathbb{R}^2} (J\tau, F + Gv)\, d\tau = 0. \tag{6.12}$$

Here l is an arbitrary closed trajectory of the vector field $v(\tau)$, (I, φ) are the action-angle variables defined in a neighborhood of l, and $a_0(I, s)$ is a smooth function.

REMARK 1. These conditions follow from the orthogonality conditions of the right-hand sides in (6.7), (6.8) to the subspaces of the cokernel of the linearized Euler operator stated in Proposition 6.1.1. Namely, (6.10) are conditions for orthogonality to the infinite-dimensional space in assertion 1 of this proposition (which comprises, along with the obvious infinite-dimensional cokernel of the linear operator (6.6), the functions constant on the trajectories of the field v). The equalities (6.11) and (6.12) are conditions for orthogonality of the right-hand side of (6.8) to the three-dimensional space in assertions 2, 3 of Proposition 6.1.1.

REMARK 2. The function a_0 characterizes the correction v^1 to the leading term of the asymptotics (see the corresponding formula below); at the same time, the equalities (6.10) form three relations for three unknown functions B_0, w, a_0. Thus the system obtained becomes "closed".

PROOF. First we prove (6.10). For that we take the scalar product of the first (vector) equation in (6.8) by the vector v and then integrate the resulting equality, as well as the second equality in (6.8) and equation (6.7), along an arbitrary closed trajectory l of the field v. We obtain

$$\oint_l (F, v)\, d\varphi + \oint_l (v, (v, \nabla_\tau)v^1)\, d\varphi + \oint_l (v, (v^1, \nabla_\tau)v)\, d\varphi + \oint_l (v, \nabla_\tau)\pi_1\, d\varphi = 0.$$

Obviously,

$$\oint_l (v, \nabla_\tau)\pi_1\, d\varphi = \omega \oint_l \frac{\partial \pi_1}{\partial \varphi}\, d\varphi = 0$$

(here $\omega(I)$ is the frequency on the trajectory l). Next, the Euler equations (6.5) imply that

$$
\begin{aligned}
\big(v, (v, \nabla_\tau)v^1 + (v^1, \nabla_\tau)v\big) &= (v^1, \nabla_\tau)\frac{v^2}{2} + (v, \nabla_\tau)(v, v^1) - \big(v^1, (v, \nabla_\tau)v\big)\\
&= (v, \nabla_\tau)(v, v^1) + (v^1, \nabla_\tau)\frac{v^2}{2} + (v^1, \nabla_\tau)\pi_0 = \omega\frac{\partial}{\partial \varphi}(v, v^1) + (v^1, \nabla_\tau)B_0.
\end{aligned}
$$

Set $v^1(\tau) = a_I\partial/\partial I + a_\varphi \partial/\partial\varphi$. Since B_0 is constant on the trajectories of v, we obtain from the above identity

$$\oint_l \big(v, (v, \nabla_\tau)v^1 + (v^1, \nabla_\tau)v\big)\, d\varphi = \oint_l (v^1, \nabla_\tau)B_0\, d\varphi = \frac{\partial B_0}{\partial I}\oint_l a_I\, d\varphi,$$

which implies the first equality in (6.10) if we let

$$a_0(I, s) = \frac{1}{2\pi} \oint_l a_I \, d\varphi.$$

The second equality follows by integration of the last equation in (6.8) along l, taking the equality

$$(\nabla_\tau, v^1) = \frac{\partial a_I}{\partial I} + \frac{\partial a_\varphi}{\partial \varphi}$$

into account. Finally, integrating (6.7) and using the fact that $\partial w/\partial \varphi = 0$ (see (6.6)), we obtain the last equality in (6.10).

Now we prove (6.11). We multiply the last equation (6.8) by the vector v and add to the first. We obtain

$$\begin{aligned} -(F + Gv)_i &= (v, \nabla_\tau) v_i^1 + (v^1, \nabla_\tau) v_i + v_i (\nabla_\tau, v^1) + \frac{\partial \pi_1}{\partial \tau_i} \\ &= (\nabla_\tau, v_i^1 v + v^1 v_i) + \frac{\partial \pi_1}{\partial \tau_i}, \qquad i = 1, 2. \end{aligned}$$

Taking the decay of v at infinity into account, we integrate these equalities over the plane (τ_1, τ_2) to obtain (6.11).

For the proof of (6.12) we take the scalar product of the last relation by $J\tau$. Taking into account the identities

$$\begin{aligned} &(\nabla_\tau, J\tau) = 0, \quad (v, \nabla_\tau) J\tau = Jv, \quad (v^1 \nabla_\tau) J\tau = Jv^1, \quad (v, Jv^1) + (v^1, Jv) = 0, \\ &(\nabla_\tau, J\tau\pi_1) = (J\tau, \nabla_\tau)\pi_1 + \pi_1(\nabla_\tau, J\tau) = (J\tau, \nabla_\tau)\pi_1, \\ &(J\tau, (v, \nabla_\tau) v^1) = (\nabla_\tau, v(v^1, J\tau)) - (v^1, (v, \nabla_\tau) J\tau) = (\nabla_\tau, v(v^1, J\tau)) - (v^1, Jv), \end{aligned}$$

we obtain

$$\begin{aligned} -(F + Gv, J\tau) &= \big(J\tau, (v, \nabla_\tau) v^1 + (v^1, \nabla_\tau) v + v(\nabla_\tau, v^1) + \nabla_\tau \pi_1\big) \\ &= \big(\nabla_\tau, \pi_1 J\tau + (v^1, J\tau) v + (v, J\tau) v^1\big). \end{aligned}$$

On integrating over the plane τ we get (6.12). □

6.2. Equations of stretched vortex stated on the Reeb graph. The Kirchhoff conditions, integral identities, and conservation laws

6.2.1. Reduction of solvability conditions for the first approximation system to equations stated on the Reeb graph. The Kirchhoff conditions. Now we consider the system of equalities (6.10). The left-hand sides in this system are functions of two variables, I, s, with I running over the graph $\mathcal{G}$ (i.e., actually over a set of joined intervals). In this section we reduce (6.10) to a form that recalls the Prandtl equations of the boundary layer theory and equations in [**146, 147, 154–159**] describing "coherent" periodic pulsations. Namely, in this subsection we prove the following assertion.

THEOREM 6.2.1. *The equalities* (6.10) *are equivalent to the following system of equations:*

$$
\left\{
\begin{aligned}
&\theta\frac{\partial\theta}{\partial s} + a\frac{\partial\theta}{\partial I} + \langle\pi_s\rangle = \nu_0\frac{\partial}{\partial I}\mathcal{D}^2\frac{\partial}{\partial I}\theta,\\
&\frac{\partial\theta}{\partial s} + \frac{\partial a}{\partial I} = 0,\\
&\theta\frac{\partial B}{\partial s} + a\frac{\partial B}{\partial I} = \nu_0\Big(\frac{\partial}{\partial I}\mathcal{D}^2\frac{\partial}{\partial I}B - \Lambda^2\Big),
\end{aligned}
\right.
\tag{6.13}
$$

where

$$
\theta = W + w, \quad a = a_0 - I\frac{\partial W}{\partial s}, \quad B = B_0 + \frac{1}{2}\big((W+w)^2 - W^2\big),
$$

$$
\langle\pi_s\rangle = \frac{1}{2\pi}\oint_l \frac{\partial}{\partial s}\Big(\pi_0 - \frac{1}{2}W^2\Big)\,d\varphi, \quad \mathcal{D}^2 = \frac{1}{2\pi}\oint_l\Big(\frac{\partial\tau}{\partial\varphi}\Big)^2\,d\varphi,
$$

$$
\Lambda^2 = (\operatorname{curl}_\tau v)_3^2 + \mathcal{D}^2\Big(\frac{\partial w}{\partial I}\Big)^2 = \Big(\frac{\partial v_2}{\partial\tau_1} - \frac{\partial v_1}{\partial\tau_2}\Big)^2 + \mathcal{D}^2\Big(\frac{\partial w}{\partial I}\Big)^2,
$$

and the parameter I runs over the edges of the graph $\mathcal{G}$. The functions θ and B are continuous on $\mathcal{G}$, and if, in addition, the conditions of Proposition 6.1.2 *are fulfilled, then the function a vanishes at each vertex of order* 1*, while at each vertex of order* 3 *the Kirchhoff conditions of the theory of electric circuits hold:*

$$
\begin{aligned}
&a_1 + a_2 = a_3, \quad \Big(\mathcal{D}^2\frac{\partial B}{\partial I}\Big)_1 + \Big(\mathcal{D}^2\frac{\partial B}{\partial I}\Big)_2 = \Big(\mathcal{D}^2\frac{\partial B}{\partial I}\Big)_3,\\
&\Big(\mathcal{D}^2\frac{\partial\theta}{\partial I}\Big)_1 + \Big(\mathcal{D}^2\frac{\partial\theta}{\partial I}\Big)_2 = \Big(\mathcal{D}^2\frac{\partial\theta}{\partial I}\Big)_3,
\end{aligned}
\tag{6.14}
$$

where the subscripts 1, 2 *denote the limits of the coresponding functions at a given vertex over the incoming edges, while the subscript* 3 *means the limit over the outgoing edge* (*recall that the edges are directed according to growth of* I).

REMARK 1. The relations (6.13) are to be treated as equations for θ, B, a (or w, B_0, a_0), with the field v being determined by B_0. Note that the function $\mathcal{D}^2$, in general, also depends on B; in this sense it may be viewed as an analog of the coefficient of turbulent viscosity.

REMARK 2. From the viewpoint of the theory of differential equations on graphs (see, e.g., [**106, 207**]) the Kirchhoff conditions (6.14) are a particular case of the so-called α-smoothness conditions, which are specified at the vertices of the graph and relate the derivatives of the unknown function over the edges incident with the vertex. Note that these conditions enable us to exclude the function a from the system (6.13) by expressing it in terms of θ successively on each edge of the graph with the aid of the second equation of the system. In the same way the analogous function is excluded from the classical Prandtl equations and the equations in [**146, 147, 154, 159**].

REMARK 3. The equations (6.13) are local with respect to the variable s, which is the parameter varying along the curve Γ. If this curve is closed, conditions (6.14) are to be completed with a periodicity condition for the functions a, θ, B with respect to s. Thus in this case the equations (6.13) are defined on the direct product $\mathcal{G} \times S^1$ of the graph by a circle.

PROOF OF THEOREM 6.2.1. Consider first the second equality in (6.10):

$$\frac{1}{2\pi}\oint_l \Big(\frac{\partial w}{\partial s} - k(s)(e(s), v)\Big)\, d\varphi + \frac{\partial a_0}{\partial I} = 0.$$

Since the field v, and hence the related canonical transformation $(\tau_1, \tau_2) \to (I, \varphi)$, depend, in general, on s, the derivative $\partial w/\partial s$ has the form

$$\frac{\partial w}{\partial s}(\tau, s) = \frac{\partial}{\partial s} w(I, \varphi, s) + \frac{\partial w}{\partial I}\frac{\partial I}{\partial s} + \frac{\partial w}{\partial \varphi}\frac{\partial \varphi}{\partial s}.$$

By (6.6) $\partial w/\partial \varphi = 0$, and hence

$$\frac{1}{2\pi}\oint_l \frac{\partial w}{\partial s}\, d\varphi = \frac{\partial}{\partial s}\frac{1}{2\pi} w \oint_l d\varphi + \frac{\partial w}{\partial I}\frac{1}{2\pi}\oint_l \frac{\partial I}{\partial s}\, d\varphi.$$

The last integral is zero (this is the well-known theorem on adiabatic invariance of the action variable in one-dimensional Hamiltonian systems, see, e.g., [**10**]). Thus

$$\frac{1}{2\pi}\oint_l \frac{\partial w}{\partial s}\, d\varphi = \frac{\partial w(I, s)}{\partial s}.$$

Now consider the integral of $(e(s), v)$ over the trajectory l. We have

$$\oint_l v\, d\varphi = \oint_l \frac{d\tau}{dt}\, d\varphi = \omega \oint_l \frac{d}{dt}\tau\, dt = 0, \tag{6.15}$$

where t is the time on the trajectory l of the field v, and ω is the frequency on this trajectory. In particular,

$$\oint_l (e(s), v)\, d\varphi = 0.$$

Thus the second equation in (6.10) transforms itself into

$$\frac{\partial w}{\partial s} + \frac{\partial a_0}{\partial I} = 0$$

or

$$\frac{\partial \theta}{\partial s} + \frac{\partial a}{\partial I} = 0.$$

Consider the last equality in (6.10). Quite similarly to the above, we deduce from it the equation

$$\frac{1}{2}\frac{\partial}{\partial s}(w + W)^2 + \frac{1}{2\pi}\oint_l (\mathcal{A}\tau, \nabla_\tau) w\, d\varphi + \langle \pi_s \rangle + a_0 \frac{\partial w}{\partial I} = \nu_0 \frac{1}{2\pi}\oint_l \Delta_\tau w\, d\varphi. \tag{6.16}$$

Consider the integral $\oint (\mathcal{A}\tau, \nabla_\tau) w\, d\varphi$. Denote by ζ the vector field $(\mathcal{A}\tau, \nabla_\tau)$ and write it as $\zeta = \zeta_I \partial/\partial I + \zeta_\varphi \partial/\partial \varphi$. We obtain

$$\oint_l (\mathcal{A}\tau, \nabla_\tau) w\, d\varphi = \frac{\partial w}{\partial I}\oint_l \zeta_I\, d\varphi.$$

Since $\tau \to (I, \varphi)$ is the canonical transformation, we have $\partial I/\partial \tau = -J \partial \tau/\partial \varphi$, and hence

$$\zeta_I = \Big(\mathcal{A}\tau, \frac{\partial I}{\partial \tau}\Big) = -\Big(\mathcal{A}\tau, J\frac{\partial \tau}{\partial \varphi}\Big).$$

Taking into account the equalities

$$\oint_l \tau_i \frac{\partial \tau_i}{\partial \varphi}\, d\varphi = 0, \quad \frac{1}{2\pi}\oint_l \tau_1 \frac{\partial \tau_2}{\partial \varphi}\, d\varphi = -\frac{1}{2\pi}\oint_l \tau_2 \frac{\partial \tau_1}{\partial \varphi}\, d\varphi = \frac{1}{2\pi}\oint_l \tau_1\, d\tau_2 = I,$$

we obtain

$$\frac{1}{2\pi}\oint_l \zeta_I\, d\varphi = I \operatorname{tr} \mathcal{A} = I\left(\left(\frac{\partial V}{\partial y_1}, n_1\right)\Big|_\Gamma + \left(\frac{\partial V}{\partial y_2}, n_2\right)\Big|_\Gamma\right) = -I\frac{\partial W}{\partial s}$$

(the last equality follows from the relation $(\nabla, V) = 0$, see (6.3), written down for the trajectory l). Now (6.16) and the first equation in (6.13) have the same left-hand sides. Consider the right-hand side of (6.16), i.e., the integral $(2\pi)^{-1}\oint \Delta_\tau w d\varphi$. We write down the Laplace operator in coordinates I, φ:

$$\Delta_\tau = \frac{\partial}{\partial I} g^{11}\frac{\partial}{\partial I} + \frac{\partial}{\partial I} g^{12}\frac{\partial}{\partial \varphi} + \frac{\partial}{\partial \varphi} g^{12}\frac{\partial}{\partial I} + \frac{\partial}{\partial \varphi} g^{22}\frac{\partial}{\partial \varphi},$$

where

$$g^{11} = \left(\frac{\partial \tau}{\partial \varphi}, \frac{\partial \tau}{\partial \varphi}\right), \quad g^{12} = -\left(\frac{\partial \tau}{\partial I}, \frac{\partial \tau}{\partial \varphi}\right), \quad g^{22} = \left(\frac{\partial \tau}{\partial I}, \frac{\partial \tau}{\partial I}\right)$$

are the components of the Euclidean metric in the coordinates I, φ. Using the fact that w does not depend on φ, on integrating with respect to φ we obtain the right-hand side of the first equality (6.13).

Now we transform the first equality in (6.10) into the last equation (6.13). Taking the scalar product of F by v and integrating along the trajectory l, we obtain

$$\begin{aligned}\frac{1}{2\pi}\oint_l &\left[(W + w)\frac{\partial}{\partial s}\frac{v^2}{2} + (v, \mathcal{A}v) + (\mathcal{A}\tau, \nabla_\tau)\frac{v^2}{2}\right] d\varphi \\ &+ k(s)(w^2 + 2Ww)\frac{1}{2\pi}\oint_l (e(s), v)\, d\varphi + a_0\frac{\partial B_0}{\partial I} = \nu_0\frac{1}{2\pi}\oint_l (v, \Delta_\tau v)\, d\varphi.\end{aligned} \tag{6.17}$$

The integral of (e, v) in the left-hand side vanishes by (6.15). Let us transform the function $(\mathcal{A}\tau, \nabla_\tau)v^2/2$. We have

$$\begin{aligned}(\mathcal{A}\tau, \nabla_\tau)\frac{v^2}{2} &= (\mathcal{A}\tau, \nabla_\tau)B_0 - (\mathcal{A}\tau, \nabla_\tau)\pi_0 = (\mathcal{A}\tau, \nabla_\tau)B_0 + (\mathcal{A}\tau, (v, \nabla_\tau)v) \\ &= (\mathcal{A}\tau, \nabla_\tau)B_0 + (v, \nabla_\tau)(\mathcal{A}\tau, v) - (v, \mathcal{A}v).\end{aligned}$$

Thus (6.17) can be rewritten as

$$\frac{1}{2\pi}\oint_l \left[(W + w)\frac{\partial}{\partial s}\frac{v^2}{2} + (\mathcal{A}\tau, \nabla_\tau)B_0\right] d\varphi + a_0\frac{\partial B_0}{\partial I} = \nu_0\frac{1}{2\pi}\oint_l (v, \Delta_\tau v)\, d\varphi.$$

Now we add this equality to the first equation in (6.13), which has already been proved, multiplied by $(W + w)$. Taking into account the equalities

$$\begin{aligned}\frac{1}{2\pi}\oint_l &(W + w)\frac{\partial}{\partial s}\frac{v^2}{2}\, d\varphi + \langle \pi_s\rangle(W + w) \\ &= \frac{1}{2\pi}\oint_l (W + w)\frac{\partial}{\partial s}\left(\frac{v^2}{2} + \pi_0\right) d\varphi - (W + w)\frac{\partial}{\partial s}\frac{W^2}{2} \\ &= (W + w)\frac{\partial}{\partial s}\left(B_0 - \frac{W^2}{2}\right),\end{aligned}$$

we obtain

$$
\begin{aligned}
&(W+w)\frac{\partial}{\partial s}\Big(B_0+\frac{(W+w)^2-W^2}{2}\Big)+a\frac{\partial}{\partial I}\Big(B_0+\frac{(W+w)^2}{2}\Big)\\
&\quad=\Big(\theta\frac{\partial}{\partial s}+a\frac{\partial}{\partial I}\Big)B=\nu_0\frac{1}{2\pi}\oint_l\big[(v,\Delta_\tau v)+(w+W)\Delta_\tau(W+w)\big]\,d\varphi.
\end{aligned}
\tag{6.18}
$$

Consider the right-hand side of the last equality. We have

$$
\begin{aligned}
\oint_l(v,\Delta_\tau v)\,d\varphi&=\oint_l\Big(\Delta_\tau\frac{v^2}{2}-\operatorname{tr}\Big(\frac{\partial v}{\partial\tau}\frac{\partial v^*}{\partial\tau}\Big)\Big)\,d\varphi\\
&=\oint_l\Big(\Delta_\tau(B_0-\pi_0)-\operatorname{tr}\Big(\frac{\partial v}{\partial\tau}\frac{\partial v^*}{\partial\tau}\Big)\Big)\,d\varphi.
\end{aligned}
\tag{6.19}
$$

We can find $\Delta_\tau\pi_0$ from equations (6.5):

$$
\Delta_\tau\pi_0+(\nabla_\tau,(v,\nabla_\tau)v)=0,
$$

whence

$$
\Delta_\tau\pi_0=-(v,\nabla_\tau)(\nabla_\tau,v)-\operatorname{tr}\Big(\frac{\partial v}{\partial\tau}\Big)^2.
$$

Putting this equality into (6.19), we obtain

$$
\begin{aligned}
\oint_l(v,\Delta_\tau v)\,d\varphi&=\oint_l\Big(\Delta_\tau B_0+\operatorname{tr}\Big(\frac{\partial v}{\partial\tau}\Big(\frac{\partial v}{\partial\tau}-\frac{\partial v^*}{\partial\tau}\Big)\Big)\Big)\,d\varphi\\
&=\oint_l\Big(\Delta_\tau B_0+(\operatorname{curl}_\tau v)_3\operatorname{tr}\Big(\frac{\partial v}{\partial\tau}J\Big)\Big)\,d\varphi=\oint_l\big(\Delta_\tau B_0-(\operatorname{curl}_\tau v)_3^2\big)\,d\varphi.
\end{aligned}
$$

Finally, the last term in (6.18) can be written as

$$
\begin{aligned}
\oint_l(w+W)\Delta_\tau(w+W)\,d\varphi&=2\pi(w+W)\frac{\partial}{\partial I}\mathcal{D}^2\frac{\partial}{\partial I}(w+W)\\
&=2\pi\Big(\frac{\partial}{\partial I}\mathcal{D}^2\frac{\partial}{\partial I}\frac{\theta^2}{2}-\mathcal{D}^2\Big(\frac{\partial\theta}{\partial I}\Big)^2\Big).
\end{aligned}
$$

Putting the last two relations into (6.18), we obtain the last equality in (6.13). To complete the proof of our assertion, it remains to prove the properties of the functions a, B, θ at the vertices of the graph $\mathcal{G}$, which will be done in the following propositions. □

PROPOSITION 6.2.1. *Suppose there exists a smooth solution* (v^1, w_1, π_1) *of the system* (6.7), (6.8). *Then the function* $a(I)$ *vanishes at each vertex of order* 1 *of the graph* $\mathcal{G}$.

PROOF. Recall that the vertices in question correspond to the singular points of type "center" of the vector field v, and $I=0$ at these points. Since $a=a_0-I\partial W/\partial s$, it suffices to prove our assertion for the function $a_0(I)$. By definition,

$$
a_0=\frac{1}{2\pi}\oint a_I\,d\varphi,
$$

where $v^1=a_I\partial/\partial I+a_\varphi\partial/\partial\varphi$ and integration is performed along a closed trajectory of the field v. In the last formula the vector field is identified with the differentiation operator in the direction of the field. Consider the coordinates I,φ in a small neighborhood of a singular point of type "center". Let r be the distance to the singular point. It is clear that

$$I = O(r^2), \quad \left|\frac{\partial \tau}{\partial I}\right| = O\Big(\frac{1}{r}\Big), \quad \left|\frac{\partial \tau}{\partial \varphi}\right| = O(r)$$

as $r \to 0$. Hence $a(I) = O(r)$ as $r \to 0$. □

PROPOSITION 6.2.2. *Suppose there exists a smooth solution (v^1, w_1, π_1) of the system (6.7), (6.8) with $v^1 = \text{const} + v_0^1$, where $|v_0^1| = o(|\tau|^{-1})$ as $\tau \to \infty$. Then at each vertex of order 3 of the graph $\mathcal{G}$ the functions $a(I)$, $\mathcal{D}^2 \partial\theta/\partial I$, $\mathcal{D}^2$, $\partial B/\partial I$ satisfy the relations (6.14).*

PROOF. Recall that each vertex of $\mathcal{G}$ of order 3 corresponds to a separatrix of the vector field v, and there is a single singularity point on this separatrix, which is a saddle-point (see Figure 6.2). The two incoming edges of the graph $\mathcal{G}$ at this vertex correspond to the interior domains of the separatrix, and the outgoing edge corresponds to the exterior domain. Denote the interior domains by D_1, D_2, and the exterior domain by D_3 (Figure 6.2).

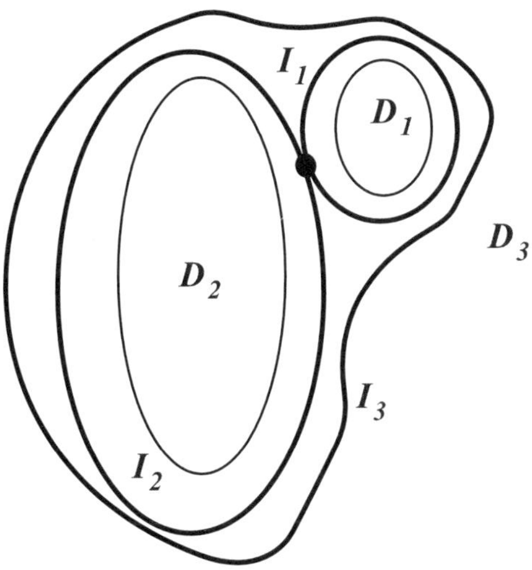

FIGURE 6.2

Let us prove the Kirchhoff equality (6.14) for the function a. Similarly to Proposition 6.2.1, it suffices to prove this equality for a_0 because it is obviously fulfilled for I. Fix a sufficiently small $\mu > 0$, and consider three closed trajectories l_1, l_2, l_3 of the field v lying close to the separatrix in the domains D_1, D_2, D_3 respectively and such that the measure of the domain encircled by these trajectories is less than μ (Figure 6.2). Denote by D_1', D_2', and D_3' the interior domains relative to l_1, l_2 and the exterior domain relative to l_3 respectively. Integrating (∇_τ, v^1) over these domains, we obtain

$$\int_{D_1'} (\nabla_\tau, v^1)\, d\tau + \int_{D_2'} (\nabla_\tau, v^1)\, d\tau + \int_{D_3'} (\nabla_\tau, v^1)\, d\tau$$
$$= -\int_{\mathbb{R}^2} (\nabla_\tau, v^1)\, d\tau + O(\mu) = O(\mu).$$

On the other hand, by the Green formula we have

$$\int_{D'_j} (\nabla_\tau, v^1)\, d\tau = \oint_{l_j} a_I\, d\varphi = a_0(l_j), \quad j = 1, 2,$$

$$\int_{D'_3} (\nabla_\tau, v^1)\, d\tau = -\oint_{l_3} a_I\, d\varphi = -a_0(l_3)$$

(recall that $v^1 = a_I \partial/\partial I + a_\varphi \partial/\partial\varphi$, $(\nabla_\tau, v^1) = \partial a_I/\partial I + \partial a_\varphi/\partial\varphi$). Putting the last two equalities into the previous one and letting μ tend to zero, we immediately obtain (6.14) for a_0. The Kirchhoff equalities for $\mathcal{D}^2 \partial B/\partial I$ and $\mathcal{D}^2 \partial\theta/\partial I$ are proved in a similar manner. Rather than (∇_τ, v^1), in these cases one has to integrate $\Delta_\tau B$ and $\Delta_\tau \theta$ respectively. □

6.2.2. Integral identities and conservation laws for equations of the stretched vortex. In spite of the complicated form of the Prandtl type equations (6.13), they turn out to possess some interesting properties. Namely, the solutions to these equations satisfy infinitely many integral identities (of energy balance type), which for $\nu_0 = 0$ (i.e., for the ideal fluid) become the conservation laws. Moreover, (6.13) possesses two more conservation laws valid for any ν_0. In this subsection we prove the following assertion.

THEOREM 6.2.2. *Suppose the functions v, π_0, w satisfy the Euler equations* (6.5), (6.6), *and the corresponding functions θ, B and some function a, smooth on the edges of $\mathcal{G}$, satisfy the Prandtl type equations* (6.13) *on $\mathcal{G}$. Further, let a, B, and θ satisfy the Kirchhoff conditions* (6.14), *and let the functions w, v, π_0, a_0 decrease at a rate of $o(I^{-1})$ as $I \to \infty$. Then the following identities hold:*

$$\frac{\partial}{\partial s} \int_{\mathcal{G}} (\theta - W)\, dI = 0, \tag{6.20}$$

$$\frac{\partial}{\partial s} K(s) \overset{\text{def}}{=} \frac{\partial}{\partial s} \int_{\mathcal{G}} \left(\theta^2 - W^2 - \left\langle \frac{v^2}{2} \right\rangle \right) dI = 0, \tag{6.21}$$

$$\frac{\partial}{\partial s} \int_{\mathcal{G}} \theta Q(B) dI = -\nu_0 \int_{\mathcal{G}} \left[Q''(B) \mathcal{D}^2 \left(\frac{\partial B}{\partial I} \right)^2 + \Lambda^2 Q'(B) \right] dI. \tag{6.22}$$

Here $Q(t)$ is an arbitrary smooth function vanishing at zero, and

$$\left\langle \frac{v^2}{2} \right\rangle = \frac{1}{2\pi} \oint_l \frac{v^2}{2}\, d\varphi.$$

REMARK. The equalities (6.20) and (6.21) are conservation laws for the Prandtl type equations. The equalities (6.22) are similar to the energy balance equations, since their right-hand sides describe the viscous dissipation. For $\nu_0 = 0$ (i.e., for the ideal fluid or for the case where viscosity is much less than ε^2) the equations (6.22) also turn into conservation laws:

$$\frac{\partial}{\partial s} \int_{\mathcal{G}} \theta Q(B)\, dI = 0;$$

in particular,

$$\frac{\partial}{\partial s} \int_{\mathcal{G}} \theta B^m\, dI = 0.$$

PROOF OF THEOREM 6.2.2. The equality (6.20) has actually been obtained (and used) in the course of the proof of Proposition 6.2.2; it is derived by integrating the second equation (6.13) over the graph $\mathcal{G}$ with (6.14) taken into account. For the proof of (6.21) we will need the following simple proposition.

PROPOSITION 6.2.3 [**56, 57**]. *Let $v(\tau), \pi_0(\tau)$ be a solution to the Euler equations* (6.5), *and let $f(\tau)$ be a smooth function satisfying the following relations as $|\tau| \to \infty$:*

$$|(\nabla_\tau f, (v, \nabla_\tau)v)| = o(|\tau|^{-2}), \qquad |(\nabla_\tau f, \nabla_\tau \pi_0)| = o(|\tau|^{-2}),$$
$$\left|\left(v, \frac{\partial^2 f}{\partial \tau^2} v\right)\right| = o(|\tau|^{-2}), \qquad |\pi_0 \nabla_\tau f| = o(|\tau|^{-1}).$$

Then

$$\int_{\mathbb{R}^2} \left(v, \frac{\partial^2 f}{\partial \tau^2} v\right) d\tau = -\int_{\mathbb{R}^2} \pi_0 \Delta_\tau f \, d\tau. \tag{6.23}$$

We omit the proof of this proposition, which can be found, e.g., in [**57**]; we only point out that the identity (6.23) is obtained by multiplying the Euler equations by $\nabla_\tau f$ and integrating.

Rewrite the first equation in (6.13) as

$$(W + w)\frac{\partial w}{\partial s} + w\frac{\partial W}{\partial s} + a\frac{\partial w}{\partial I} + \frac{1}{2\pi}\oint_l \frac{\partial \pi_0}{\partial s}\, d\varphi = \nu_0 \frac{\partial}{\partial I}\mathcal{D}^2 \frac{\partial}{\partial I} w$$

and integrate it over the graph $\mathcal{G}$. We obtain

$$\frac{\partial}{\partial s}\int_{\mathcal{G}} \left(\frac{w^2}{2} + wW\right) dI + \int_{\mathcal{G}} a\frac{\partial w}{\partial I}\, dI + \int_{\mathbb{R}^2} \frac{\partial \pi_0}{\partial s}\, d\tau = \int_{\mathcal{G}} \nu_0 \frac{\partial}{\partial I}\mathcal{D}^2 \frac{\partial}{\partial I} w \, dI. \tag{6.24}$$

Let us transform the second term in the left-hand side. Integrating by parts and taking Proposition 6.2.1 into account, we obtain

$$\int_{\mathcal{G}} a\frac{\partial w}{\partial I}\, dI = \sum_k \big(a_1(k) + a_2(k) - a_3(k)\big) w_k - \int_{\mathcal{G}} w\frac{\partial a}{\partial I}\, dI,$$

where the summation is extended over all vertices of $\mathcal{G}$ of order 3, the $a_j(k)$ are the limits of a at the kth vertex taken over the incoming ($j = 1, 2$) and outgoing ($j = 3$) vertices, and w_k is the value of w at the kth vertex. According to the Kirchhoff equations (6.14) all the summands vanish; hence

$$\int_{\mathcal{G}} a\frac{\partial w}{\partial I}\, dI = -\int_{\mathcal{G}} w\frac{\partial a}{\partial I}\, dI = \int_{\mathcal{G}} w\frac{\partial}{\partial s}(w + W)\, dI.$$

In a similar way one can check that the right-hand side of (6.24) vanishes. Taking (6.20), which has been proved before, into account, we obtain from (6.24)

$$\frac{\partial}{\partial s}\int_{\mathcal{G}} (w^2 + 2wW)\, dI + \int_{\mathbb{R}^2} \frac{\partial \pi_0}{\partial s}\, d\tau = 0.$$

Now we use the identity (6.23) (see Proposition 6.2.3), where we set $f(\tau) = |\tau|^2/4$. This identity and the last equality immediately imply (6.21). Finally, we prove (6.22). Differentiating the function

$$\int_{\mathcal{G}} \theta Q(B)\, dI$$

with respect to s, applying the second and the last equalities (6.13), and using the Kirchhoff conditions (6.14), we obtain

$$\begin{aligned}\frac{\partial}{\partial s}\int_{\mathcal{G}}\theta Q(B)\,dI &= \int_{\mathcal{G}}\frac{\partial\theta}{\partial s}Q(B)\,dI + \int_{\mathcal{G}}\theta Q'(B)\frac{\partial B}{\partial s}\,dI\\ &= -\int_{\mathcal{G}}\Big(\frac{\partial a}{\partial I}Q(B) + Q'a\frac{\partial B}{\partial I}\Big)\,dI + \nu_0\int_{\mathcal{G}}Q'(B)\Big(\frac{\partial}{\partial I}\mathcal{D}^2\frac{\partial}{\partial I}B - \Lambda^2\Big)\,dI\\ &= -\nu_0\int_{\mathcal{G}}\Big(Q''(B)\mathcal{D}^2\Big(\frac{\partial B}{\partial I}\Big)^2 + Q'(B)\Lambda^2\Big)\,dI. \qquad \square\end{aligned}$$

6.3. Additional conditions on the parameters and the integral of the vortex equations. Determination of the twisting angle.

Here we obtain additional conditions on the parameters involved in the leading term U of the asymptotic solution to the Navier–Stokes equations that describes the stretched vortex. These conditions again arise from the requirement of solvability of the first approximation system (6.7), (6.8). Part of them was discussed in §6.1; these are conditions (6.11) and (6.12) related to the symmetries of the two-dimensional Euler equations and to the presence of the corresponding three-dimensional subspace in the cokernel of the linearized operator. It turns out that these conditions determine the integral K of the stretched vortex equations and the twisting angle α of the asymptotic solution; in 6.3.1 and 6.3.2 they are reduced to the form we need. In 6.3.3 we obtain conditions on the remaining free parameters (the values of I at the vertices and the boundary values w, B) of the Prandtl type equations (6.13). These are also necessary conditions for solvability of the system (6.7), (6.8); they are related to the presence of a singular component in the cokernel of the linearized Euler operator. Roughly speaking, they arise from the condition of "orthogonality of the right-hand side to delta-functions".

6.3.1. Conditions on the integral K. We begin with the transformation of condition (6.11).

PROPOSITION 6.3.1. *The equalities* (6.11) *are equivalent to the single relation*

$$k(s)K = 0, \tag{6.25}$$

where $k(s)$ is the curvature of the curve Γ, and

$$K = \int_{\mathbb{R}^2}\Big(\theta^2 - W^2 - \frac{1}{2}v^2\Big)\,d\tau = \int_{\mathbb{R}^2}\Big(w^2 - \frac{1}{2}v^2\Big)\,d\tau$$

is the integral (6.21) *of the Prandtl type equations* (6.13).

PROOF. Consider the left-hand side of (6.11). Since for any trajectory l of the field v

$$\oint_l v\,d\varphi = 0$$

(see (6.15)), we obtain

$$\int_{\mathbb{R}^2}\frac{\partial w}{\partial s}v\,d\tau + \int_{\mathbb{R}^2}\frac{\partial v}{\partial s}(w+W)\,d\tau = \frac{\partial}{\partial s}\Big(\int_{\mathbb{R}^2}wv\,d\tau + W\int_{\mathbb{R}^2}v\,d\tau\Big) = 0.$$

Thus the equalities (6.11) are equivalent to

$$k(s)\int_{\mathbb{R}^2}\big[(w^2+2wW)e(s)-(e(s),v)v\big]\,d\tau=0. \tag{6.26}$$

We complete the vector $e(s)$ to an orthonormal basis $e,\tilde{e}$ on the plane and decompose the vector v in this basis, $v=v_e e+v_n\tilde{e}$. Obviously, the equalities (6.26) are equivalent to the following two equalities:

$$k(s)\int_{\mathbb{R}^2}(\theta^2-W^2-v_e^2)\,d\tau=0,$$

$$k(s)\int_{\mathbb{R}^2}v_e v_n\,d\tau=0.$$

Now we employ Proposition 6.2.3. Taking the functions $(e,\tau)(\tilde{e},\tau)$ and $(e,\tau)^2-(\tilde{e},\tau)^2$ for $f(\tau)$, we obtain from (6.23) the equalities

$$\int_{\mathbb{R}^2}v_e v_n\,d\tau=0,\qquad \int_{\mathbb{R}^2}v_e^2\,d\tau=\int_{\mathbb{R}^2}v_n^2\,d\tau=\frac{1}{2}\int_{\mathbb{R}^2}v^2\,d\tau,$$

which immediately imply (6.25). □

Thus we get the following conclusion.

If the trajectory Γ of the outer flow in not rectilinear, then the flow of the stream is possible only if the integral K of the Prandtl type equations (6.13) *vanishes.*

REMARK 1. As we pointed out in §6.1, when describing solutions localized in a neighborhood of a point, it is natural to consider the problem on extension of the vortex, which is similar to the problem on extension of a laminary boundary layer [**193**]. This means that the additional conditions for the Navier–Stokes equation (6.1) are set on a surface $\mathcal{M}$ which intersects the curve Γ transversally (suppose that the natural parameter s on Γ is counted starting from the intersection point). Namely, there must be a vector field $\overset{\circ}{v}(y/\varepsilon)$ and a scalar function $\overset{\circ}{w}(y/\varepsilon)$ specified on this surface, where $y=\{y_1,y_2\}$ are coordinates in the normal plane to the curve Γ. These data cannot be specified arbitrarily: the field $\overset{\circ}{v}$ must satisfy the two-dimensional Euler equations (6.5), the function $\overset{\circ}{w}$ must be an integral of this field, and, finally, the integral K corresponding to the field $\overset{\circ}{v}$ must vanish, unless Γ is a straight line. Specification of such data determines the initial data $\overset{\circ}{B}$, $\overset{\circ}{w}$ for the Prandtl type equations.

REMARK 2. The integral K describes the distribution of the energy between transverse circulations (v) and translational motion (w) inside the vortex. The condition that this integral vanishes means that the flow is balanced, viz., the vortex velocity has equal projections on the three vectors of the Frenet frame (or the tuple $e,\tilde{e},dR/ds$).

6.3.2. Determination of the twisting angle α. Now we consider the equality (6.12), which is the last condition for solvability of the system (6.7), (6.8) from among those discussed in 6.1.2.

PROPOSITION 6.3.2. *The equality* (6.12) *is equivalent to the equation*

$$\begin{aligned}\frac{\partial}{\partial s}&\int_{\mathbb{R}^2}(w+W)(J\tau,v)\,d\tau\\ &=k(s)\int_{\mathbb{R}^2}\big[(\tau,\tilde e(s))(\theta^2-W^2)+(e(s),v)(J\tau,v)\big]\,d\tau,\end{aligned}\tag{6.27}$$

where $\tilde e=(-e_2,e_1)$.

PROOF. We can write (6.12) as

$$\begin{aligned}\int_{\mathbb{R}^2}\Big(J\tau,(W+w)\frac{\partial v}{\partial s}&+k(s)e(s)(w^2+2wW)+\mathcal{A}v\\ &+(\mathcal{A}\tau,\nabla_\tau)v+v\frac{\partial w}{\partial s}-k(s)(e(s),v)v-\nu_0\Delta_\tau v\Big)\,d\tau=0.\end{aligned}\tag{6.28}$$

Consider the integral

$$\int_{\mathbb{R}^2}\big(J\tau,\mathcal{A}v+(\mathcal{A}\tau,\nabla_\tau)v\big)\,d\tau.$$

Since we consider the vector field v decreasing at a rate of $o(|\tau|^{-3})$ as $|\tau|\to\infty$, we obtain

$$\begin{aligned}\int_{\mathbb{R}^2}(J\tau,(\mathcal{A}\tau,\nabla_\tau)v)\,d\tau&=-\int_{\mathbb{R}^2}\Big((J\tau,v)(\nabla_\tau,\mathcal{A}\tau)+\big(v,(\mathcal{A}\tau,\nabla_\tau)J\tau\big)\Big)\,d\tau\\ &=-\int_{\mathbb{R}^2}\big((J\tau,v)\operatorname{tr}\mathcal{A}+(v,J\mathcal{A}\tau)\big)\,d\tau.\end{aligned}\tag{6.29}$$

Now we observe that for any smooth divergence-free vector field v on the plane decreasing at infinity sufficiently fast the following obvious identities hold for all $i,j=1,2$:

$$\int_{\mathbb{R}^2}v_i\tau_j\,d\tau=-\int_{\mathbb{R}^2}v_j\tau_i\,d\tau.$$

Using them, we obtain

$$\begin{aligned}\int_{\mathbb{R}^2}(\mathcal{A}v,J\tau)d\tau&=\frac12\operatorname{tr}\mathcal{A}\int_{\mathbb{R}^2}(v,J\tau)\,d\tau,\\ \int_{\mathbb{R}^2}(v,J\mathcal{A}\tau)d\tau&=\frac12\operatorname{tr}\mathcal{A}\int_{\mathbb{R}^2}(v,J\tau)\,d\tau.\end{aligned}$$

Putting these relations into (6.29) yields

$$\int_{\mathbb{R}^2}\big(J\tau,\mathcal{A}v+(\mathcal{A}\tau,\nabla_\tau)v\big)\,d\tau=-\operatorname{tr}\mathcal{A}\int_{\mathbb{R}^2}(v,J\tau)\,d\tau=\frac{\partial W}{\partial s}\int_{\mathbb{R}^2}(v,J\tau)\,d\tau$$

(as was pointed out in 6.2.1, when deriving the stretched vortex equations, the equality $\operatorname{tr}\mathcal{A}+\partial W/\partial s=0$ holds because the outer flow, i.e., the three-dimensional vector field $V(x)$, is divergence-free, see (6.3)). Finally, putting the equality just obtained into (6.28) and taking the equalities $(J\tau,e)=-(\tau,Je)=-(\tau,\tilde e)$ into account, after obvious calculations we obtain (6.27). □

REMARK. The equation (6.27) can be naturally interpreted as the condition which determines the rotation angle α twisting the stream. Namely, as was pointed out in §6.2, along with any solution v, w, the equations (6.5), (6.6) possess a family of solutions v_α, w_α of the form $v_\alpha(\tau) = T(\alpha)v(T^*(\alpha)\tau)$, $w_\alpha(\tau) = w(T^*(\alpha)\tau)$, where $\alpha(s)$ is a smooth function and $T(\alpha)$ is the matrix of rotation at the angle α. Putting v_α, w_α into (6.27) and making the change of variables $\tau \to \eta = T^*(\alpha)\tau$ in the integrals, we obtain after elementary calculations that this equation is satisfied if and only if the angle α is determined by the equality

$$\cos(\alpha - \sigma) = \frac{C(s)}{k(s)} \frac{\partial}{\partial s} \int_{\mathbb{R}^2} (w + W)(\eta)(v(\eta), J\eta)\, d\eta,$$

where

$$\tan\sigma = \frac{\int_{\mathbb{R}^2} \big((\eta, e)(w^2 + 2wW) - (v(\eta), \tilde{e})(v(\eta), J\eta)\big)\, d\eta}{\int_{\mathbb{R}^2} \big((\eta, \tilde{e})(w^2 + 2wW) + (v(\eta), e)(v(\eta), J\eta)\big)\, d\eta},$$

$$C(s) = \Big[\Big(\int_{\mathbb{R}^2} \big((\eta, \tilde{e})(w^2 + 2wW) + (v(\eta), e)(v(\eta), J\eta)\big)\, d\eta\Big)^2 + \Big(\int_{\mathbb{R}^2} \big((\eta, e)(w^2 + 2wW) - (v(\eta), \tilde{e})(v(\eta), J\eta)\big)\, d\eta\Big)^2\Big]^{-\frac{1}{2}}.$$

6.3.3. Conditions on free parameters of the vortex equations. Note that the Prandtl type equations (6.13) involve free parameters related, first, to the freedom in parametrization of the graph $\mathcal{G}$ and, secondly, to the freedom in the choice of the boundary conditions for equations (6.13). Let us describe these parameters explicitly. Specification of the parametrization of the graph $\mathcal{G}$ is equivalent to specification of the values of I at the end-points of the edges of $\mathcal{G}$; at all vertices of order 1 this value must be zero, wheweas at each vertex of order 3 the values corresponding to the incident edges must satisfy the Kirchhoff relation $I_3 = I_1 + I_2$ (see §6.1). Thus specification of the parametrization involves $2m$ free parameters I_1^j, I_2^j, $j = 1, \ldots, m$, where m is the number of vertices of order 3 in the graph $\mathcal{G}$ (or the number of saddle-point singularities of the divergence-free field v). Moreover, for equations (6.13) on the graph we can arbitrarily specify the boudary conditions w^j, B^j at its vertices of order 1 (see [**106, 207**]), i.e., at the singular points of type "center" of the field v; since $\mathcal{G}$ is a planar graph, there are $m+1$ such vertices. Thus we get a set of $4m + 2 = 2M$ parameters, where $M = 2m + 1$ is the total number of singular points of the field v. It turns out that these parameters satisfy certain relations, which again follow from the solvability conditions for the first approximation equations (6.7), (6.8).

PROPOSITION 6.3.3. *Suppose equations* (6.7), (6.8) *possess a* C^2*-smooth solution* v^1, w^1. *Then at each singular point* r_j, $j = 1, \ldots, M$, *of the vector field* v *the following relations hold:*

$$(6.30) \qquad H(r_j) = 0, \quad (\operatorname{curl}_\tau F(r_j))_3 - G(r_j)(\operatorname{curl}_\tau v(r_j))_3 = 0.$$

PROOF. Consider the equality (6.7) at a singular point r_j. Since $v(r_j) = 0$, the first term in the left-hand side of this equality vanishes. Next, the function w is constant on the trajectories of the vector field v, and hence the singular points of v are critical points of w. Therefore the second term in the left-hand side of (6.7) also vanishes, which proves the first equality (6.30). For the proof of the second equality we apply the operation curl_τ to the first (vector) equation (6.8) and substitute the singular point r_j into the equality thus obtained. After elementary calculations we obtain at this point

$$\begin{aligned}(\operatorname{curl}_\tau F)_3 &+ \big(\operatorname{curl}_\tau((v,\nabla_\tau)v^1)\big)_3 + \big(\operatorname{curl}_\tau((v^1,\nabla_\tau)v)\big)_3 \\ &= (\operatorname{curl}_\tau F)_3 + (v,\nabla_\tau)(\operatorname{curl}_\tau v^1)_3 + (v^1,\nabla_\tau)(\operatorname{curl}_\tau v)_3 \\ &\quad + (\nabla_\tau, v)(\operatorname{curl}_\tau v^1)_3 + (\nabla_\tau, v^1)(\operatorname{curl}_\tau v)_3 = 0.\end{aligned}$$

The second through the fourth terms here are equal to zero because r_j is a singular point of v, the function $(\operatorname{curl}_\tau v)_3$ is constant of the trajectories of v, and the field v is divergence-free. Using the relation $(\nabla_\tau, v^1) = -G$, we immediately obtain the second equality in (6.30). □

REMARK. The second equality in (6.30) can be interpreted as a consequence of the presence of delta-functions in the cokernel of the linearized Euler operator. Indeed, consider the linearized Euler equations

$$(v,\nabla_\tau)\xi + (\xi,\nabla_\tau)v + \nabla_\tau\chi = 0, \quad (\nabla_\tau, \xi) = 0.$$

We introduce the stream functions ψ, ϕ of the fields v, ξ ($v = \operatorname{sgrad}\psi$, $\xi = \operatorname{sgrad}\phi$) and apply the operator curl_τ to the equations. Then we obtain

$$\{\Delta_\tau\phi, \psi\} + \{\Delta_\tau\psi, \phi\} = 0,$$

where $\{\cdot,\cdot\}$ is the Poisson bracket, i.e.,

$$\{f, g\} = \frac{\partial f}{\partial \tau_1}\frac{\partial g}{\partial \tau_2} - \frac{\partial f}{\partial \tau_2}\frac{\partial g}{\partial \tau_1}.$$

Since v satisfies the Euler equations, $\Delta_\tau\psi$ is locally a function of ψ. Suppose that this function is defined everywhere, i.e., $\Delta_\tau\psi = f(\psi)$. Then the above equation for ϕ can be rewritten as

$$\{\psi, (\Delta_\tau - f'(\psi))\phi\} = 0,$$

and the adjoint equation as

$$(\Delta_\tau - f'(\psi))\{\psi, \hat\phi\} = 0.$$

Since v is divergence-free, we get that $\hat\phi = \delta(x - r_j)$ satisfies this equation, where r_j is a singular point of v.

Now we rewrite the conditions (6.30) in terms of the functions B, θ. Let $B_j(s), \theta_j(s)$ denote the values of these functions at the singular points r_j of the vector field v.

PROPOSITION 6.3.4. *The equalities* (6.30) *are equivalent to the relations*

$$\left\{\begin{aligned}&\frac{\partial B_j(s)}{\partial s} = \nu_0\Delta_\tau\theta(r_j),\\ &q_j(s)\frac{\partial\theta_j(s)}{\partial s} - \theta_j(s)\frac{\partial q_j(s)}{\partial s} = -\nu_0\Delta_\tau(\operatorname{curl}_\tau v)_3(r_j).\end{aligned}\right. \tag{6.31}$$

Here $q_j(s)$ denotes the value of the vorticity function $q_j(s) = (\operatorname{curl}_\tau v(r_j))_3$ of the field v at the point r_j (recall that $\operatorname{curl}_\tau v$ is constant on the trajectories of the field v and, in particular, the r_j are its critical points).

REMARK. The equalities (6.31) constitute $2M$ relations for the $2M$ free parameters mentioned above which are involved in the solutions of the stretched vortex equations (6.13).

PROOF. We begin with the first equality in (6.30). Putting $H(\tau, s)$ as in (6.7) into this equality, we obtain

$$\begin{aligned}\theta\frac{\partial\theta}{\partial s}(r_j) - \frac{1}{2}\frac{\partial}{\partial s}W^2 - k(s)\big(e(s), v(r_j)\big)\theta(r_j) + (b, v(r_j))\\ + \big(\mathcal{A}\tau(r_j), \nabla_\tau\big)\theta(r_j) + \frac{\partial\pi_0}{\partial s}(r_j) - \nu_0\Delta_\tau\theta(r_j) = 0.\end{aligned}\tag{6.32}$$

Since $v(r_j) = 0$ and $\nabla_\tau\theta(r_j) = 0$ (because θ is the integral of the field v), the third through the fifth terms in (6.32) vanish. Consider the function $\partial\pi_0/\partial s(r_j)$. We have

$$\frac{\partial\pi_0}{\partial s}(r_j) = \frac{\partial}{\partial s}\big(\pi_0(r_j)\big) - \Big(\nabla_\tau\pi_0(r_j), \frac{\partial r_j}{\partial s}\Big).$$

The Euler equations (6.5) imply that $\nabla_\tau\pi_0(r_j) = 0$; moreover, by definition of the Bernoulli integral,

$$B_0(r_j) = \frac{1}{2}v^2(r_j) + \pi_0(r_j) = \pi_0(r_j).$$

Therefore $\partial\pi_0/\partial s(r_j) = \partial/\partial s\big(B_0(r_j)\big)$. Finally, since r_j is a critical point of the function θ,

$$\theta(r_j)\frac{\partial\theta}{\partial s}(r_j) = \theta_j\left(\frac{\partial\theta_j}{\partial s} - \Big(\nabla_\tau\theta(r_j), \frac{\partial r_j}{\partial s}\Big)\right) = \frac{\partial}{\partial s}\Big(\frac{1}{2}\theta_j^2\Big).$$

Putting the equalities thus obtained into (6.32), we obtain the first equality in (6.31).

Now we consider the second equality (6.30). Putting F and G as in (6.8) into this equality, we obtain

$$\begin{aligned}\operatorname{curl}_\tau\Big(\theta\frac{\partial v}{\partial s}\Big)\Big|_{\tau=r_j} + k(s)\nabla_\tau(W^2 + 2wW)\big|_{\tau=r_j} \times e(s)\\ + \operatorname{curl}_\tau\big(\mathcal{A}v + (\mathcal{A}\tau, \nabla_\tau)v\big)\big|_{\tau=r_j}\\ - \nu_0\Delta_\tau\operatorname{curl}_\tau v\big|_{\tau=r_j} - q_j\Big(\frac{\partial w}{\partial s}(r_j) - k(s)\big(e, v(r_j)\big)\Big) = 0.\end{aligned}\tag{6.33}$$

All the terms in this formula containing $k(s)$ vanish (recall that r_j is a critical point of w). Next, the first term is of the form

$$\operatorname{curl}_\tau\Big(\theta\frac{\partial v}{\partial s}\Big)\Big|_{\tau=r_j} = \theta_j\frac{\partial\operatorname{curl}_\tau v}{\partial s}\Big|_{\tau=r_j} + \nabla_\tau\theta_j \times \frac{\partial v}{\partial s}\Big|_{\tau=r_j} = \theta_j\frac{\partial q_j}{\partial s}.$$

Finally, consider the expression $\operatorname{curl}_\tau\big(\mathcal{A}v + (\mathcal{A}\tau, \nabla_\tau)v\big)(r_j)$. Applying standard formulas of the vector calculus, we obtain

$$\begin{aligned}&\operatorname{curl}_\tau\big(\mathcal{A}v + (\mathcal{A}\tau, \nabla_\tau)v\big)\big|_{\tau=r_j}\\ &\quad = \Big((\mathcal{A}, \nabla_\tau)\operatorname{curl}_\tau v + \operatorname{tr}\mathcal{A}\operatorname{curl}_\tau v + \big(a_{12} - a_{21}(\nabla_\tau, v)\big)\Big)\Big|_{\tau=r_j} = q_j\operatorname{tr}\mathcal{A},\end{aligned}$$

where a_{ij} are the elements of the matrix $\mathcal{A}$. Thus formula (6.33) becomes

$$\theta_j \frac{\partial q_j}{\partial s} + q_j \operatorname{tr} \mathcal{A} - q_j \frac{\partial}{\partial s} w(r_j) - \nu_0 \Delta_\tau \operatorname{curl}_\tau v(r_j) = 0.$$

Recall that $\operatorname{tr} \mathcal{A} = -\partial W/\partial s$ and $\theta = w + W$. Then we obtain the second equality in (6.31). □

6.4. Radially-symmetric stretched vortex

6.4.1. General formulas. In this section we consider the vector fields v with the simplest phase portrait. Namely, these will be the fields with a single singular point on the plane τ, which is necessarily a point of type "center". In this case the corresponding graph $\mathcal{G}$ has no vertices of order 3 and exactly one vertex of order 1; thus it can be represented by the half-line $I = [0, \infty)$. There is a well-known class of such solutions to the Euler equations (6.5), namely, radially-symmetric fields of the form $v = (-\partial\Psi(r)/\partial\tau_2, \partial\Psi(r)/\partial\tau_1)$, where $r = \sqrt{\tau_1^2 + \tau_2^2}$ and $\Psi(r)$ is a smooth function. Any such vector field satisfies equations (6.5), and any function $w(r)$ satisfies (6.6). No other solutions to the Euler equations with this topology of the phase curves are known; there is a conjecture [**177**] that none exist. Anyway, we will restrict our consideration to the radially-symmetric solutions.

In the radially-symmetric case the action-angle variables are defined on the entire plane τ with the only singularity at zero; they have the form $I = r^2/2$, $\varphi = \phi$ (the angle in polar coordinates). To simplify the calculations, we will use the frequency $\omega(I)$ on a trajectory rather then the Bernoulli integral B_0. The solution v, π_0 and the Bernoulli integral are expressed in terms of ω by the formulas

$$v = \omega \frac{\partial}{\partial \varphi}; \quad v^2 = 2I\omega^2; \quad \frac{\partial \pi_0}{\partial I} = \omega^2; \quad B_0 = I\omega^2 + \int \omega^2 \, dI,$$
$$(v, \Delta_\tau v) = 4I\omega \frac{\partial^2}{\partial I^2}(I\omega).$$

We will write down for this case the Prandtl type equations (6.13) describing the stretched vortex. On subtracting the first equation in (6.13) multiplied by θ from the last we obtain

$$\left(\theta \frac{\partial}{\partial s} + a \frac{\partial}{\partial I}\right)(I\omega^2) + a\omega^2 = 4\nu_0 I\omega \frac{\partial^2}{\partial I^2}(I\omega),$$

or

$$\left(\theta \frac{\partial}{\partial s} + a \frac{\partial}{\partial I}\right)(I^2\omega^2) = 4\nu_0 I^2\omega \frac{\partial^2}{\partial I^2}(I\omega). \tag{6.34}$$

The second equation in (6.13) yields

$$\theta = \frac{\partial \psi}{\partial I}, \quad a = -\frac{\partial \psi}{\partial s},$$

where $\psi(I, s)$ is a smooth function (the stream function). Differentiating the first equation in (6.13) with respect to I and writing down (6.34) in terms of ψ, we finally obtain the following system of equations, which in this setting is equivalent to the Prandtl type equations:

$$
\left\{
\begin{aligned}
&\{\psi, I^2\omega^2\} = 4\nu_0 I^2 \omega \frac{\partial^2}{\partial I^2}(I\omega),\\
&\frac{\partial}{\partial s}\omega^2 + \left\{\psi, \frac{\partial^2\psi}{\partial I^2}\right\} = 2\nu_0 \frac{\partial^2}{\partial I^2} I \frac{\partial^2\psi}{\partial I^2},
\end{aligned}
\right.
\tag{6.35}
$$

where $\{f, g\} = (\partial f/\partial I)(\partial g/\partial s) - (\partial f/\partial s)(\partial g/\partial I)$ is the Poisson bracket. The condition (6.25) on the integral K of the Prandtl type equations becomes

$$
k(s) \int_0^\infty \left(\left(\frac{d\psi}{dI}\right)^2 - W^2 - I\omega^2 \right) dI = 0.
\tag{6.36}
$$

Finally, there is no condition on the twisting angle α whatsoever — it is not hard to check that the equalities (6.27) in the radially-symmetric case reduce to a trivial identity. This is quite natural, because in this case the very parameter α becomes meaningless since in view of the radial symmetry the solution to the Euler equations transforms into itself under rotation. We will consider two particular cases of equations (6.35).

6.4.2. Radially-symmetric vortex in the ideal fluid. Let $\nu_0 = 0$; then the system of the stretched vortex equations (6.35) becomes

$$
\left\{
\begin{aligned}
&\{\psi, I^2\omega^2\} = 0,\\
&\tfrac{\partial}{\partial s}\omega^2 + \left\{\psi, \tfrac{\partial^2\psi}{\partial I^2}\right\} = 0.
\end{aligned}
\right.
$$

Obviously, the first equality is fulfilled if $I^2\omega^2 = f_0(\psi)$ for some smooth function f_0. The substitution $\omega^2 = f_0(\psi)/I^2$ transforms the second equality into

$$
\left\{\psi, \frac{\partial^2\psi}{\partial I^2}\right\} + \frac{1}{I^2} f_0'(\psi) \frac{\partial\psi}{\partial s} = 0,
$$

or

$$
\left\{\psi, \frac{\partial^2\psi}{\partial I^2}\right\} - f_0'(\psi) \frac{\partial\psi}{\partial s} \frac{\partial}{\partial I} \frac{1}{I} = 0.
$$

We rewrite the expression $-f_0'(\psi)(\partial\psi/\partial s)(\partial(I^{-1})/\partial I)$ as

$$
f_0'(\psi)\left\{\psi, \frac{1}{I}\right\} = \left\{\psi, \frac{f_0'(\psi)}{I}\right\}.
$$

Then we obtain

$$
\left\{\psi, \frac{\partial^2\psi}{\partial I^2} + \frac{f_0'(\psi)}{I}\right\} = 0.
$$

Obviously, this equality will be fulfilled if

$$
\frac{\partial^2\psi}{\partial I^2} + \frac{f_0'(\psi)}{I} = -g(\psi)
$$

for some smooth function g. Thus in the radially-symmetric case the Prandtl type equations reduce to the "ordinary differential equation"

$$
I\frac{d^2\psi}{dI^2} + f(\psi) + Ig(\psi) = 0,
$$

where f, g are arbitrary smooth functions ($f = f_0'$). Of course, this equation has to be completed with regularity conditions on ψ at zero, with the condition

$$\psi = W(s)I + \xi(s) + o(I^{-1})$$

at infinity, as well as with condition (6.36).

6.4.3. The Moffatt–Kida–Ohkitani vortex. Now we consider the radially-symmetric stretched vortex in the particular case of a rectilinear outer flow $V = (\lambda_1 x_1, \lambda_2 x_2, \lambda_3 x_3)$. We take the trajectory Γ to be the straight line $x_1 = x_2 = 0$. Then the equality (6.36) is fulfilled automatically, and the system (6.35) admits a simple solution, for which

$$\psi = I\lambda_3 s, \quad \frac{\partial \omega}{\partial s} = 0, \quad -I\lambda_3 \frac{\partial}{\partial I}(I^2\omega^2) = 4\nu_0 I^2 \omega \frac{\partial^2}{\partial I^2}(I\omega).$$

The last equality can be rewritten as

$$\left(\lambda_3 + 2\nu_0 \frac{\partial}{\partial I}\right) \frac{\partial}{\partial I}(I\omega) = 0,$$

whence we find that

$$\omega = \text{const}\, \frac{1 - \mathrm{e}^{-\beta I}}{I} = \text{const}\, \frac{1 - \mathrm{e}^{-\beta r^2/2}}{r^2},$$

where $\beta = \lambda_3 (2\nu_0)^{-1}$. This solution (along with the first correction) was obtained and thoroughly investigated by Moffatt, Kida, and Ohkitani [**177**].

APPENDIX

List of Equations

1. The Korteweg–de Vries equation (§1.1)

$$\frac{\partial u}{\partial t} + 3u\frac{\partial u}{\partial x} + \varepsilon^2 \frac{\partial^3 u}{\partial x^3} = 0.$$

2. The Korteweg–de Vries equation with variable coefficients (§§1.3 and 5.2)

$$\frac{\partial u}{\partial t} + (\rho_1(x,t) + 3\rho_2(x,t)u)\frac{\partial u}{\partial x} + \varepsilon^2 \rho_3(x,t)\frac{\partial^3 u}{\partial x^3} + \rho_4(x,t)u = 0.$$

The coefficients $\rho_j(x,t)$ are assumed to be smoth with $\rho_2 > 0$ and $\rho_3 > 0$. The following particular examples are considered:

$$\rho_1 = \sqrt{gH(x)}, \quad \rho_2 = \frac{1}{2}\sqrt{\frac{g}{H(x)}}, \quad \rho_3 = \frac{1}{6}\sqrt{gH^5(x)}, \quad \rho_4 = \frac{1}{2}\frac{\partial \rho_1(x)}{\partial x},$$

where $H(x) > 0$ is the depth of the unperturbed fluid;

$$\rho_1 = 0, \quad \rho_4 = 0, \quad \rho_2 = \rho_2(t), \quad \rho_3 = \rho_3(t);$$

and the "spherical" Korteweg–de Vries equation

$$\rho_1 = 0, \quad \rho_2 = 1, \quad \rho_3 = 1, \quad \rho_4 = 1/t, \quad t > 0.$$

3. Korteweg–de Vries–Burgers equation (§1.2)

$$\frac{\partial u}{\partial t} + 3u\frac{\partial u}{\partial x} + \varepsilon_1 \frac{\partial^3 u}{\partial x^3} = \varepsilon_2 \frac{\partial^2 u}{\partial x^2},$$

where the parameters ε_1 and ε_2 are related either by

$$\varepsilon_1 \approx \varepsilon_2 \ll 1$$

or by

$$\varepsilon_1 \approx (\varepsilon_2)^{1/3} \ll 1.$$

4. The nonlinear Schrödinger equation with variable coefficient (§5.1)

$$i\varepsilon\frac{\partial \psi}{\partial t} + \varepsilon^2 \Delta\psi - \alpha(x,t)|\psi|^2\psi = 0,$$

where $\alpha(x,t) \geq \text{const} > 0$ is a smooth function

5. The Boussinesq equation (§1.4)

$$\frac{\partial^2 u}{\partial t^2} - \frac{\partial^2 u}{\partial x^2} = \frac{3}{2}\frac{\partial^2}{\partial x^2}u^2 + \varepsilon^2 \frac{\partial^4 u}{\partial x^4}.$$

6. The sine-Gordon equation with variable coefficients (§§1.5, 5.2, 5.3)

$$\varepsilon^2\frac{\partial^2 u}{\partial t^2} - \varepsilon^2\frac{\partial^2 u}{\partial x^2} + \frac{\alpha^2(x,t)}{\gamma(x,t)}\sin(\gamma(x,t)u) = 0.$$

7. The sine-Gordon equation in a medium with dissipation and energy influx

$$\varepsilon^2\frac{\partial^2 u}{\partial t^2} - \varepsilon^2\frac{\partial^2 u}{\partial x^2} + \alpha^2(x)\sin u$$
$$= -\varepsilon^2(\varepsilon_1\rho_1(x) + 3\varepsilon\rho_2(x)\cos u)\left|\frac{\partial u}{\partial t}\right|\frac{\partial u}{\partial t} + \varepsilon^4\eta(x)\frac{\partial^3 u}{\partial x^2\partial t} + \varepsilon_1\delta(x,t).$$

The case $\varepsilon_1 = \varepsilon$ corresponds to weak dissipation (§1.5) and the case $\varepsilon_1 = 1$ to strong dissipation (§1.6).

8. The Benjamin, Bona, and Mahogany equation (also called the PBBM equation or the regularized equation of long waves) (§1.5)

$$\frac{\partial u}{\partial t} + \frac{\partial u}{\partial x} + 3u\frac{\partial u}{\partial x} = \varepsilon^2\frac{\partial^3 u}{\partial x^2\partial t}.$$

9. The multidimensional sine-Gordon equation (§2.2)

$$\varepsilon^2\frac{\partial^2 u}{\partial t^2} - \varepsilon^2\Delta u + \frac{\alpha^2(x,t)}{\gamma(x,t)}\sin(\gamma(x,t)u) + \varepsilon F = 0, \quad x\in\mathbb{R}^n, \quad n = 2,3.$$

The following particular cases are considered: $F = 0$,

$$F = \varepsilon^2(\rho_1(x) + 3\rho_2(x)\cos u)\left|\frac{\partial u}{\partial t}\right|\frac{\partial u}{\partial t} - \eta(x)\varepsilon^3\frac{\partial}{\partial t}\Delta u + \delta(x,t),$$

and

$$F = f(x,t)(1-\cos(\gamma(x,t)u)).$$

10. The Kadomtsev–Petviashvili equation (§§2.1 and 5.2)

$$\frac{\partial}{\partial x}\left(4\frac{\partial u}{\partial t} + 6u\frac{\partial u}{\partial x} + \varepsilon^2\frac{\partial^3 u}{\partial x^3}\right) + 3\sigma\frac{\partial^2 u}{\partial y^2} = 0, \quad \sigma = \pm 1.$$

11. The Kadomtsev–Petviashvili equation with variable coefficients (§2.1)

$$\frac{\partial u}{\partial t} + \left(H^{1/2} + \frac{3}{2}\alpha H^{-3/4}u\right)\frac{\partial u}{\partial x} + \frac{\varepsilon^2}{6}H^{3/2}\frac{\partial}{\partial x}\left(H^{1/2}\frac{\partial}{\partial x}\left(H^{1/2}\frac{\partial u}{\partial x}\right)\right)$$
$$+\frac{\kappa}{2}H^{1/4}\frac{\partial}{\partial y}\left(H\frac{\partial w}{\partial y}\right) = 0,$$
$$H^{3/4}\frac{\partial w}{\partial x} = u,$$

where $H = H(x,y) > 0$ is the depth of the unperturbed fluid, and α, κ are constants.

12. The Boussinesq system of equations I (§2.3)

$$\frac{\partial u}{\partial t} + \nabla\left\{\frac{1}{2}|u|^2 + g\rho\right\} = 0,$$
$$\frac{\partial \rho}{\partial t} + \langle\nabla, \{(h(x)+\rho)u + \varepsilon^2 h^3(x)\Delta u\}\rangle = 0,$$

where $x\in\mathbb{R}^2$, $u = (u_1, u_2)$, and $h(x) > 0$ is the depth of the unperturbed fluid.

13. The Boussinesq system of equations II (§2.4)

$$\frac{\partial u}{\partial t} + \nabla\Big\{\frac{\alpha}{2}|u|^2 + \rho + \varepsilon^2\Delta\rho\Big\} = 0,$$
$$\frac{\partial \rho}{\partial t} + \big\langle \nabla, (1+\alpha\rho)u\big\rangle = 0,$$

where $x \in \mathbb{R}^2$, $u = (u_1, u_2)$, $\alpha = \mathrm{const} > 0$.

14. The Whitham equation (also referred to as Joseph's equation or the "intermediate" equation) (§3.1)

$$\frac{\partial u}{\partial t} + (\rho_1 + \rho_2 u)\frac{\partial u}{\partial x} + \rho_3 V.P.\int_{-\infty}^{\infty}\Big(\coth\frac{\pi(x-y)}{\varepsilon H} - \operatorname{sign}(x-y)\Big)\frac{\partial^2 u(y,t)}{\partial y^2}\,dy = 0,$$

where $\rho_1, \rho_2, \rho_3 > 0$, $H > 0$ are constants.

15. The infinite Toda chain with slowly varying coefficients (§3.2)

$$m_k\frac{d^2 q_k}{dt^2} = a_{k+1}\big(e^{r_{k+1}(q_{k+1}-q_k)} - 1\big) - a_k\big(e^{r_k(q_k - q_{k-1})} - 1\big),$$

where $k = 0, \pm1, \pm2, \dots$, $q_k = O(\varepsilon)$ with $\varepsilon \ll 1$ being a small parameter. It is assumed that there exist smooth functions α, μ, ρ such that

$$m_k = \varepsilon\mu(x)\big|_{x=k\varepsilon}, \qquad r_k = \frac{1}{\varepsilon}\rho\Big(x + \frac{\varepsilon}{2}\Big)\Big|_{x=k\varepsilon},$$
$$a_k = \alpha\Big(x + \frac{\varepsilon}{2}\Big)\Big|_{x=k\varepsilon}, \qquad \alpha(x)\rho(x) \geq \mathrm{const} > 0.$$

Moreover, the case $m_k = m_k(t)$, $r_k = r_k(t)$, $a_k = a_k(t)$, and $q_k = O(1)$ is considered.

16. The infinite Langmuir chain with slowly varying coefficients (§3.3)

$$M_k\frac{\partial N_k}{\partial t} + N_k(N_{k+1} - N_{k-1}) = 0, \quad k = 0, \pm1, \pm2, \dots.$$

It is assumed that there exists a smooth function $\mu(x,t) > 0$ such that

$$M_k(t) = 2\varepsilon\mu(x,t)\big|_{x=k\varepsilon}, \quad \varepsilon \ll 1.$$

17. The system of the phase field equations (§4.1)

$$\frac{\partial}{\partial t}(\theta + \psi) = \langle \nabla, k(x,\psi)\nabla\theta\rangle,$$
$$\varepsilon^2\frac{\partial \psi}{\partial t} = \varepsilon^2\Delta\psi + \psi - \psi^3 + \varepsilon\kappa(\psi^2 - 1)\theta,$$

where $x \in \mathbb{R}^n$, $n \leq 3$, $\kappa = \mathrm{const}$,

$$k(x,\psi) = \frac{k^+(x) + k^-(x)}{2} + \frac{k^+(x) - k^-(x)}{2}\eta(\psi),$$

$k^{\pm}(x) > 0$, and $\eta(\psi) \to \pm1$ as $\psi \to \pm1$.

18. The conserved phase field system (§§4.2 and 4.3)

$$\frac{\partial}{\partial t}(\theta+\psi)=\Delta\theta+f(x),$$
$$\kappa\frac{\partial\psi}{\partial t}+\Delta(\varepsilon^2\Delta\psi+\psi-\psi^3+\varepsilon\kappa_1\theta)=0,$$

where $x\in\Omega\subset\mathbb{R}^n$, $n\le 3$, $\kappa=\text{const}>0$, $\kappa_1=\text{const}$.

19. The gas dynamics equations for large Reynolds numbers

$$\frac{\partial\rho}{\partial t}+\langle\nabla,\rho u\rangle=0,$$
$$\frac{\partial}{\partial t}(\rho u_i)+\langle\nabla,\rho u u_i\rangle+\frac{\partial p}{\partial x_i}=\varepsilon_1\varepsilon\frac{\partial}{\partial x_j}\sigma_{ij},\quad i=1,2,3,$$
$$\frac{\partial}{\partial t}\left(\frac{1}{2}\rho u^2+\frac{p}{\gamma-1}\right)+\left\langle\nabla,\left(\frac{1}{2}\rho u^2+\frac{\gamma}{\gamma-1}p\right)u+\varepsilon_1\varepsilon\sigma\cdot u-\varepsilon_2\kappa\nabla T\right\rangle=0.$$

Here $u=(u_1,u_2,u_3)$, $x\in\mathbb{R}^3$, $\gamma>1$, the tensor σ has elements

$$\sigma_{ij}=\eta\left(\frac{\partial u_i}{\partial x_j}+\frac{\partial u_j}{\partial x_i}-\frac{2}{3}\delta_{ij}\frac{\partial u_k}{\partial x_k}\right)+\zeta\delta_{ij}\frac{\partial u_k}{\partial x_k},$$

$\eta>0$, $\zeta>0$, δ_{ij} is Kronecker's delta, and repeated indices indicate summation.

A solution of smoothed shock wave type is constructed in the following cases:

large Prandtl numbers (§2.5), i.e.,

$$\varepsilon_1=1,\quad \varepsilon_2=\varepsilon^2,\quad \varepsilon=1/\operatorname{Re}\ll 1,\quad \operatorname{Pr}\sim\operatorname{Re};$$

finite Prandtl numbers (§2.6), i.e.,

$$\varepsilon_1=1,\quad \varepsilon_2=\varepsilon,\quad \varepsilon=1/\operatorname{Re}\ll 1,\quad \operatorname{Pr}\sim 1;$$

small Prandtl numbers (§§2.6 and 5.4), i.e.,

$$\varepsilon_1=\varepsilon,\quad \varepsilon_2=\varepsilon,\quad \varepsilon=1/\sqrt{\operatorname{Re}}\ll 1,\quad \operatorname{Pr}\sim 1/\sqrt{\operatorname{Re}}.$$

A solution of boundary layer type (§2.7) is constructed in the case

$$\varepsilon_1=\varepsilon,\quad \varepsilon_2=\varepsilon^2,\quad \varepsilon=1/\sqrt{\operatorname{Re}}\ll 1,\quad \operatorname{Pr}\sim 1.$$

20. The stationary Navier–Stokes equations for large Reynolds numbers (Chapter 6)

$$\langle u,\nabla\rangle u+\nabla p=\varepsilon^2\Delta u,$$
$$\langle\nabla,u\rangle=0.$$

Here $u=(u_1,u_2,u_3)$, $x\in\mathbb{R}^3$, $\varepsilon\sim 1/\sqrt{\operatorname{Re}}\ll 1$.

Bibliography

1. M. J. Ablowitz, *Applications of slowly varying nonlinear dispersive wave theories*, Stud. Appl. Math. **50** (1971), 329–344.
2. M. J. Ablowitz and H. Segur, *Solitons and the inverse scattering transfrom*, SIAM, Philadelphia, 1981.
3. G. N. Abramovich, T. A. Girshovich, S. Yu. Krasheninnikov, A. N. Sekundov, and I. P. Smirnova, *Theory of turbulent jets*, 2nd ed., Nauka, Moscow, 1984; English transl. of 1st ed., MIT Press, Cambridge, MA, 1963.
4. S. M. Allen and J. W. Cahn, *A microscopic theory for antiphase boundary motion and its application to antiphase domain coarsening*, Acta Metall. **27** (1979), 1085–1095.
5. H. W. Alt and I. Pawlow, *A mathematical model of dynamics of non-isothermal phase separation*, Physica D **59** (1992), 389–416.
6. ———, *On the entropy principle of phase transition models with a conserved order parameter*, Adv. Math. Sci. Appl. **6** (1996), 291–376.
7. V. K. Andreev, O. V. Kaptsov, V. V. Pukhnachev, and A. A. Rodionov, *Application of group-theoretic methods in hydrodynamics*, Nauka, Novosibirsk, 1994; English transl., Kluwer, Dordrecht, 1998.
8. A. A. Andronov, E. A. Leontovich, I. I. Gordon, and A. G. Maier, *Qualitative theory of second-order dynamic systems*, Nauka, Moscow, 1966; English transl., Halsted Press, New York, 1973.
9. J. Aragona and F. Villareal, *Colombeau's theory and shock waves in a problem of hydrodynamics*, J. Anal. Math. **61** (1993), 113–144.
10. V. I. Arnold, *Mathematical methods of classical mechanics*, Nauka, Moscow, 1974; English transl., Springer-Verlag, Berlin, 1978.
11. ———, *The asymptotic Hopf invariant and its applications*, Materials All-Union School on Differential Equations with an Infinite Number of Variables and Dynamical Systems with an Infinite Number of Degrees of Freedom (Dilizhane, 1973), Izdat. Akad. Nauk Armyan. SSR, Erevan, 1974, pp. 229–256; English transl., Selecta Math. Sovietica **5** (1986), 327–345.
12. V. I. Arnold and B. A. Khesin, *Topological methods in hydrodynamics*, Springer-Verlag, Berlin, 1998.
13. V. M. Babich and V. S. Buldyrev, *Asymptotic methods in short-wave diffraction theory*, Nauka, Moscow, 1972; English transl., Springer-Verlag, Berlin, 1991.
14. A. O. Barut (ed.), *Nonlinear equations in physics and mathematics*, Reidel, Dordrecht, 1978.
15. G. K. Batchelor, *An introduction to fluid dynamics*, Cambridge University Press, Cambridge, 1970.
16. J. H. Batteh and J. D. Powell, *Soliton propagation in a one-dimensional lattice under shock compression*, in **[129]**, pp. 257–275.
17. T. B. Benjamin, J. L. Bona, and J. J. Mahony, *Model equations for long waves in nonlinear dispersive systems*, Philos. Trans. Roy. Soc. London Ser. A **272** (1972), 47–78.
18. H. A. Biagioni, *A nonlinear theory of generalized functions*, Springer-Verlag, Berlin, 1992.
19. H. A. Biagioni and M. Oberguggenberger, *Generalized solutions to Burgers' equation*, J. Diff. Equations **97** (1992), 263–287.
20. ———, *Generalized solutions to the Korteweg–de Vries and the regularized long-wave equations*, SIAM J. Math. Anal. **23** (1992), 923–940.
21. V. N. Bogaevskii and A. Ya. Povzner, *Algebraic methods in nonlinear perturbation theory*, Nauka, Moscow, 1987; English transl., Springer-Verlag, Berlin, 1991.

22. N. N. Bogolyubov and Yu. A. Mitropol′skii, *Asymptotic methods in the theory of non-linear oscillations*, 4th ed.,Nauka, Moscow, 1974; English transl. of 2nd ed., Hindustan, Delhi, and Gordon and Breach, New York, 1961.
23. A. V. Bolsinov and A. T. Fomenko, *Introduction to the topology of integrable Hamiltonian systems*, Nauka, Moscow, 1997. (Russian)
24. K. A. Brakke, *The motion of a surface by its mean curvature*, Princeton Univ. Press, Princeton, NJ, 1978.
25. V. V. Bulatov, Yu. V. Vladimirov, V. G. Danilov, and S. Yu. Dobrokhotov, *On motion of the point algebraic singularity for two-dimensional nonlinear equations of hydrodynamics*, Mat. Zametki **55** (1994), no. 3, 11–20; English transl., Math. Notes **55** (1994), 243–250.
26. B. K. Bullough and P. J. Caudrey (eds.), *Solitons*, Topics in Current Physics, vol. 17, Springer-Verlag, Berlin, 1980.
27. V. F. Butuzov, *Asymptotic properties of solutions of singularly perturbed elliptic equations in rectangular regions*, Differentsial′nye Uravneniya **11** (1975), 1030-1041; English transl., Differential Equations **11** (1975), 780-787.
28. ———, *The angular boundary layer in mixed singularly perturbed problems for hyperbolic equations*, Mat. Sb. **104** (1977), 460-485; English transl., Math. USSR–Sb. **33** (1977), 403-425.
29. G. Caginalp, *An analysis of a phase field model of a free boundary*, Arch. Rat. Mech. Anal. **92** (1986), 205–245.
30. ———, *A conserved phase system: implication for kinetic undercooling*, Phys. Rev. B **38** (1988), 789–791.
31. ———, *The dynamics of a conserved phase field system: Stefan like, Hele–Shaw, and Cahn–Hillard models as asymptotic limits*, IMA J. Appl. Math. **44** (1990), 77–94.
32. ———, *Stefan and Hele–Shaw type models as asymptotic limits of the phase field equations*, Phys. Rev. A **39** (1990), 101–111.
33. G. Caginalp and X. Chen, *Convergence of the phase model to its sharp interface limits*, European J. Appl. Math. **9** (1998), 417–445.
34. J. W. Cahn and S. M. Allen, *A microscopic theory for domain wall motion and its experimental verification in Fe-Al alloy domain growth kinetics*, Journal de Physique, Coll. C7 suppl. (12) **38** (1977), 51–54.
35. J. W. Cahn and J. E. Hilliard, *Free energy of a nonuniform system. I. Interfacial free energy*, J. Chem. Phys. **28** (1958), 258–267.
36. ———, *Spinodal decomposition: A reprise*, Acta Metall. **19** (1971), 151–161.
37. F. Calogero and A. Degasperis, *Spectral transform and solitons.* Vol 1: *Tools to solve and investigate nonlinear evolution equations*, North-Holland, Amsterdam, 1982.
38. F. E. Campfield and R. L. Street, *Shoaling of solitary waves on small beach. Water ways and harbor devices*, Proc. Amer. Soc. Civil. Engrs. **95** (1969), 1–22.
39. J. J. Cauret, J. F. Colombeau, and A. Y. Le Roux, *Discontinuous generalized solutions of nonlinear nonconservative hyperbolic equations*, J. Math. Anal. Appl. **139** (1989), 552–573.
40. S. Chapman and T. G. Cowling, *The mathematical theory of non-uniform gases*, Cambridge University Press, Cambridge, 1952.
41. Y.-G. Chen, Y. Giga, and S. Goto, *Uniqueness and existence of viscosity solutions of generalized mean curvature flow equations*, J. Diff. Geom. **33** (1991), 749–786.
42. J. D. Cole, *Perturbation methods in applied mathematics*, Blaisdell, Waltham, MA, 1968.
43. J. F. Colombeau, *Elementary introduction to new generalized functions*, North-Holland, Amsterdam, 1985.
44. R. Courant, *Partial differential equations*, Interscience, New York, 1962.
45. V. G. Danilov, V. P. Maslov, and V. M. Shelkovich, *The algebra of singularities of singular solutions to quasilinear strictly hyperbolic systems of first order*, Teoret. Mat. Fiz. **114** (1998), 3–55; . English transl., Theoret. and Math. Phys. **114** (1998), 1–42.
46. V. G. Danilov, V. P. Maslov, and K. A. Volosov, *Mathematical modelling of heat and mass transfer processes*, Kluwer, Dordrecht, 1995.
47. V. G. Danilov and G. A. Omel′yanov, *Propagation of singularities and related problems of solidification*, Matematichki Vesnik (Belgrade) **49** (1997), 143–150.
48. ———, *Calculation of the singularity dynamics for quadratic nonlinear hyperbolic equations. Example: the Hopf equation*, Nonlinear Theory of Generalized Functions (Vienna,

1997; M. Grosser *et al.*, eds.), Res. Notes in Math., vol. 401, Chapman and Hall, London, 1999, pp. 63–74.

49. V. G. Danilov, G. A. Omel′yanov, and E. V. Radkevich, *Asymptotics of the solution to the phase field system and the modified Stefan problem*, Differentsial′nye Uravneniya **31** (1995), 483-491; English transl., Differential Equations **31** (1995), 446-454.
50. ———, *Weak solutions to the phase field system*, Integral Transforms and Special Functions **6** (1997), 27–35.
51. V. G. Danilov and V. M. Shelkovich, *Propagation and interaction of nonlinear waves*, Eighth International Conference on Hyperbolic Problems. Theory–Numerics Applications (Magdeburg, Feb. 28–March 3, 2000), 2000, pp. 326–328.
52. R. I. Demchenko and M. I. Zheleznyak, *Nonlinear-dispersion effects in the travel of surface waves over irregularities of the bottom topography*, Gidromekhanika **48** (1983), 17–22; English transl., Fluid Mech., Soviet Res. **12** (1983), 15–22.
53. S. Yu. Dobrokhotov and V. P. Maslov, *Multiphase asymptotics of nonlinear partial differential equations with small parameter*, in Soviet Sci. Rev. Ser. C: Math. Phys. Rev., vol. 3, Harwood Academic Publ., Chur, 1982, pp. 221–280.
54. S. Yu. Dobrokhotov and A. I. Shafarevich, *Parametrix and asymptotics of localized solutions of Navier–Stokes equations in R^3 linearized on a smooth flow*, Mat. Zametki **51** (1992), no. 1, 72–82; English transl., Math. Notes **51** (1992), 47–54.
55. ———, *Asymptotic solutions of linearized Navier-Stokes equations*, Mat. Zametki **53** (1993), 25-35; English transl., Math. Notes **53** (1993), 19-26.
56. ———, *Some integral identities and remarks on the decay at infinity of the solutions to the Navier–Stokes equations in the entire space*, Russian J. Math. Phys. **2** (1994), 133–135.
57. ———, *On the behavior of an incompressible fluid velocity field at infinity*, Izv. Akad. Nauk, Mekh. Zhidk. Gaza **1996**, no. 4, 38-42; English transl., Fluid Dyn. **31** (1996), 511-514.
58. B. A. Dubrovin, V. B. Matveev, and S. P. Novikov, *Non-linear equations of Korteweg-deVries type, finite-zone linear operators, and Abelian varieties*, Uspekhi Mat. Nauk **31** (1976), no. 1, 55–136; English transl., Russian Math. Surveys **31** (1976), no. 1, 59-146.
59. Yu. V. Egorov, *A contribution to the theory of generalized functions*, Russian Math. Surveys **45** (1990), no. 5, 1–49.
60. C. M. Elliott and J. R. Ockendon, *Weak and variational methods for moving boundary problems*, Pitman, Boston, 1982.
61. L. C. Evans, H. M. Soner, and P. E. Souganidis, *Phase transtions and generalized motion by mean curvature*, Comm. Pure Appl. Math. **45** (1992), 1097–1123.
62. L. C. Evans and J. Spruck, *Motion of level sets by mean curvature.* I, J. Diffefential Geom. **33** (1991), 635–681.
63. ———, *Motion of level sets by mean curvature.* II, Trans. Amer. Math. Soc. **330** (1992), 321–332.
64. ———, *Motion of level sets by mean curvature.* III, J. Geom. Anal. **2** (1992), 121–150.
65. P. C. Fife and V. Greenlee, *Interior transition layers for elliptic boundary-value problems with a small parameter*, Uspekhi Mat. Nauk **29** (1974), no. 4, 103–130; English transl., Russian Math. Surveys **29** (1974), no. 4, 103–131.
66. H. Flaschka, *The Toda lattice.* I. *Existence of integrals*, Phys. Rev. B **9** (1974), 1924–1925.
67. ———, *The Toda lattice.* II. *Inverse scattering solution*, Progr. Theoret. Phys. **51** (1974), 703–706.
68. H. Flaschka, M. Forest, and D. McLaughlin, *Multiphase averaging and the inverse spectral solution of the Korteweg-de Vries equation*, Comm. Pure Appl. Math. **33** (1980), 739–784.
69. L. E. Fraenkel, *On steady vortex rings of small cross-secton in an ideal fluid*, Proc. Roy. Soc. London Ser. A **316** (1970), 29–62.
70. ———, *Examples of steady vortex rings of small cross-secton in an ideal fluid*, J. Fluid Mech. **51** (1972), 119–135.
71. S. Friedlander and M. M. Vishik, *Instability criteria for the flow of an inviscid incompressible fluid*, Phys. Rev. Lett. **17(66)** (1991), 2204–2206.
72. C. S. Gardner, J. M. Green, M. D. Kruskal, and R. M. Miura, *A method for solving the Korteweg–de Vries equation*, Phys. Rev. Lett. **19** (1967), 1095–1097.
73. I. M. Gel′fand, *Some problems in the theory of quasilinear equations*, Uspekhi Mat. Nauk **14** (1959), no. 2, 87–158; English transl., Amer. Math. Soc. Transl. (2) **29** (1963), 295–381.

74. P. Germain and R. Bader, *Unicité des écoulements avec chocs dans la mécanique de Burgers*, Office National d'Etudes et de Recherches Aeronautiques (ONERA), Paris, 1953.
75. K. A. Gorshkov and L. A. Ostrovsky, *Interactions of solitons in nonintegrable systems: direct perturbation method and applications*, Physica D **182** (1981), 428–438.
76. R. Grimshaw, *Slowly varying solitary waves.* I. *Korteweg–de Vries equation*, Proc. Roy. Soc. London Ser. A **368** (1979), 359–375.
77. ———, *Slowly varying solitary waves.* II. *Nonlinear Schrödinger equation*, Proc. Roy. Soc. London Ser. A **368** (1979), 377–388.
78. V. V. Grushin, *On a class of elliptic pseudodifferential operators degenerating on a submanifold*, Mat. Sb. **84** (1971), 163-195; English transl., Math. USSR–Sb. **13** (1971), 155–185.
79. V. Guillemin and S. Sternberg, *Geometric asymptotics*, Amer. Math. Soc., Providence, RI, 1977.
80. M. E. Gurtin, *Thermomechanics of evolving phase boundaries in the plane*, Clarendon Press, Oxford, 1993.
81. Ph. Hartman, *Ordinary differential equations*, Wiley, New York, 1964.
82. R. Hirota, *Direct methods of finding exact solutions of nonlinear evolution equations*, Bäcklund Transformations (R. M. Miura, ed.), Lecture Notes in Math., vol. 515, Springer-Verlag, Berlin, 1976, pp. 40–68.
83. ———, *Direct methods in soliton theory*, in **[26]**, pp. 157–176.
84. E. H. Hirschel and W. Kordulla, *Shear flow in surface-oriented coordinates*, Notes on Numer. Fluid Mech., vol. 4, Vieweg, Braunschweig, 1981.
85. E. Hopf, *The partial differential equation* $u_t + uu_x = \mu u_{xx}$, Comm. Pure Appl. Math. **3** (1950), 201–230.
86. ———, *On the right weak solution of the Cauchy problem for a quasi-linear equation of first order*, J. Math. Mech. **19** (1969), 483–487.
87. A. M. Il′in, *Matching of asymptotic expansions of solutions of boundary value problems*, Nauka, Moscow, 1989; English transl., Amer. Math. Soc., Providence, RI, 1992.
88. A. M. Il′in and L. A. Kalyakin, *Perturbation of finite-soliton solutions to the Korteweg-de Vries equation*, Dokl. Ross. Akad. Nauk **336** (1994), 595-598; English transl., Phys. Dokl. **40** (1995), 603–606.
89. A. M. Il′in and T. N. Nesterova, *Asymptotics of the solution of the Cauchy problem for a quasi-linear parabolic equation with a small parameter*, Dokl. Akad. Nauk SSSR **240** (1978), 11–13; English transl., Soviet Math. Dokl. **19** (1978), 529–532.
90. S. V. Iordanskii, *The asymptotics of an axially symmetric diverging wave in a heavy fluid*, Dokl. Akad. Nauk SSSR **125** (1959), 1211–1214; English transl., Soviet Phys. Dokl. **4** (1959), 284–287.
91. V. K. Ivanov, *An associate algebra of the simplest generalized functions*, Siberian Math. J. **20** (1979), 509–516.
92. R. S. Johnson, *On the development of a solitary wave moving over an uneven bottom*, Proc. Cambridge Philos. Soc. **73** (1973), 183–203.
93. R. I. Joseph, *Solitary waves in a finite depth fluid*, J. Phys. A: Math. Gen. **10** (1977), L225–L227.
94. B. B. Kadomtsev, *Collective effects in a plasma*, Nauka, Moscow, 1976; French transl., Mir, Moscow, 1979.
95. B. B. Kadomtsev and V. I. Petviashvili, *On the stability of solitary waves in weakly dispersing media*, Dokl. Akad. Nauk SSSR **192** (1970), 753–756; English transl., Soviet Phys. Dokl. **15** (1970), 539–541.
96. A. S. Kalashnikov, *The construction of generalized solutions to quasilinear first order equations without convexity condition as limits of solutions to parabolic equations with a small parameter*, Dokl. Akad. Nauk SSSR **127** (1959), 27–30. (Russian)
97. L. A. Kalyakin, *Long wave asymptotics. Integrable equations as asymptotic limits of nonlinear systems*, Uspekhi Mat. Nauk **44** (1989), no. 1, 5–34; English transl., Russian Math. Surveys **44** (1989), no. 1, 3–42.
98. ———, *The asymptotics of an integral occurring in the theory of soliton perturbations*, Mat. Zametki **50** (1991), no. 5, 32–42; English transl., Math. Notes **50** (1991), 1114–1122.
99. ———, *Perturbation of the Korteweg–de Vries soliton*, Teoret. Mat. Fiz. **92** (1992), 62–76; English transl., Theoret. and Math. Phys. **92** (1992), 736–747.

100. ———, *On the problem of first correction in the theory of soliton perturbations*, Mat. Sb. **186** (1995), no. 7, 51–76; English transl., Sb. Math. **186** (1995), 977–1002.
101. ———, *Weak interaction of the KdV solitons under perturbation*, Physica D **87** (1995), 193–200.
102. ———, *Asymptotic well-posedness for the problem of soliton perturbation*, Russian J. Math. Phys. **4** (1998), no. 3, 3–27.
103. ———, *Weak interaction of perturbed KdV solitons*, Fundam. i Prikl. Mat. **4** (1998), 1179–1205. (Russian)
104. ———, *Perturbation of a singular solution to the Liouville equation*, Teoret. Mat. Fiz. **118** (1999), 390–397; English transl., Theoret. and Math. Phys. **118** (1999), 307–314.
105. ———, *Asymptotic decay of the solution to the perturbed Liouville equation*, Mat. Zametki **68** (2000), 195–209; English transl. in Math. Notes **68** (2000).
106. I. G. Karelina and Yu. V. Pokornyi, *On the Green function of the Dirichlet problem on a graph*, Dokl. Akad. Nauk SSSR **318** (1991), 542–544; English transl., Soviet Math. Dokl. **43** (1991), 732–734.
107. V. I. Karpman, *Soliton evolutions in the presence of perturbations*, Physica Scripta **20** (1979), 462–478.
108. V. I. Karpman and E. M. Maslov, *The structure of tails formed under effect of perturbations on solitons*, Zh. Eksper. Teoret. Fiz. **75** (1978), 504–517; English transl., Soviet Phys. JEPT **48** (1978), 252–259.
109. ———, *On the soliton propagation in slowly varying media*, Preprint of IZMIRAN (Inst. of Terrestial Magnetism, Ionsphere, and Radiowave Propagation), no. 39a, 1981.
110. D. J. Kaup, *A perturbation theory for inverse scattering transforms*, SIAM J. Appl. Math. **31** (1976), 121–123.
111. D. J. Kaup and A. C. Newell, *Solitons as particles, oscillators and in slowly changing media: a singular perturbation theory*, Proc. Roy. Soc. London Ser. A **361** (1978), 413–446.
112. J. P. Keener and D. W. McLaughlin, *Solitons under perturbations*, Phys. Rev. A **16** (1977), 777–790.
113. N. Kenmochi and M. Niezgódka, *A nonlinear system for thermodynamical phase separation*, Techn. Rep. Math. Sci. no. 7, Chiba Univ., 1992.
114. Yu. S. Kivshar′, *On the perturbation theory for solitons: the sine–Gordon equation*, preprint no. 21–84, Fiz.-Tekhn. Inst. Nizkikh Temperatur Akad. Nauk Ukrain. SSR, Kharkov, 1984.
115. C. J. Knickerbocker and A. C. Newell, *Internal solitary waves near a turning point*, Phys. Lett. **75A** (1980), 326.
116. K. Ko and H. H. Kuehl, *Korteweg–de Vries soliton in a slowly varying medium*, Phys. Rev. Lett. **40** (1978), 233–236.
117. Y. Kodama and T. Taniuti, *Higher order corrections to the soliton velocity and the linear dispersion relation*, Physica Scripta **20** (1979), 486–489.
118. M. N. Kogan, *Rarefied gas dynamics*, Nauka, Moscow, 1967; English transl., Plenum Press, New York, 1969.
119. Yu. A. Kravtsov and Yu. I. Orlov, *The geometric optics of inhomogeneous media*, Nauka, Moscow, 1980; English transl., Springer-Verlag, Berlin, 1990.
120. I. M. Krichever, *Method of averaging for two-dimensional "integrable" equations*, Funktsional. Anal. i Prilozh. **22** (1988), no. 3, 37–52; English transl., Functional Anal. Appl. **22** (1988), 200–213.
121. A. G. Kulikovskii, *On the structure of shock waves*, Prikl. Mat. Mekh. **26** (1962), 631–641; English transl., J. Appl. Math. Mech. **26** (1962), 950–964.
122. G. E. Kuzmak, *Asymptotic solutions of nonlinear second order differential equations with variable coefficients*, Prikl. Mat. Mekh. **23** (1959), 515–526; English transl., J. Appl. Math. Mech. **23** (1959), 730–744.
123. G. L. Lamb, Jr., *Elements of soliton theory*, Wiley, New York, 1980.
124. G. L. Lamb, Jr. and D. W. McLaughlin, *Aspects of soliton physics*, in **[26]**, pp. 65–106.
125. L. D. Landau and E. M. Lifshits, *Course of theoretical physics.* Vol. 6: *Fluid mechanics*, 3rd ed., Nauka, Moscow, 1986; English transl., Pergamon Press, Oxford, 1987.
126. P. D. Lax, *Integrals of nonlinear equations of evolution and solitary waves*, Comm. Pure Appl. Math. **21** (1968), 467–490.

127. S. Leibovich and A. R. Seebass (eds.), *Nonlinear waves*, Cornell University Press, Ithaca, NY, 1974.
128. J. Leray, *Analyse lagrangienne et méchanique quantique*, Séminaire sur les Équations aux Dérivées Partielles (1976–1977), Vol. 1, Exposé 1, Collège de France, Paris, 1977; English transl., MIT Press, Cambridge, MA, 1981.
129. K. Lonngren and A. Scott (eds.), *Solitons in action*, Academic Press, New York, 1978.
130. S. Luckhaus, *Solutions of the two-phase Stefan problem with the Gibbs–Thomson law for the melting temperature*, European J. Appl. Math. **1** (1990), 101–111.
131. S. Luckhaus and L. Modica, *The Gibbs–Thomson relation within the gradient theory of phase transitions*, Arch. Rat. Mech. Anal. **107** (1989), 71–83.
132. S. Luckhaus and A. Visintin, *Phase transition in multicomponent systems*, Manuscripta Math. **43** (1983), 261–288.
133. J. C. Luke, *A perturbation method for nonlinear dispersive wave problems*, Proc. Roy. Soc. London Ser. A **292** (1966), 403–412.
134. R. K. Lunenburg, *Mathematical theory of optics*, Univ. of California Press, Berkeley and Los Angeles, 1964.
135. O. S. Madsen and S. S. Mei, *The transformation of a solitary wave over an uneven bottom*, J. Fluid Mech. **39** (1969), 781–791.
136. S. A. Manakov, *On complete integrability and stochastization in discrete dynamical systems*, Zh. Eksp. Teoret. Fiz. **67** (1974), 543–555; English transl., Soviet Phys JETP **40** (1974), 269–274.
137. S. V. Manakov, V. E. Zakharov, L. A. Bogard, and A. B. Matveev, *Two-dimensional solitons of the Kadomtsev–Petviashvili equation and their interaction*, Phys. Lett. **63A** (1977), 205–206.
138. J.-A. Marti, *Fundamental structures and asymptotic microlocalization in sheaves of generalized functions*, Integral Transforms and Special Functions **6** (1998), 223–228.
139. E. M. Maslov, *On the theory of perturbations of solitons in second approximation*, Teoret. Mat. Fiz. **42** (1980), 362–373; English transl., Theoret. and Math. Phys. **42** (1980), 237–245.
140. ———, *Dynamics of rotationally symmetric solitons in near-sg field models with applications to large-area Josephson junctions and ferromagnets*, Physica D **15** (1985), 433–443.
141. V. P. Maslov, *The theory of perturbations and asymptotic methods*, Moscow Univ. Press, Moscow, 1965; French transl., *Théorie des perturbations et méthodes asymptotiques. Suivi de deux notes complémentaires de V. I. Arnol'd et V. C. Bouslaev*, Dunod, Paris, 1972.
142. ———, *Asymptotic methods and the theory of perturbations*, Nauka, Moscow, 1988. (Russian)
143. ———, *Propagation of shock waves in an isentropic, nonviscous gas*, Itogi Nauki i Tekhn.: Sovremennye Probl. Mat., vol. 8, VINITI, Moscow, 1977, pp. 199–271; English transl., J. Soviet Math. **13** (1980), 119–163.
144. ———, *The complex WKB method for nonlinear equations*, Nauka, Moscow, 1977; English transl. of Part I, *The complex WKB method for nonlinear equations. I: Linear theory*, Progress in Physics., vol. 16, Birkhäuser, Basel, 1994.
145. ———, *Three algebras corresponding to nonsmooth solutions of systems with quasilinear hyperbolic equations*, Uspekhi Mat. Nauk **35** (1980), no. 2, 252–253. (Russian)
146. ———, *Coherent structures, resonances, and asymptotic nonuniqueness of solutions to the Navier–Stokes equations for large Reynolds numbers*, Uspekhi Mat. Nauk **41** (1986), no. 6, 19–35; English transl., Russian Math. Surveys **41** (1986), no. 6, 19–35.
147. ———, *Violation of causality principle for nonstationary equations of two-dimensional and three-dimensional gas dynamics at sufficiently large Reynolds numbers*, Teoret. Mat. Fiz. **69** (1986), 361–378; English transl., Theoret. and Math. Phys. **69** (1986), 1202–1212.
148. V. P. Maslov and V. E. Nazaikinskii, *Asymptotic methods for the solution of pseudo-differential equations*, Nauka, Moscow, 1987; English transl., *Asymptotics of operator and pseudodifferential equations*, Consultants Bureau, New York, 1988.
149. V. P. Maslov and M. V. Fedoryuk, *Semi-classical approximation in quantum mechanics*, Nauka, Moscow, 1970; English transl., Math. Phys. and Appl. Math., vol. 7, Reidel, Dordrecht, 1981.
150. V. P. Maslov and G. A. Omel′yanov, *Asymptotic soliton-form solutions of equations with small dispersion*, Uspekhi Mat. Nauk **36** (1981), no. 3, 63–126; English transl., Russian Math. Surveys **36** (1981), no. 3, 73–149.

151. ———, *Hugoniot-type conditions for infinitely narrow solutions of the equation for simple waves*, Sibirsk. Mat. Zh. **24** (1983), no. 5, 172–182; English transl., Siberian Math. J. **24** (1983), 787–795.
152. ———, *Soliton-type asymptotic behavior of interior waves in a stratified liquid with weak dispersion*, Differentsial'nye Uravneniya **21** (1985), 1766–1775; English transl., Differential Equations **21** (1985), 1197–1204.
153. ———, *Three-wave interaction including frequency doubling effects*, Izv. Vyssh. Uchebn. Zaved. Fiz. **29** (1986), no. 3, 3–23; English transl., Soviet Phys. J. **29** (1986), 157–175.
154. ———, *Single-phase asymptotics for magnetic hydrodynamic equations with large Reynolds numbers*, Sibirsk. Mat. Zh. **29** (1988), no. 5, 172–180; English transl., Siberian Math. J. **29** (1988), 824–830.
155. ———, *Lin–Lees equation for boundary layers in domains with curvilinear boundary*, Physica D **33** (1988), 266–280.
156. ———, *Rapidly oscillating asymptotic solution of magnetohydrodynamic equations in the tokamak approximation*, Teoret. Mat. Fiz. **92** (1992), 269–292; English transl., Theoret. and Math. Phys. **92** (1992), 879–895.
157. ———, *Three-scale expansion of the solution of the magnetohydrodynamic equations and the Reynolds equation for a tokamak*, Teoret. Mat. Fiz. **98** (1994), 297–311; English transl., Theoret. Math. Phys. **98** (1994), 202–211.
158. ———, *Fluctuation-generated Tokamak pinch instabilities*, Russian J. Math. Phys. **2** (1995), 463–485.
159. ———, *Nonlinear evolution of fluctuations in a tokamak plasma and the dynamics of the boundary of a pinch*, Fizika Plazmy **21** (1995), 684–696; English transl., Plasma Phys Rep. **21** (1995), 646–658.
160. ———, *The turbulent dynamo problem*, Mat. Zametki **58** (1995), 936–939; English transl., Math. Notes **58** (1995), 1352–1355.
161. V. P. Maslov, G. A. Omel'yanov, and V. A. Tsupin, *The behavior of a solitary wave taking a small dispersion into account*, Processes of Excitation and Propagation of Tsunami, Institute of Oceanology, Moscow, 1982, pp. 58–97. (Russian)
162. ———, *Asymptotics of some differential and pseudodifferential equations, and dynamical systems with small dispersion*, Mat. Sb. **122(164)** (1983), 197–219; English transl., Math. USSR-Sb. **50** (1985), 191–212.
163. ———, *Quasi-linear nonstationary equations and evolutions of interior boundary layers: asymptotic soliton-like solutions*, BAIL-IV, Proc. IV Internat. Conf. Boundary and Interior Layers, Boole Press, Dublin, 1986, pp. 362–367.
164. V. P. Maslov and A. I. Shafarevich, *Asymptotic laminar solutions to the Navier-Stokes equations localized in a small neighborhood of a curve and their small-scale disturbances*, Dokl. Ross. Akad. Nauk **351** (1996), 52–55; English transl., Phys. Dokl. **41** (1996), 540–543.
165. ———, *Localized asymptotic solutions to the Navier-Stokes equations and laminar traces in incompressible fluid*, Prikl. Mat. Mekh. **62** (1998), 424–432; English transl., J. Appl. Math. Meck. **62** (1998), 389–396.
166. V. P. Maslov and V. A. Tsupin, *Propagation of a shock wave in an isentropic gas with small viscosity*, Itogi Nauki i Tekhn.: Sovremennye Probl. Mat., vol. 8, VINITI, Moscow, 1977, pp. 273–308; English transl., J. Soviet Math. **13** (1980), 163–185.
167. ———, *Delta-shaped Sobolev generalized solutions of quasi-linear equations*, Uspekhi Mat. Nauk **34** (1979), no. 1, 235–236; English transl., Russian Math. Surveys **34** (1979), no. 1, 231–232.
168. Y. Matsuno, *Exact multi-soliton solution for nonlinear waves in a stratified fluid of finite depth*, Phys. Lett. **74A** (1979), 233–235.
169. D. W. McLaughlin and A. C. Scott, *A multisoliton perturbation theory*, in **[129]**, pp. 201–256.
170. D. W. McLaughlin, G. C. Papanicolaou, and O. R. Pironneau, *Convection of microstructures and related problems*, SIAM J. Appl. Math. **45** (1985), 780–797.
171. A. M. Meirmanov, *The Stefan problem*, Nauka, Novosibirsk, 1986; English transl., de Gruyter, Berlin, 1992.
172. J. W. Miles, *The Korteweg–de Vries equation: a historical essay*, J. Fluid Mech. **106** (1981), 131.

173. A. S. Mischenko, B. Yu. Sternin, and V. E. Shatalov, *Lagrangian manifolds and the method of canonical operator*, Nauka, Moscow, 1978; English transl, *Lagrangian manifolds and the Maslov operator*, Springer-Verlag, Berlin, 1990.
174. R. M. Miura, *Korteweg–de Vries equation: a model equation for nonlinear dispersive waves*, in [**127**], pp. 212–234.
175. H. K. Moffatt, *Magnetostatic equilibria and analogous Euler flows of arbitrary complex topology. Part 1. Fundamentals*, J. Fluid Mech. **159** (1985), 359–378.
176. ______, *Generalized vortex rings with and without swirl*, Fluid Dynamics Research **3** (1988), 22–30.
177. H. K. Moffatt, S. Kida, and K. Ohkitani, *Stretched vortices – the sinews of turbulence; large-Reynolds-number asymptotics*, J. Fluid Mech. **259** (1994), 241–264.
178. I. A. Molotkov and S. A. Vakulenko, *Nonlinear longitudinal waves in inhomogeneous rods*, Zap. Nauchn. Sem. Leningrad. Otdel. Mat. Inst. Steklov. **99** (1980), 64–73; English transl., J. Soviet Math. **20** (1982), 2434–2441.
179. ______, *Concentrated nonlinear waves*, Leningrad. Univ., Leningrad, 1988. (Russian)
180. P. I. Naumkin and I. A. Shishmarev, *On the breaking of waves for the Whitham equation*, Dokl. Akad. Nauk SSSR **265** (1982), 809–811; English transl., Soviet Math. Dokl. **26** (1982), 150–152.
181. ______, *On nonlinear nonlocal equations in wave theory. I: The Whitham equation*, Vestnik Moskov. Univ. Ser. 3 Phys. Astr. **31** (1990), no. 5, 3–16; English transl., Moscow Univ. Phys. Bull. **45** (1990), no. 5, 5–17.
182. A. H. Nayfeh, *Perturbation methods*, Wiley, New York, 1973.
183. A. C. Newell, *The inverse scattering transform*, in [**26**], pp. 177–242.
184. J. Norbury, *A family of steady vortex rings*, J. Fluid Mech. **57** (1973), 417–431.
185. A. Novick-Cohen, *The nonlinear Cahn–Hillard equation: transition from spinodal decomposition to nucleation behavior*, J. Stat. Phys. **38** (1985), 707–722.
186. ______, *The Cahn–Hillard equation: mathematical and modelling perspectives*, Adv. Math. Sci. Appl. **8** (1998), 968–985.
187. M. Oberguggenberger, *Multiplication of distributions and applications to partial differential equations*, Pitman. Res. Notes Math. 259, Longman Sci. Techn., Harlow, Essex, 1992.
188. O. A. Oleinik, *The Cauchy problem for nonlinear equations in a class of discontinuous functions*, Dokl. Akad. Nauk SSSR **95** (1954), 451–455; English transl., Amer. Math. Soc. Transl. (2) **42** (1964), 7–12.
189. ______, *Discontinuous solutions of non-linear differential equations*, Uspekhi Mat. Nauk **12** (1957), no. 3, 3–73; English transl., Amer. Math. Soc. Transl. (2) **26** (1963), 95–172.
190. ______, *On the construction of a generalized solution to the Cauchy problem for a quasilinear equation by introducing the "vanishing viscosity"*, Uspekhi Mat. Nauk **14** (1959), no. 2, 159–164; English transl., Amer. Math. Coc. Transl. (2) **23** (1963), 277–283.
191. ______, *On uniqueness and stability of the generalized solution to the Cauchy problem for a quasilinear equation*, Uspekhi Mat. Nauk **14** (1959), no. 2, 165–170; English transl., Amer. Math. Soc. Transl. (2) **23** (1963), 285–290.
192. ______, *Mathematical problems of boundary layer theory*, Uspekhi Mat. Nauk **23** (1968), no. 3, 3–65; English transl., Russian Math. Surveys **23** (1968), no. 3, 1–66.
193. O. A. Oleinik and V. N. Samokhin, *Mathematical models in boundary layer theory*, Nauka, Moscow, 1997; English transl., Chapman and Hall, Boca Raton, FL, 1999.
194. G. A. Omel′yanov, *Boundary value problems for elliptic systems of nonlinear differential equations with a small parameter*, Differentsial′nye Uravneniya **18** (1982), 1829-1831. (Russian)
195. ______, *Asymptotic description of two metastable processes of solidification for the case of large relaxation time*, preprint IC/95/162, International Centre for Theoretical Physics, Trieste, 1995.
196. G. A. Omel′yanov, V. G. Danilov, and E. V. Radkevich, *On regularization of initial data of the modified Stefan problem*, Mat. Zametki **57** (1995), 793–802; English transl., Math. Notes **57** (1995), 559–564.
197. ______, *Soliton-type asymptotic solutions of the conserved phase field system*, Portugaliae Mathematica **53** (1996), 471–501.
198. ______, *Asymptotic solution of the conserved phase field system in the fast relaxation case*, European J. Appl. Math. **9** (1998), 1–21.

199. ———, *Tanh-type asymptotic solution of the conserved phase field system*, Adv. Math. Sci. Appl. **8** (1998), 663–689.
200. G. A. Omel′yanov and V. V. Trushkov, *A geometric correction in the problem on the motion of a free boundary*, Mat. Zametki **63** (1998), 151–153; English transl., Math. Notes **63** (1998), 137–139.
201. ———, *Dynamics of a free boundary in a binary medium with variable thermal conductivity*, Mat. Zametki **66** (1999), 231–241; English transl., Math. Notes **66** (1999), 181–189.
202. L. A. Ostrovskii, *Approximate methods in the theory of nonlinear waves*, Izv. Vyssh. Uchebn. Zaved. Radiofizika **17** (1974), 454–476; nglish transl. in Radiophys. and Quantum Electronics.
203. L. A. Ostrovskii and E. N. Pelinovskii, *Nonlinear evolution of tsunami-type waves*, in Theoretical and Experimental Study of Tsunami Waves, Nauka, Moscow, 1977, pp. 52–60. (Russian)
204. L. A. Ostrovskii and V. N. Shrira, *Instability and self-refraction of solitons*, Zh. Eksper. Teoret. Fiz. **71** (1976), 1412–1420; English transl., Soviet Phys. JETP **44** (1976), 738–743.
205. R. D. Parmentier, *Fluxons in long Josephson junctions*, in **[129]**, pp. 173–199.
206. E. N. Pelinovskii and Yu. A. Stepanyan, *Cylindrical tsunami waves*, in Propagation and Shoaling of Tsunami Waves, Nauka, Moscow, 1981, pp. 104–117. (Russian)
207. O. M. Penkin and Yu. V. Pokornyi, *On some qualitative properties of equations on a one-dimensional CW-complex*, Mat. Zametki **59** (1996), 777–780; English transl., Math. Notes **59** (1996), 562–565.
208. O. Penrose and P. C. Fife, *Thermodynamically consistent models of phase-field type for the kinetics of phase transitions*, Physica D **43** (1990), 44–62.
209. D. H. Peregrine, *Calculations of the development of an undular bore*, J. Fluid Mech. **25** (1966), 321–330.
210. S. Pilipović, *Colombeau's generalized functions and the pseudo-differential operators*, Lectures in Mathematical Sciences, The University of Tokyo, Tokyo, 1994.
211. P. I. Plotnikov and V. N. Starovoitov, *Stefan problem as the limit of the phase field system*, Differential Equations **29** (1993), 461–471.
212. L. Prandtl, *Führer durch die Strömungslehre*, 3rd ed., Vieweg, Braunschweig, 1949.
213. *Proceedings of a conference on the theory and applications of solitons*, 1978, Rocky Mountain J. Math. **8** (1978), no. 1–2.
214. V. I. Pryazhinskii and V. G. Sushko, *Small parameter asymptotics of some solutions of the Cauchy problem for a quasilinear parabolic equation*, Dokl. Akad. Nauk SSSR **247** (1979), 283–285; English transl., Soviet Math. Dokl. **20** (1979), 698–700.
215. E. V. Radkevich, *On conditions for the existence of a classical solution of the modified Stefan problem (the Gibbs–Thomson law)*, Mat. Sb. **183** (1992), no. 2, 77-101; English transl., Math. USSR-Sb. **75** (1993), 221-246.
216. ———, *On the asymptotic solutions of a phase field system*, Differentsial′nye Uravneniya **29** (1993), 487-500; English transl., Differential Equations **29** (1993), 418-429.
217. R. Ravindran and P. Prasad, *A new theory of shock dynamics*, Appl. Math. Lett. **3** (1990), 107–109.
218. V. A. Reutov, *The motion of a solitary wave over an underwater ridge*, Izv. Akad. Nauk SSSR, Mekh. Zhidk. Gaza **4** (1975), 79–85. (Russian)
219. B. L. Rozhdestvenskii and N. I. Yanenko, *Systems of quasilinear equations and their applications to gas dynamics*, Nauka, Moscow, 1978; English transl., Amer. Math. Soc., Providence, RI, 1983.
220. A. V. Rybin and M. A. Sall′, *Solitons of the Korteweg-de Vries equation on the background of a known solution*, Teoret. Mat. Fiz. **63** (1985), 333–339; English transl., Theoret. and Math. Phys. **63** (1985), 545–550.
221. A. M. Samsonov, *Solitons in nonlinear elastic rods with variable characteristics*, in Nonlinear and Turbulent Processes in Physics (R. Z. Sagdeev, ed.), vol. 2, Gordon and Breach, New York, 1984, pp. 1029–1035.
222. H. Segur, *Solitons as approximate descriptions of physical phenomena*, Rocky Mountain J. Math. **8** (1978), 15–24.
223. A. I. Shafarevich, *Behavior of rapidly decreasing asymptotic solutions of linearized Navier–Stokes equations as $t \to \infty$*, Mat. Zametki **55** (1994), no. 6, 124–145; English transl., Math. Notes **55** (1994), 632–647.

224. ———, *Differential equations on graphs describing localized asymptotic solutions to the Navier–Stokes equations, and stretched jets in an incompressible fluid*, preprint no. 604, Keldysh Inst. Prikl. Mat. Ross. Akad. Nauk, Moscow, 1997. (Russian)
225. ———, *Differential equations on graphs describing asymptotic solutions of the Navier-Stokes equations localized in a small neighborhood of a curve*, Differentsial′nye Uravneniya **34** (1998), 1119–1130; English transl., Differential Equations **34** (1998), 1124–1134.
226. ———, *Generalized Prandtl–Maslov equations on graphs describing stretched vortices in an incompressible fluid*, Dokl. Ross. Akad. Nauk **358** (1998), 752–756; English transl., Dokl. Math. **57** (1998), 152–155.
227. ———, *Asymptotic description of vortical threads in incompressible fluid*, Prikl. Mat. Mekh. **64** (2000), 264–275; English transl., J. Appl. Math. Mech. **64** (2000), 255–265.
228. ———, *Localised asymptotic solutions of the Navier–Stokes equations and topological invariants of vector fields. Prandtl–Maslov equations on Reeb graphs and Fomenko invariants*, Russian J. Math. Phys. **7** (2000), 401–448.
229. ———, *Asymptotic solutions for Navier–Stokes equations describing smoothed tangent dicontinuities*, Mat. Zametki **67** (2000), 938–949; English transl., Math. Notes **67** (2000).
230. A. N. Shiryaev, *Essentials of stochastic finance. Vol. 2: Theory*, FAZIS, Moscow, 1998; English transl. of Vols. 1, 2, combined into one volume, World Sci. Publ, Singapore, 1999.
231. H. Schlichting, *Boundary layer theory*, 7th ed., McGraw–Hill, New York, 1979.
232. A. I. Shnirelman, *Lattice theory and flows of ideal incompressible fluid*, Russian J. Math. Phys. **1** (1993), 105–113.
233. V. I. Shrira, *The influence of cylindrical divergence and refraction on the propagation of nonlinear tsunami waves*, Propagation and Shoaling of Tsunami Waves, Nauka, Moscow, 1981, pp. 146–156. (Russian)
234. H. M. Soner, *Convergence of the phase field equations to the Mullins–Sekerka problem with kinetic undercooling*, Arch. Rat. Mech. Anal. **131** (1995), 139–197.
235. J. Sprekels and S. Zheng, *Global solutions to the equations of a Ginzburg–Landau theory for structural phase transitions in shape memory alloys*, Physica D **39** (1989), 59–76.
236. R. Srinivasan and P. Prasad, *On the propagation of a multi-dimensional shock of arbitrary strength*, Proc. Indian Acad. Sci. (Math. Sci.) **94** (1985), 27–42.
237. B. Stoth, *Convergence of the Cahn–Hillard equation to the Mullins–Sekerka problem in spherical symmetry*, J. Differential Equations **125** (1996), 154–183.
238. V. Yu. Subochev and V. A. Tsupin, *Asymptotic soliton-like solutions to the nonlinear Schrödinger equation with variable coefficients*, Teoret. Mat. Fiz. **56** (1983), 42–53; English transl., Theoret. and Math. Phys. **56** (1983), 661–668.
239. V. G. Sushko and L. A. Chudov, *Asymptotic behavior of some solutions to quasi-linear parabolic equations with small parameter at a higher derivative*, Vychisl. Metody i Programmirovanie **7** (1967), 113–141. (Russian)
240. V. G. Sushko and E. A. Lapshin, *Asymptotic expansions of solutions to some problems related to nonlinear acoustics*, in Interaction of One-Dimensional Waves in Media without Dispersion, Moscow Univ. Press, Moscow, 1983, pp. 167–181. (Russian)
241. M. E. Taylor, *Pseudodifferential operators*, Princeton Univ. Press, Princeton, NJ, 1981.
242. ———, *Vibration of a chain with nonlinear interaction*, J. Phys. Soc. Japan **22** (1967), 431–436.
243. ———, *Wave propagation in unharmonic lattices*, J. Phys. Soc. Japan **23** (1967), 501–506.
244. ———, *On a nonlinear lattice (the Toda lattice)*, in **[26]**, pp. 143–155.
245. J. C. Tolédano and P. Tolédano, *The Landau theory of phase transitions*, World Scientific, Singapore, 1987.
246. F. Treves, *Introduction to pseudodifferential and Fourier integral operators*, Plenum Press, New York, 1982.
247. G. E. Uhlenbeck and G. W. Ford, *Lectures in statistical mechanics*, Lectures in Appl. Math., vol. 1, Amer. Math. Soc., Providence, RI, 1963.
248. B. R. Vainberg, *Asymptotic methods in equations of mathematical physics*, Gordon and Breach, New York, 1989.
249. M. Van Dyke, *Perturbation methods in fluid mechanics*, Academic Press, New York, 1964.
250. A. B. Vasil′eva and V. F. Butuzov, *Asymptotic expansion of solutions for singularly pertubed equations*, Nauka, Moscow, 1973. (Russian)

251. A. L. Velikovich and M. A. Liberman, *Physics of shock waves in gases and plasma*, Nauka, Moscow, 1987; English transl. (of the ms.), Springer-Verlag, Berlin, 1986.
252. M. I. Vishik and L. A. Lyusternik, *Regular degeneration and boundary layer for linear differential equations with small parameter*, Uspekhi Mat. Nauk **12** (1957), no. 5, 3–122; English transl., Amer. Math. Soc. Transl. (2) **20** (1962), 239–364.
253. ———, *The solution of some perturbation problems for matrices and selfadjoint differential equations. I*, Uspekhi Mat. Nauk **15** (1960), no. 3, 3–80; English transl., Russian Math. Surveys **15** (1960), no. 3, 1–73.
254. W. Wasow, *Asymptotic expansions for ordinary differential equations*, Interscience, New York, 1965.
255. G. B. Whitham, *A new approach to problems of shock dynamics. Part I. Two-dimensional problems*, J. Fluid Mech. **2** (1957), 146–171.
256. ———, *On the propagation of shock waves through regions of non-uniform area of flow*, J. Fluid Mech. **4** (1958), 337–360.
257. ———, *A new approach to problems of shock dynamics. Part* II. *Three-dimensional problems*, J. Fluid Mech. **5** (1959), 369–386.
258. ———, *Some comments on wave propagation and shock wave structure with application to magnetohydrodynamics*, Comm. Pure Appl. Math. **12** (1959), 113–158.
259. ———, *A general approach to linear and nonlinear dispersive waves using a Lagrangian*, J. Fluid Mech. **22** (1965), 273–283.
260. ———, *Nonlinear dispersive waves*, Proc. Roy. Soc. London Ser. A **283** (1965), 283–291.
261. ———, *Two-timing, variational principles and waves*, J. Fluid Mech. **44** (1970), 373–395.
262. ———, *Linear and nonlinear waves*, Wiley, New York, 1974.
263. V. E. Zakharov, *The inverse scattering method*, in **[26]**, pp. 243–285.
264. ———, *Instability and nonlinear oscillations of solitons*, Pis′ma Zh. Eksper. Teoret. Fiz. **22** (1975), 364–367; English transl., JETP Letters **22** (1975), 172–173.
265. V. E. Zakharov, S. V. Manakov, S. P. Novikov, and L. P. Pitaevskii, *Theory of solitons. The inverse problem method*, Nauka, Moscow, 1980; English transl., Plenum Press, New York, 1984.
266. V. E. Zakharov, S. L. Musher, and A. M. Rubenchik, *On the nonlinear stage of excitation of waves in a plasma*, Pis′ma Zh. Eksper. Teoret. Fiz. **19** (1974), 249–251; English transl., JETP Letters **19** (1974), 151–152.
267. V. E. Zakharov and A. M. Rubenchik, *Instability of waveguides and solitons in nonlinear media*, Zh. Eksper. Teoret. Fiz. **65** (1973), 997–1011; English transl., Soviet Phys. JETP **38** (1973), 494–500.
268. V. E. Zakharov and A. B. Shabat, *Exact theory of two-dimensional self-focusing and one-dimensional automodulation of waves in nonlinear media*, Zh. Eksper. Teoret. Fiz. **61** (1971), 118–134; English transl., Soviet Phys. JETP **34** (1971), 62–69.
269. ———, *A scheme for integrating the nonlinear equations of mathematical physics by the method of the inverse scattering problem. I*, Funktsional. Anal. i Prilozhen. **8** (1974), no. 3, 43–53; English transl., Functional Anal. Appl. **8** (1974), 226–235.
270. V. E. Zakharov, L. A. Tahtadzjan, and L. D. Faddeev, *Complete description of solutions of the 'sine-Gordon' equation*, Dokl. Akad. Nauk SSSR **219** (1974), 1334–1337; English transl., Soviet Phys. Dokl. **19** (1974), 824–826.
271. Ya. B. Zel′dovich and Yu. P. Raizer, *The physics of shock waves and high-temperature hydrodynamical phenomena*, Nauka, Moscow, 1966; English transl., Vols. 1, 2, Academic Press, New York, 1966, 1967.

Selected Titles in This Series

(*Continued from the front of this publication*)